ALCOHOLISM
New Directions in Behavioral Research and Treatment

NATO CONFERENCE SERIES

I   Ecology
II   Systems Science
III Human Factors
IV   Marine Sciences
V   Air—Sea Interactions
VI   Materials Science

III HUMAN FACTORS

ALCOHOLISM
New Directions in Behavioral Research and Treatment

Edited by
Peter E. Nathan
Rutgers University
New Brunswick, New Jersey

G. Alan Marlatt
University of Washington
Seattle, Washington

and
Tor Løberg
University of Bergen
Bergen, Norway

Published in coordination with NATO Scientific Affairs Division

PLENUM PRESS · NEW YORK AND LONDON

Library of Congress Cataloging in Publication Data

Nato Conference on Experimental and Behavioral Approaches to Alcoholism, Os, Norway, 1977.
  Alcoholism: new directions in behavioral research and treatment.

  (NATO conference series: III, Human factors; v. 7)
  "Proceedings of the NATO Conference on Experimental and Behavioral Approaches to Alcoholism, held in Os, Norway, August 28—September 1, 1977, and sponsored by the NATO Special Panel on Human Factors."
  Includes index.
  1. Alcoholism — Treatment — Congresses. 2. Behavior therapy — Congresses. I. Nathan, Peter E. II. Marlatt, G. Alan. III. Løberg, Tor. IV. Nato Special Program Panel on Human Factors. V. Title. VI. Series.
RC565.N3 1977   616.8'61   78-11876
ISBN-13: 978-1-4613-2876-6   e-ISBN-13: 978-1-4613-2874-2
DOI: 10.1007/978-1-4613-2874-2

Proceedings of the NATO Conference on Experimental and Behavioral Approaches to Alcoholism, held in Os, Norway, August 28—September 1, 1977, and sponsored by the NATO Special Program Panel on Human Factors

© 1978 Plenum Press, New York
Softcover reprint of the hardcover 1st edition 1978
A Division of Plenum Publishing Corporation
227 West 17th Street, New York, N.Y. 10011

For L.A.C. and J.A.E.

Data non sententia

PREFACE

This book contains the fifteen invited papers delivered at the
NATO International Conference on Experimental and Behavioral Approaches to Alcoholism, held August 28 through September 1, 1977, at
the Solstrand Fjord Hotel, Os, Norway. The editors of the book were
Co-Directors of that conference. As well, 65 other scientists from
12 countries in the free world presented scientific papers on experimental and behavioral topics of relevance to alcoholism at the
meeting. A most receptive audience of almost 200 persons also participated actively in the discussions which followed every invited
and contributed paper.

The beauty of Norway, the hospitality of the proprietors of the
Solstrand Fjord Hotel, the aura of Grieg and Troldhaugen, the enthusiasm of the speakers and participants - all combined to make
the conference most memorable for those who attended it.

Many persons and institutions deserve special thanks for their
part in the success of the conference. Among these persons are Dr.
J.C. Brengelmann, European Co-Director of the Conference, Professor
M.N. Özdas, Assistant Secretary General for Scientific and Environmental Affairs, NATO, Dr. B.A. Bayraktar, Executive Officer, Human
Factors Programme, Scientific Affairs Division, NATO, and Professor
H. Ursin, member of NATO's Special Programme Panel on Human Factors
and a faculty member of the Institute of Psychology, University of
Bergen, the conference's host institution. All three Co-Directors
were delighted with the commitment of these individuals to a successful conference; Dr. Bayraktar deserves special commendation for
the efficiency and good graces with which he helped us organize and
carry out the meeting.

While we have chosen and edited these 15 papers with care, it is
impossible to convey the very special quality of the interactions
that took place at the meeting. If critics of NATO's scientific
programs had attended the conference, they could not have helped
realizing that the good will, international understanding, and invaluable scientific exchange that characterized it made NATO's support for the endeavor a most tangible and effective undertaking.

July, 1978

Peter E. Nathan
G. Alan Marlatt
Tor Løberg

CONTENTS

COMMENTS ON THE MANY FACES OF ALCOHOLISM

John L. Horn

University of Denver

WHY THIS CHAPTER: SOME POINTS OF STRATEGY AND PHILOSOPHY

The purpose of this chapter is to provide an overview of results from studies of the personalities of people who have entered
treatment in the Alcoholism Division of the Fort Logan Mental Health
Center in Denver, Colorado. To considerable extent the chapter
will deal with scale construction that has resulted from these
studies. Such work is not in the mainstream of research presented
elsewhere in this volume. For this reason, it is desirable to do
a bit more stage management than otherwise would be necessary to
indicate how the work is relevant in the present context. A major
question must be: "What are the values (if any), the uses and
possible uses, of scale construction for researchers and practition-
ers who are concerned primarily with studies and applications of
treatment techniques?" Several useful observations can be made in
considering these questions.

Debates about Process Should be Based Upon Differential Diagnosis

In an article concerned primarily with the recent revival of
old arguments about whether or not controlled drinking, in contrast
to abstinence, can be a realistic goal of therapy for people who
are said (often rather glibly) to be alcoholics, Miller and Caddy
(1977) point out that these arguments are likely to generate more
heat than light until such time that they involve diagnosis that
is sufficiently refined to permit distinction between problem drinking is contraindicated. This illustrates a fundamental difficulty
in solving what is referred to as the alcoholism problem, and a

fundamental reason for using the kind of measurement that is represented by the research described in the present chapter. For it is
almost certainly true that there is no singular alcoholism problem;
instead, there are a number of problems--psychological/sociological/
medical problems they might be called--that are rather closely
associated with the use of alcohol.[2]

For some of these problems, at some stages in their development, abstinince may be a necessary part of a realistic goal of
therapy. This is suggested not only by the rhetoric of AA, but also
by a substantial amount of research and clinical experience (e.g.,
Glatt, 1976). It seems, indeed, that perhaps 20 to 30 percent of
our Fort Logan patients require abstinence as a therapy goal, at
least in the short run of 2 to 3 years. But it is clear that abstinence should not be a major objective of therapy for a substantial number of people who enter alcoholism treatment under conditior
similar to those that bring patients to the Fort Logan program. An
insistence on abstinence forces some of these patients out of therap
altogether, and it is evident that they need help; for others a requirement of abstinence results in a string of personal failures
ending in loss of confidence in oneself, dissolution of the integrit
of one's personality, self-destructive rebellion, and similar identifiable losses in the power of the individual's drive to realize his
potential; for still other patients, the attainment of abstinence
leads to maladjustment and/or maladaptation that is more serious
than the problems associated with use of alcohol. As Miller and
Caddy point out, it does little good to quibble about whether or
not such patients truly are alcoholics. The genuine alcoholism
problem of any nation involves recognizing that such problem drinkers exist, accurately distinguishing among the many problems they
present, and specifying ways to cope with these problems. In this
work objective diagnosis, achieved by measurement, is essential.

Treatment-by-Personality Interaction Should be Considered

Related to this point is the fact that most clinical psychologists, and perhaps most behavior therapists of any discipline,
pay at least lip service to the idea of patient-by-treatment interaction. Indeed, in the treatment of alcoholism it is usually recognized that although only a relatively small proportion of patient
can accept the treatment constraints and philosophy of AA, for those
who have the "right kind" of personality AA can be a very effective
therapy, whereas for those who do not have this kind of personality,
AA is either simply ineffective or actually harmful. This particula
treatment-by-personality interaction is rather well understood clinically. Although the interaction is less well understood scientifically, it is fairly well documented that it pertains to, in the
personality obeisance, dependency, and needs for structure, whereas
in the treatment there must be clear and firm statement of require-

ments and goals coupled with compassionate support.

This is only one example of many treatment-by-personality interactions that are of interest for the theoretical understanding
of drug use and value for treatment of problems associated with this
use. Another example, less well documented, is treatment employing
conditioned nausea. Such treatment should be related to the individual differences of personality. For one thing, there very well
can be self-selection for this kind of treatment (even as there is
with AA). For another thing, not all of those willing to try this
form of treatment continue with it. And, of course, even among those
who enter and complete treatment there are differences in the effectiveness of the treatment six months, two years, and five years down
the line. Each of these factors of input and output is, in principle, predictable (in part) from knowledge of personality, both that
which is measurable prior to treatment and that which can be measured as treatment and personality development progress.

The case for conditioned nausea is also only an example. In
the last analysis, one must understand the personality in order to
make maximum use of therapies concerned with improving behavior
management, producing behavior management, producing tension reduction, controlling craving, mitigating marital problems, instill-
ing assertiveness, making use of relaxation, and controlling cognitive precursors to drinking, to mention only a few therapy techniques considered in this volume.

This is not to say, of course, that the scales developed in
our research can provide the comprehensive picture of personality
that is needed for full clinical diagnosis to say that these scales
will always provide the kind of personality measurements a researcher needs to study treatment-by-personality interaction. For one
thing, most of the scales developed in our work depend ultimately
on the reliability and validity of self-reports. Contrary to what
is sometimes asserted by novices to the field of personality study
and by a few non-novices who have particular axes to grind (Cf.
Mischel, 1971, but see Block, 1977; Eysenk, 1977 for perspective),
self-report data often is stable and relevant (i.e., valid) for many
purposes. Certainly this is true for the study of some problems of
alcohol use and abuse (see L.C. Sobell in this volume). But the
general point here is that while self-report personality scales of
the kind to be discussed in this chapter do provide a basis for the
study of many questions pertaining to personality-by-treatment interactions, and the enterprising researcher would do well to consider these scales carefully before going on to develop his own
scales (since often his resources and samples will be less appropriate for this work than those of our research), still it must be
recognized that a full science relating personality to treatment will
involve many more, and many better, personality measures than have
been (or ever will be) provided by our work.

Practical and Theoretical Concerns Dictate Objective Measurement

Another reason why one with a treatment orientation should
consider the work of this chapter has to do with matters pertaining to objectivity, economy, transportability and historical continuity in research. Most of us know that under the right condition
expert clinicians can supply a diagnosis that is often better, even
for research purposes, than anything we can yet produce with our
scales, graphs, prediction equations and the like. But the cost for
this diagnosis should be kept well in mind. This cost is usually
several tens or hundreds of times as great as the cost for a comparable diagnosis provided by use of personality scales. There
are other disadvantages of depending on the diagnoses of expert
clincians. Some of the diagnoses and diagnosticians are not good.
The difficulties of replication are attempted across cultures (or
even subcultures) and over extended periods of time. Perhaps more
important is the fact that because so much of the expert judgment
measurement is within the clinician rather than within the items of
a scale, it is almost impossible to study just what it is that might
be changing as diagnosis is shifted from one historical period to
another. Quite simply, clinical judgments, however good they are,
intrinsically are not as objective (in the above-mentioned senses
of this term) as are self-report (i.e., subjective in this sense)
measures of the kind developed in the research to be reported here.

Research Designs Should be Based Upon Multidimensional Conceptions

One important bit of advice that emerges from both clinical
experience and research is that one should recognize that under
the generic heading "alcoholic" there are a variety of quite differen
alcoholic personalities. There are many ways to take this advice
seriously in the design of research studies. Oddly enough, however,
in most of the published research on the use of alcohol (and other
drugs) this advice has not been heeded, not even to the extent of
being pointedly rejected. One suspects that the advice, somehow,
had not been adequately comprehended by researchers.

A crass example of this neglect is the seemingly never-ending
proliferation of studies in which a group of so-called alcoholics
is contrasted with some other group and analyses are directed at
identifying the items that provide discrimination between the groups
One might have thought that this kind of largely wasted activity
would have ceased following critiques of such work by Dawes (1972),
Horn and Wanberg (1969), Horn, Wanberg and Appel (1973), and Mac-
Andrew and Geertsma (1963), among others, but no such salubrious
denouement has yet been realized. This is illustrated by Knox's
(1976) recent review of measurement issues in relation to alcohol-
use problems. A major portion of this review is devoted to studies

of "alcoholics versus nonalcoholics."

Other failures to heed advice to distinguish different kinds of alcohol-related problems are found in ever-popular efforts to find the causes of alcoholism (singular) in the peer group. The family, the society, the physiological structure, the genes or somewhere similar. Again the hallmark of this work is an a priori contrast between alcoholics and "the others" or problem drinkers and "the others." Frequently in such work a rather elaborate theory is stated and sampling restrictions are imposed, as in matching, it being suggested that these features make the study scientific. However, usually the theory stems more from general considerations and graduate training than from careful observations of drinkinkg behavior or problem drinkers and, of course, matching as a means for simulating genuine experimental control creates more design and inference problems than it solves (see Humphreys and Dachler, 1969, for discussion of this in another context, also Horn, 1967). This kind of effort is well illustrated by the work of Jessor and colleagues (Jessor, Graves, Hansen and Jessor, 1968), but in fact a major portion of the alcohol-use research involving human subjects is of this ilk (as indicated in reviews and compilations such as those of Braucht et al, 1973; Knox, 1976; Maddox, 1970; and Tarter, 1975).

The fundamental problems with such research are formally the same as the problems with scale construction based on contrasting a priori groupings, as discussed at length by Horn, Wanberg and Appel (1973). The problem can be conceptualized as one of multiple prediction in which the criterion is not unidimensional, but instead is a melange of categories. Under such conditions, almost any result can be found depending on the categories actually sampled and the frequencies of subjects sampled within these categories. Each category of the heterogeneous dependent variable presents the possibility for a main effect and interactions with each category of treatment variable. If these effects do indeed occur, they can cancel each other, augment each other in peculiar ways and, in general, produce results that are extremely difficult to understand and replicate.

A straight-forward example of this kind of interaction, and of the manner in which an effect can be suppressed, is provided by Hodgson's paper in this volume; had Hodgson not made the personality measurements of severity in this study, no craving effect would have been found. Many examples of this kind of thing have been produced in psychology and education (e.g., Cronbach and Snow, 1977; Eysenck, 1977). There is no reason to suppose that studies of alcohol use should yield any fewer examples.

As concerns issues of identifying different forms of alcoholism or even alcoholism itself, the problem with simple diagnoses of alcoholics versus others is that the study design begs the question; there is assumption of that (alcoholism) which is to be proved. The sampling treats a conglomerate as if it were homogeneous and no dependable analytic means are employed to differentiate the elements of the conglomerate. The research strategy is deductive, with no provisions for discovery of regularities not posited in theory or allowed for in design. It is true that sometimes in this kind of research an investigator is nudged in the direction of recognizing distinct alcohol-related problems by findings of large error and interaction terms in his analyses, but such nudging occurs in spite of a research strategy that works against it, not because of it, and more often than not the evidence of heterogeneity in the alcoholism categorization is simply ignored or forced to serve a theory about different causes for what continues to be regarded as a singular alcoholism problem.

The solution to this kind of problem is relatively simple; all one need do is obtain measures of different alcohol-use problems, as well as relevant measures of personality, and do analyses that can sensitively reveal pretest, in-therapy and dependent variable relationships among and between these variables.

Even the Search for Unitary Alcoholism should be Based on Multi-dimensional Thinking

Also running counter to a research thrust aimed at distinguishing different forms of alcoholism are scalogram studies of stages of development of a (singular) condition and analytic studies that focus on describing and measuring the first principal component (or the like) among variables pertaining to alcohol use (Overall and Patrick, 1972). Again, the problem here is that the research strategy does not well serve the needs for induction. It allows for refined descriptions of influences that work together, but it doesn't lead one to discover distinct sets of influences.

If indeed there is a unitary alcoholism process imbedded within a conglomerate of drinking problems, it is possible, but not assured, that a strategy of the kind represented by defining the first principal component will help to identify the process. For example, if the theory is that there is a "true alcoholism" (e.g., alcohol addiction) that can be distinguished from a variety of other problem drinking conditions, one cannot be assured of converging on an improved definition of this by doing principal components or scalogram analyses on collections of symptoms. This is not assured because the alcoholism process may represent only a small proportion of the total variance in the samples under study and may be correlated with other processes. Thus a procedure that searches out the major sources of variance may converge on a different factor than the alcoholism process or may do no more than bring in a weighted combination of different processes.

The Design Philosophy of the Fort Logan Research

This latter problem has been recognized for many years in the study of abilities (see Horn, 1972, 1976 for review). It was for the purpose of dealing with this problem, primarily, that Thurstone developed the nethods of simple structure factor analysis.[3] Thurstone noted also, however, that these methods help to solve problems of replication of research results and, most important for present purposes, they encourage inductive as well as deductive advancements in comprehending a complex area of study. For these various reasons the Fort Logan research group has adopted the metatherapy of simple structure as a principal guide for analysis of data.

THE SETTING, THE SUBJECTS, AND THE RAW DATA

With this brief introduction to philosophical stance, let us move to a consideration of the context within which findings were produced. The ultimate purpose of this chapter is to provide a broad view of results that appear to be stable, and an indication of some major points of theory, in the research of what might be called the Fort Logan group--namely, Wanberg, Foster, Wackwitz, Diesenhaus, Admas, Ward, Hober, and the present writer. The first published report of this work appeared in Horn and Wanberg (1969). Since that time roughly 20 published articles, 10 psychometric devices, and a number of unpublished technical reports and position papers have been produced by the group. It will not be possible to give a detailed account of this work within the bounds of time and space allowed here, but a bird's eye view of major features can be provided.

The Fort Logan Mental Health Center is one of the two major State of Colorado (population approximately 3 million) hospital facilities for treatment of psychological problems. It is located on the grounds of a former military installation southwest of the principal residential and business sections of Denver (population approximately 600,000 for the city itself and, with surrounding suburbs, a metropolitan population of about 1.6 million). To the west of the hospital are rolling hills rising into the Rocky Mountains. To the south, for about 3 miles, are suburbs comprised mainly of upper middle class whites. For 5 miles to the north and east are business, factories, and lower middle class white residences. Beyond, along a northeast line are, in order, the main concentration of Spanish surnamed people (somewhat less than 10 percent of the metropolitan populations). These features of population distribution relative to hospital location probably account in part for the fact that the proportions of persons in minority groups that

are treated at the hospital are somewhat less than the corresponding proportions in the metropolitan population.

One of the major treatment sections of the Fort Logan hospital
is the Alcoholism Division. Over the years since before our research
began, each week on Monday morning approximately 15 persons have
been admitted to an in-hospital, two-week treatment program designed and administered by the Alcoholism Division. A variety of information about the patients of this program have been gathered. This
information is the raw data of our research.

The patients of the alcoholism treatment program have come
from all parts of Colorado, but most have resided in the metropolitan Denver area. Roughly 85 percent have been white, 11 percent
Spanish surnamed, and 4 percent black. About 15 to 20 percent have
been female; that is, over the last 10 years this percent has increased from about 15 to 20. The age range has been from 18 to over
70 years, but roughly 75 percent of the patients have been between
30 and 55 years of age, the mean age being approximately 42 years.
The average educational level has been slightly greater than 11th
grade, less than high school graduation (i.e., completion of the
12th grade). Roughly 25 percent of the patients reported being un-
employed for 3 months or more at the time of admission.

Data on more than 6,000 people passing through the Fort Logan
alcoholism treatment program have been gathered and analyzed. The
major portion of these data have been obtained at the time the patient was admitted to the program, but in-therapy and follow-up information has been gathered in some years. The intake data have
been obtained primarily from questionnaires (self-administering
after general instructions have been given, but administered as a
scheduled interview to patients who can't read); however, interviews, as in a physical examination, and behavior rating forms
have been used also. In some years it has been possible to accumulate information about the in-hospital program in comparison with
an in-community program, in which the patient was not asked to take
residency at the hospital, but instead was visited in his home or
other parts of his community by therapists trained to help solve
problems in the settings in which they arise (Wanberg, Horn, and
Fairchild, 1974). For some patients it has been possible to obtain
information on progress (and retrogression) immediately following
therapy and 6 months, 1 year, 18 months, and 2 years after leaving
the program. In very recent years Diesenhaus, Wackwitz, and Foster
have obtained data on the use of drugs other than alcohol, and they
have obtained this information not only from patients at Fort Logan
but also from people in the community and clients being treated at
a large outpatient treatment center.

Analyses here deal with drinking behavior, current conditions, characteristic behavior and attitudes, physical/medical symptoms, and retrospections about one's background. For practical as well as theoretical reasons, it has been assumed that the separate questionnaires curcumscribe logically distinct, although overlapping, domains of behavior. Within each domain multiple factor analyses, usually for a simple structure model, have been used to help specify the number and nature of independent dimensions that might be reliably measured. Scales have been constructed to provide operational definition of the concepts thus specified. These scales have been used in clinical practice, as in planning a course of therapy for a particular patient, and in further research work, as in studies of relationships between factors of different domains (Horn, Wangerg and Adams, 1974). The effort of the present paper is to bring together experience derived from both of these uses in an integrated account of what we now think we know about the variety of people and problems presented to an alcoholism treatment facility of the kind that is represented by the Fort Logan Center.

SCALES TO MEASURE ALCOHOL USE AND ADJUSTMENT/ADAPTATION

FOLLOWING THERAPY

How do people describe their use of alcohol, its benefits and harms? In particular, how do they describe problems that can be closely related to alcohol use? There is, of course, an infinity of such descriptions. But can one sample from this infinity in a manner that will provide a basis for identifying major recurrent themes—themes that point to basic reasons for problem drinking? It has been an assumption of the Fort Logan research team that one can sample in this way, albeit crudely, by following two basic prescriptions:

1) Attend to the self reports patients volunteer when they seek entry to, are referred to, and in fact enter a facility for treatment of problem drinking, and

2) Sample these reports comprehensively and in accordance with a metatheory of simple structure.

In developing our present base for talking about major themes in descriptions of alcohol use, these prescriptions have been used in an iterative manner. In one phase the clinical and research staff cooperated in the writing of questionnaire items, the responses to which were believed (according to hypothesis) to indicate a stable order that is psychologically (i.e., theoretically) interesting.[4] In the next phase factoring and scale construction analyses of the kind described earlier were conducted (with N's ranging from over 300 to over 3,000) to test hypotheses already indicated and to develop an empirical basis for new hypotheses. These results were

then used, along with insights gleaned from clinical use of the
scales, in another round of item writing, whence the entire procedure was repeated. In constructing what is presently being called
the Alcohol Use Inventory (AUI)[5] there were three major rounds of
this process and an unrecorded number of minor rounds (in which only
a few new items were written and only a few old items were rewrit-
ten).

Considerable attention has been given to the issue of ensur-
ing that measurement scales developed to represent factors can represent truly independent concepts (Horn and Cattell, 1965). Specifically, we have required that the internal consistency reliability, r^2_{xt}, representing the proportion of variance that is common among the elements of a scale (factor), be substantially larger
than the squared multiple correlation, SMR, of that scale with all
other scales of the domain, this indicating the proportion of variance a scale has in common with other, supposedly different, scales
(factors). Thus the concern has been to ensure that the proportion
of reliable unique variance, or PUV (see Table 1), is large.

This kind of requirement for independence has been neglected
in much work in the behavioral sciences, perhaps particularly that
pertaining to alcohol abuse, and yet is very important. When operational identifiers of concepts are not independent, there can be
much redundant reworking of the same concepts with new labels. For
example, investigator A may talk (perhaps at great length) about
quality B in alcoholics, using his particular scale to identify B,
while investigator G, using what appears to be a different set of
observational techniques, talks (perhaps again at great length)
about a quality H that is supposed to be quite different from B;
but when a test of independence is finally run, it is found that B
and H present virtually the same (except for errors of measurement)
order of subjects. Thus, although different words may abound in two
theories about alcoholism, in fact no essentially different information is conveyed by one theory in distinction from the other.

While this kind of floundering may be characteristic of early
scientific formulations, nevertheless it is extremely wasteful of
the resources of scientists. Just as measurement is the foundation
of science, so it is that independence in measurement is necessary
for efficient, architectonic growth of scientific theory.

The AUI Scales

Details of construction and description of the AUI are provided in Wanberg, Horn and Foster (1977), and need not be repeated
here. However, to indicate the concepts, and labels for concepts,
to be used in further discussion, and to suggest some ideas about
what, and how accurately, the scales measure, the summaries of

TABLE 1

INFORMATION ABOUT AUI SCALES*

	G Load	2nd Load	Avg 2 r_{xt}	Avg PUV	Avg SMR	Chronic Mean N = 154	1st Admit Mean N = 154	Out Patient Mean N=150	F Stat	Signif .01?	6 Mos Not Sober vs Sober N=70	Signif @ ? Level
I. "Core" Primary Scales												
14. Daily Average Quantity Alcohol — QUNT	55	D,C	41			54	50	46	33.5	yes	16	
10. Social-Role Maladaptation — SOMA	52	D,C	66	31	35	56	50	44	59.0	yes	36	.01
11. Perceptual Withdrawal Symptoms — DTS	78	D,C,B	87	35	52	54	49	47	22.6	yes	27	.05
12. Psychophysiological Withdrawal — HANG	80	D,C,B	72	16	56	55	50	44	50.1	yes	27	.05
9. Drinking Control Loss — CTRL	72	D,C,B	79	32	47	54	50	46	24.2	yes	22	
4. Compulsive Drinking — COMP	70	D,C,B	76	24	52	55	51	44	50.0	yes	22	
5. Daily, Habitual Drinking — HABT	22		67	49	16	52	50	48	4.6	no	16	
II. "Concerned" Primary Scales												
6. Drinking Worry Guilt — WOGT	61	C	75	27	48	54	52	44	41.7	yes	25	.05
8. Sought External Help — HELP	32	C	68	46	22	53	48	48	14.7	yes	26	.05
III. Change State Primary Scales												
7. Drink to Change Mood — MOOD	56	C,A	76	32	44	53	50	47	12.6	yes	29	.05
1. Drink to Improve Sociability — IMSO	46	A,B	80	30	50	52	51	46	14.4	yes	21	
2. Drink to Improve Cognition — IMCG	33	A	70	30	40	52	51	47	9.3	yes	02	
3. Gregarious Drinking — GREG	09	A	68	48	20	50	50	49	0.9	no	22	
IV. Marital Involvement Primary Scales												
16. Drink Provokes Marital Problems — DMRP	35		72									
15. Marital Problems Provoke Drink — MRPD	19		70									
V. Expansion Primary Scale												
13. Use Other Drugs — DRUG	22	D,C	71	58	13	51	48		2.3	no	14	
VI. Broad Secondary Scales												
D. Bottom Level Apathetic Deterioration — BOTM	85		87			55	50	45	45.8	yes	32	.01
C. Anxious Concerned Deterioration — ACOD	68		86			53	51	45	27.7	yes	25	.05
B. Obsessive Ambivalent Deterioration — OBAD	51		78			54	50	46	27.1	yes	26	.05
A. Commitment to Drinking — COMT	27		77			52	51	47	10.4	yes	21	

*Decimal points omitted in correlation, reliabilities and PUV's.

Table 1 are provided.

The purpose of Table 1 is to provide some information that will help the reader attempt to understand just what the AUI scales measure. The columns of the table contain the following information:

Scale number, the order of listing in profiles, etc.

Scale name.

Abbreviation, acronym for scale.

G: Correlation of scale with a scale to measure first principal component among all primary scales (not corrected for item overlap).

2nd Ord: Second order factors with which first order scales are correlated most highly

r^2_{xt}: Internal consistency reliability (r_{xx}).

PUV: Percent (proportion) of reliable unique variance.

SMR: Squared multiple correlation of factor with all remaining primary factors.

Chronic: T-score mean in a sample of Alcoholism Division patients judged by the clinical staff (independently of AUI information) to be "chronic alcoholics."

1st Admission: T-score mean in a sample of first admission patients in the Alcoholism Division.

Outpatient: T-score mean in a sample of patients recommended by Alcoholism Division staff (independently of AUI information) for outpatient care only.

F: F-value for testing the main effect significance of difference among the preceding three means.

Signif: Indication of significance of F-value at the .01 level.

Not Sober: Point biserial correlation for contrast between a sample of patients who had not remained abstinent after 6 months (N = 49) and a sample of patients who had remained abstinent (N = 29).

Signif: Indication of the significance of the point-biserial
correlation at either the .01 or .05 level.

Although it can be a bit difficult to extract meaning from the numbers, as such, in Table 1, in fact the table indicates several important things about problem drinkers.

Profiles Viewed from Within and Without

If one examines the columns numbered 6 through 10 in Table 1,
he or she will find that the mean for persons judged by counselors
(operating without the AUI) to be "chronic alcoholics" is notably
above the means for other groups of patients on almost all scales.
Similarly, most of the means for the group of patients admitted to
the hospital are above the corresponding means for the group for
which outpatient care was recommended. These results indicate that
the clinician is observing patient behavior that contains information that is similar to, or at least consistent with, the information summarized in the AUI scales, and he/she is using this information in a somewhat interchangeable manner to make an overall,
unidimensional assessment of extent of alcoholism (ranging from,
say, "chronic" to "insipient"). The interchangeability of use is
suggested by the fact that although there is a considerable amount
of independence among the AUI scales, the order of the means for
the "chronic," "first admissions" and "outpatients" groups is the
same on all scales. It's as if the clinician judges a patient to
be "chronic" if the behavior of scale 1 or the behavior of scale 2
or the behavior of any other scale is extreme. However, since all
the scales are positively correlated with a general factor, it is
possible that the clinician's judgments are based on the information in the conjunction of all scales. Some clues about the nature
of the conjunction may be suggested by Table 2.

The rank order for G in this table reflects an order of scales
with respect to extremeness of alcoholism as patients themselves
tend to report it. That is, patients who report (in the AUI) a
large number of the symptoms (and thus score high on the general
factor) tend to report the withdrawal symptoms of HANG and STS, as
well as the control-loss symptoms of CTRL and COMP and the "lesser"
symptoms of worry (WOGT) and social role maladaptation (SOMA). But
patients who report an intermediate number of symptoms, and typically do not report the "extreme" symptoms of withdrawal and control
loss, will report the SOMA symptoms. The rank order for the ANOVA
F-values, on the other hand, can be regarded as a paramorphic
(Hoffman, 1960) representation of clinicians' judgments. That is,
this suggests an order of importance (of the behavior reflected by
scales) in the clinicians' judgments about extremeness of alcoholism. For example, the behavior represented by the social role
maladaptation factor seems to be weighted heavily, whereas the

TABLE 2

RANK ORDER OF THE G-LOADINGS AND F-STATISTICS
SHOWN FOR 14 PRIMARY SCALES IN TABLE 1

Rank	High	13	12	11	10	9	8	7	6	5	4	3	2	Low
G	80	78	72	70	61	56	55	52	46	33	32	22	22	09

For G HANG   CTRL   WOGT   QUNT   IMSO   HELP   DRUG
           DTS   COMP   MOOD   SOMA   IMCG   HABT   GREG

For F SOMA   COMP   QUNT   DTS   IMSO   IMCG   DRUG
           HANG   WOGT   CTRL   HELP   MOOD   HABT   GREG

| F | 59 | 50 | 50 | 42 | 34 | 24 | 23 | 15 | 14 | 13 | 9 | 5 | 2 | 1 |

TABLE 3

INTERCORRELATIONS AMONG ADJUSTMENT/ADAPTATION FACTORS

		Factor				
	1	2	3	4	5	6
A. Self Ratings						
1. Abstinence	-					
2. Control Drinking	48	-				
3. Employment/Productivity	10	10	-			
4. Decrease Sociopathy	24	04	03	-		
5. Intrapersonal Adjustment	28	23	14	26	-	
6. Social Involvement	01	26	21	(26)	(24)	-
7. Adjustment/Adaptation(by Clinician)	57	56	30	33	34	07
$r_j7.1$(Part Abstin.)	-	40	30	24	23	06
$r_j7.2$(Part Control)	41	-	30	37	26	(09)

withdrawal symptom behavior of DTS is weighted only moderately, in the clinicians' assessments. More concretely this suggests that if the clinician perceives the patient as having been jailed, charged with a driving offense, moving from town to town, not working, living alone, and separated from spouse, all as a consequence of drinking, then he/she (the clinician) is likely to regard the patient as a "chronic alcoholic" even when the patient himself reports (in the AUI) relatively few (or none) of the symptoms of perceptual withdrawal--fuzzy thinking, visual-auditory-kinaesthetic misperception, weird and frightening sensations, delirium tremens. As noted, these latter symptoms do tend to be reported by patients who report many of the other symptoms of alcoholism--i.e., are "chronic alcoholics" when so assessed by the general factor of the AUI.

Less extreme disjunctions of this kind are indicated for MOOD, HELP, and CTRL. Patients reporting relatively many symptoms tend to report those of MOOD (drink to relieve tension, forget, let down, relieve depression, etc.) and CTRL (pass out, get belligerent, blackout, stumble, weave, attempt suicide when drunk), but clinicians do not seem to give these kinds of symptoms as much weight, or use them as often, as other symptoms in their judgments about alcoholism. By contrast, for clinicians the symptoms of HELP (have used tranquilizers, disulfirim, medical help, counselors, AA, and religion to help stop drinking) seem to be regarded as important in judgments about alcoholism--relative, that is, to the symptoms of mood and relative to the order of extremeness of HELP symptoms in patients' reports.

Here, then, are two faces of alcoholism that contrast a profile sketched by clinicians (or perhaps persons generally who are external to the patient) and a profile presented by the patient under the conditions represented when he completes a written questionnaire on his own (i.e., without someone else asking the questions, looking on, etc.).

In this respect it is well to recognize a fact that perhaps is too often neglected, namely that neither of the faces depicted in these ways is necessarily the true face of alcoholism, or any closer to this than the other. As Cattell (more than any other theorist) has repeatedly emphasized (e.g., Cattell, 1950, 1957), the self-report data of a subject himself and the behavioral rating data of those observing the subject are as windows, each with its own angle and peculiar distortions, through which an investigator may peer in an effort to see the true personality inside. When the squint is to see the true profile of alcoholism it seems that one should be aware that the window provided by clinician judgments may unduly enhance elements of deviance from the social roles most valued in the dominant culture and recurrence of cries for help, but the window provided by a patient's questionnaire reports may cloud these elements and magnify others pertaining to loss of control and dependence on alcohol for effective mood changes.

It is interesting to consider how these two faces may relate to alcohol addiction. In this volume Keller has emphasized the viewpoint that very possibly when accurate diagnosis of alcohol ad- diction is made (and he has stressed the need for good diagnosis) usually the addict must attain abstinence, at least for some (as yet unknown) period of time before any form of cure (abstinence or controlled drinking) can be obtained. This viewpoint agrees with our clinical experience. When an alcohol user is truly strung out, in the sense that he uses alcohol every day if he can and reports extreme physiological discomfort (the withdrawal symptoms of the DTS scale) for at least a couple of days when he begins to abstain, then it is our judgment that he must be encouraged to maintain ab- stinence, at least over a period of two or three years. This is not to say that we expect that he will maintain abstinence over this period--usually he does not--but that it seems most therapeutic to strongly encourage it. Our practice has been to anticipate a "fall from grace" by helping the patient to work out means for getting back on the wagon should he fall off. In this respect our clinical practice appears to be consistent with the advice from learning theory provided in this volume by Wilson, Miller and others.

In these judgments our diagnoses of alcohol addiction have been based on a combination of high scores on CTRL, COMP, DTS, and QUNT (in roughly the given order of magnitude of importance in the judgment). As indicated before (Table 2) with a few notable excep- tions, diagnoses achieved in this way seem to agree fairly well with those obtained independently of the AUI (but of course based on re- ports, in interview, of many of the same symptoms) by expert clini- cians. The question we have not answered satisfactorily, however, is how close is the agreement between our AUI diagnosis and the diagnosis Keller would require to be assured that addictive alco- holism had in fact been identified. This question is not easily answered for reasons alluded to earlier, namely because the true criterion is difficult to pin down. Clearly, it is not just any expert clinician's judgments and, indeed, no clinician is the Pope in this matter, much less the God, as such. However, there is a good case for a claim that the expert clinician provides a more valid bench mark than our AUI diagnosis. In fact, however, research is needed to relate both kinds of diagnoses, together, in construct validity studies of the kind represented by Hodgson's work in this volume. In this respect, also, it would be very useful to see the two kinds of diagnoses in research such as that of Alterman (also represented in this volume).

Other Features of the Failure-of-Abstinence Profile

Looking again for the face of alcoholism represented by failure to maintain abstinence, it can be seen in the last two columns
of Table 1 that this failure is best predicted 6 months earlier by
the SOMA and HELP symptoms that tend to be emphasized in the clinicians' assessment of alcoholism and by the MOOD symptoms that come
rather high in the patient's (relative to the clinician's) scale of
severity, as well as by the psychophysiological withdrawal (HANG)
symptoms that are prominent in both assessments. Thus it seems the
ex-patient who is not abstinent is likely to be one who 6 months
earlier displayed the face of social maladaptation and much previous
floundering for help to the clinician and showed the face of dependence on alcohol for mood change to the psychometrician, while
also manifesting the extremes of hangover symptoms for all to see.
This, it should be noted, is not the face of the alcohol addict
discussed in the previous section.

The Follow-up Scales

As is often noted, abstinence is only one of several outcomes
desired from a program designed to treat drinking problems. If
such a program is to be regarded as worthwhile, surely there must
be evidence that, as a consequence of treatment, people cope better
with all of the problems of existence, including--but not confined
to--those problems that can be more or less directly linked to the
use of alcohol. If a therapy program produces abstinence accompanied by increase in some forms of maladjustment and/or maladaptation, then it might well be concluded that the program was unsuccessful. It is important, therefore, to look closely at how the
use of alcohol is and is not enabling one to cope.

To explore coping behavior following alcoholism treatment
provided by the Fort Logan Center, the research and clinical staff
put together some 55 questions, the answers to which might indicate
personal adjustment (happiness with self) and adaptation in respect
to demands of the kind that are typically placed upon an adult in
our society. These questions were given in the form of an interview
by a specially trained clinician to 202 people who three months before had passed through either an in-hospital or in-community treatment program administered by the Fort Logan facility. In addition
to gathering responses to questions, the interviewer rated the ex-
patient on a three-point scale of adjustment/adaptation in respect
to seven areas: drinking control, personal problems, marriage and
family problems, job performance, friendship-peer relationships,
community involvement, overall adjustment/adaptation. Analyses of
these data are described in detail in Foster, Horn and Wanberg
(1972). The research of particular interest in the present context

are shown in Table 3.

One finding of interest was that the interviewer ratings were positively and substantially correlated. Most of the common variance these variables had was captured in a single rater factor. The correlations of this factor with factors involving the ex-patients' responses to questions are shown in row 7 of Table 3.

A second finding of importance was that the self-reports of abstinence, together with interviewer ratings of this (presumably involving some evaluation of the truth of an ex-patient's avowal of abstinence), stood apart in a factor that was quite separate from (although correlated with) other factors indicating success. In particular the abstinence factor was distinct from a controlled drinking factor in which the responses indicated that drinking was not interfering with one's responsibilities, was no longer a problem, and so forth. This factor correlated .37 with the interviewer' assessment that drinking was under control or no drinking was occurring, and that overall adjustment was good.

The abstinence and controlled drinking factors were intercorrelated .48. They correlated (r = .56) with the factor of clinical ratings of adjustment/adaptation. It might seem, therefore, that they were providing little independent information in the prediction of this latter. In fact, however, the partial correlation for factors when the other was controlled was quite substantial, as can be seen in the last two rows of Table 3.

Thus, there is noteworthy relationship between self-reports of controlled drinking and a clinician's judgments about adjustment/ adaptation after all that can be linearly predicted in these judgments by self-reports of abstinence has been partialled out--i.e., controlled in this sense. This was found to be true, also, for three of the other factors in Table 2. After controlling for abstinence in the manner indicated, there was still significant non-zero variance in common between the clinician's judgments and self-reports of intrapersonal adjustment, employment/productivity and decrease in sociopathy.

Related here is a finding of outcome differences associated with different treatments (Wanberg, Horn & Fairchild, 1974). The in-hospital (IH) program at Fort Logan was compared with a treatment in which the patients were counseled in their home/community (IC). The IH treatment was superior to the IC in respect to most indicants of adjustment/adaptation that were obtained in the months following treatment, but these effects were recorded principally in items pertaining to social relationships and reduction of anxiety. The differences were insignificant for between-group comparisons of self-reports of whether or not alcohol had been consumed, length of time of abstinence following treatment, and days since last drink.

(However, significantly more subjects in the IC than in the IH program reported that they had drunk continuously since leaving treatment.) Again, then, our findings suggest that abstinence, as such, is not the most sensitive indicant of adjustment/adaptation produced by theory.

A number of nuances of interpretation are possible with these results. Some of these interpretations are discussed in Foster et al. (1972) and Wanberg, Horn and Fairchild (1974) and will not be repeated here. The important points for present purposes are that: (1) in self-reports, abstinence is different from controlled drinking and other indicants of adjustment and adaptation, (2) factors in addition to abstinence weigh heavily in clinicians' judgments about how well ex-patients of an alcoholism treatment program are doing, (3) factors other than abstinence appear to be more sensitive indicants of adjustment/adaptation than is abstinence. Thus abstinence appears to be neither a necessary nor a sufficient condition of successful treatment of drinking problems.

Faces of Types and Phases

Returning now to now-familiar Table 1, the discerning eye can find yet other ways to see faces of alcoholism. For example, each of the broad secondary scales in the lower part of Table 1 represents a distinct type, or perhaps stage, of alcoholism. That is, a type can be specified by demarkation along the gradations of a scale, as by defining a Bottom Level Apathetically Deteriorated (BOTM) alcoholic as any person whose score on scale D is above the 73rd percentile (top 27 percent).[6] The distinctiveness of this type in contrast to one specified by a nonoverlapping set of scores on the same scale--e.g., the lower 27 percent of scores on scale D--is assured by the largeness of the reliability of the scale (r_{xt}^2 = .87). The distinctiveness of the type in contrast to types defined in a similar way or other scales is assured by the fact that the correlation of the scale with other scales is small (relative to scale reliability). In clinical practice at the Fort Logan center, the AUI scales, particularly the secondary scales, are used in this manner.

For example, a person classified as the BOTM type in accordance with the procedure outlined here is regarded in somewhat the same way as one classified as "chronic" using guidelines similar to the well known Jellinek (1952, 1960) set. An important feature of the BOTM type is that the defining high score is obtained without counting the self reported symptoms of WOGT. These are symptoms that can be interpreted as indicating that the rationalization system is intact and active--one admits to avoiding talking about drinking, making excuses to cover up drinking, being depressed after a drinking bout, etc. They are prominent in the ACOD dimension.

High scores on the ACOD scale thus define another type, one that is
similar in several respects to the "crucial phase" alcoholic described in Jellinek's (1952) paper.

This illustrates the use of the AUI scales for one kind of
differential diagnosis. The distinction between the BOTM and ACOD
extreme-score types is that the latter involves the symptoms of
WOGT, interpreted as indicating rationalization defense of one's
drinking, but the former does not. Both are defined by many of the
same symptoms--symptoms long recognized as characteristic of problem drinking; these include, respectively, for acronyms QUNT, SOMA,
DTS, CTRL, and COMP (compulsive, loss of control in drinking relatively large amounts of alcohol, withdrawal of both a psychophysiological and perceptual kind, and social role maladaptation). But
the BOTM type is defined exclusively by these symptoms, whereas the
ACOD type involves not only the WOGT indications of rationalization, but also a broad set of symptoms (of HELP and MOOD) indicating anxious concern about many aspects of one's drinking, particularly the compulsion to use alcohol to cope with life's circumstances.

Extreme scores on the third of the second-order dimensions
among symptoms specify an OBAD type that also is, at the level of
description of symptoms, similar to a phase in the Jellinek system,
namely, the prodromal phase. This type is similar to the BOTM and
ACOD types by virtue of involving the symptoms of DTS, HANG, CTRL
and COMP, but it is different from BOTM and ACOD partly because the
amount of drinking is less and the evidences of social maladaptation
are largely lacking (QUNT and SOMA are absent from the factor), and
mainly because it involves prominent expression of belief in the
efficacy of use of alcohol (i.e., IMSO), coupled with indications
that the patient has become obsessed with drinking. Thus the OBAD
type is characterized by preoccupation with alcohol, particularly
the idea that alcohol can be used to facilitate social interaction
(help one to relax socially, make friends, relate to the opposite
sex, etc.).

It may be Procrustean to link a high-score type on the COMT
dimension with one of the concepts of the Jellinek theory. Indeed
the COMT type might not seem to be a problem drinker at all, much
less an alcoholic. The symptoms of the "core" and "concerned"
primary factors are absent in the dimension. This means that one
can obtain a high score on the factor without reporting any of the
symptoms that are usually assumed to characterize a problem drinker
or alcoholic. A high score means simply that the person has reported that drinking helps him/her in a variety of ways (socially,
mentally, etc.) and that he/she does not drink alone, but only to
be sociable in company with others. This has some resemblance to
the pre-alcoholic symptomatic phase of the Jellinek system. Also,
of course, the type does appear among persons admitted to an

alcoholism treatment facility, so for this reason it might be regarded as a form of alcoholism. In our samples of male problem drinkers, the pattern is associated with youth, thus again suggesting a link with the prealcoholic symptomatic phase (Wanberg and Horn, 1973). The fact that it is unrelated to age in our female samples may result because of the wide spread of the ages at which women in our culture enter what might be called the drinking subculture (Horn and Wanberg, 1973; Schuckit, 1972). A woman of relatively advanced age is just as likely to be in an early phase of alcoholism as a younger woman, because so many women start drinking at a late age. In any case, there is at least some resemblance between the COMT type and the prealcoholic symptomatic phase, and this type is found in samples of people who enter treatment for alcoholism.

While the types specified here are similar to the phases described in Jellinek's near-classic paper, they are not isomorphic to these latter or to any modifications (that I have seen) of the Jellinek theory (unless our work is regarded as such a modification). This is illustrated by the fact that, depending on the interpretation given to the symptoms of our types, any one of them might be regarded as most similar to, say, the crucial phase of Jellinek's system. For example, assertions that drinking has benefits can be interpreted (correctly in some cases at least) as indication of the strong system of rationalization that Jellinek described as characteristic of the crucial phase, in which case the OBAD type might be regarded as most similar to this phase rather than to the prodromal phase. Similarly, if the clinician interprets the absence of WOGT symptoms in the BOTM type as evidence of rationalization-- that not admitting to making excuses about one's drinking is rationalization--then this type might be regarded as more nearly indicative of the crucial phase than of the critical phase. Thus one should be wary about equating the faces of alcoholism seen in the Jellinek theory (and offshoots of this) with the faces that can be objectively identified with the AUI second-order dimensions.

AGE DIFFERENCES EVIDENCE PERTAINING TO PHASE THEORIES

A major problem with linking the second-order factor types with the phases of Jellinek's theory was adumbrated above in discussing the relationship of age to the COMT type. It is reasonable to suppose that if there is transition from the COMT to the OBAD to the ACOD to the BOTM types, because they represent the phases of the Jellinek theory, there then should be a systematic increase in age from COMT to ACOD to BOTM. More simply, the secondary dimensions and the primary factors of the dimensions should be correlated with age. This does not follow necessarily from the theory, of course, because it depends on assumptions that the bulk of people enter the drinking subculture at roughly the same age and progress through the stages of alcoholism at roughly the same rate, but it is a

reasonable extrapolation from the theory because these assumptions
are plausible.

The results from our research provide little support for a
COMT-to-OBAD-to-ACOD-to-BOTM phase theory that also involves these
assumptions, or indeed for any other phase theory except perhaps one
that regards COMT as preceding other types. As noted, the COMT type
is associated with youth in males (but not in females). Also the
HELP factor of the ACOD type is positively associated with age as
required by the theory. However, the WOGT factor of this type is
negatively associated with age, contrary to the theory. More important is the fact that with an N of 1884 patients distributed through
9 categories of age (no category containing fewer than 60 cases and
most containing well over 100 cases), our analyses revealed no significant age differences (i.e., nonlinear as well as linear effects)
for any of the primary factors, except the three mentioned above
(Wanberg and Horn, 1973). This result has been replicated in an
unpublished study involving a sample of 805 Fort Logan patients.
Although such evidence cannot be regarded as conclusive (for the
reasons stated above), it does not give much comfort to one who
wishes to regard our types as phases.

Because the AUI second-order factor types are so similar descriptively to Jellinek's phases (and almost all the behavior referred to in his theory is sampled in the AUI factors), this evidence on age relationships seriously questions the phase hypothesis of the Jellinek theory. As has been pointed out in another
context (Horn, 1976), phase theories are attractive for a variety
of reasons (pedagogical, theories, etc.), and so they tend to be
accepted rather more uncritically than is warranted by supportive
evidence. They are much more easily stated than tested. This
seems to be the case with the Jellinek theory. It is certain, of
course, that the outcome witnessed in BOTM is the result of a developmental process. This is true also for ACOD and OBAD and even
COMT. Moreover, our evidence on types, as well as much clinical
experience, indicates that the behavior described by Jellinek in
his phase theory certainly is descriptive of problem drinking.
Almost anyone who has done any drinking knows, for example, the
sensation of needing a drink, of rationalizing about drinking, of
feeling an impending loss of control, or of actually losing control, in drinking, etc. Moreover, it certainly is plausible, and
quite an attractive theory, to suppose that a process of alcoholism
proceeds through phases to a bottom level, from whence it may then
proceed in phases to a recovered level (as suggested in Figure 1).
But all of these observations about drinking problems and the fact
that they must develop, plus the other information given in support
of a Jellinek's theory, needn't add up to a reality that a developmental process of the kind described does in fact occur or if it
occurs that it describes any substantial number of cases of problem

drinking. Indeed the experience of our work at the Fort Logan center is that quite young people who have not been drinking for very long may present the profile of the bottom level alcoholics, our BOTM type. In contrast, some older patients who have been drinking for many, many years look for all intents and purposes to be chronic prodromal cases (and here I use chronic in the general sense of the word, not as in Jellinek's theory). Also, it seems that many BOTM types have never been ACOD types and many of these, in turn, have never been ACOD types and many of these, in turn, have never been OBAD types. In short it seems that what our analyses with respect to age suggest is verified by clinical experience; there are few cases of problem drinkers that fit a phase syndrome that is at all similar to that created by Jellinek. His descriptions of alcoholic types are reasonably accurate (and can now be objectively identified with the AUI), but his account of how these develop one into the other is probably not valid for the lion's share of alcoholics.

SCALES TO MEASURE BACKGROUND, CURRENT CONDITIONS, AND

GENERAL PERSONALITY

As with the AUI, these scales, designed to measure background, current conditions, and general personality, have been developed (using large samples) in a manner designed to ensure scale independence within a domain (i.e., satisfactory percent reliable unique variance, as described previously). The domain of the Life Situation Questionnaire (LSQ) encompasses both current conditions and background conditions, while the domain of the Personality Assessment Survey (PAS) is represented by items of the kind that appear in many personality questionnaires, the MMPI and 16PF being typical.

It should be noted in this respect, however, that unlike the 16PF, but like the MMPI, some of the items of the PAS pertain to symptoms (self-reported behaviors, beliefs, etc.) of a kind that are regarded as indicants of abnormality. This means that the scales stretch, as it were, into the abnormal reaches of personality as well as provide measurements within the normal range. More specifically, this means that while the second-order factors of extroversion and anxiety no doubt are quite similar to, perhaps colinear with, the corresponding factors in Cattell's and Eysenck's theories (e.g., Cattell, 1973; Eysenck, 1977), they provide for a more extended range of measurement of the abnormal than is provided by Cattell's 16PF or Eysenck's EPI.

Results pertaining to the development and applications of the PAS and LSQ will not be discussed in any detail here, but the scales that have come out of this work will be mentioned from time to time in descriptions of patterns of alcohol use. Thus it is desirable to have at least a glimpse at the factors of Charts 3 and 4.

Further Issues Pertaining to Development

The factors of retrospection about conditions of childhood, as indicated in the LSQ, are of particular interest in considering theories about the causes of alcoholism. Self reports cannot be taken strictly at face value, of course. But even in the absence of complete veridicality, these reports can be of use in understanding the patient and how he came to be what he is. For example, whether or not one is in fact from an economically depressed background, one's report that this is the case can be of value in indicating values, satisfaction with life, etc.

Parental Drinking Related to Problem Drinking

Braucht, Brakarsh, Follergstand and Barry (1973) reviewed a considerable body of research that pointed to the importance of parental drinking in the development of adolescent drug use generally, and alcohol use in particular. The background factor 4 identified and some of its correlates, may speak to a few of the issues in this area. The factor measure is obtained from questions about whether one's mother, father, or other relative had a drinking problem. Thus in our samples it provides a measure of an adult's assertion that a drinking problem existed in his family of origin.

The Braucht et al. review indicated that adolescent drinking typically began in the home with parents and/or relatives present, that drinking was perceived as adult behavior and that heavy drinking among adolescents and young adults was often associated with heavy drinking, or at least reported (by the young person) heavy drinking, on the part of the parents. Maddox (1970) reported that knowledge of the parents' drinking pattern is the single most accurate tool for predicting adolescent drinking behavior.

But as Braucht et al. pointed out, while Maddox may be correct this does not mean that we now know how to use this "most accurate tool." The relationship between parental drinking and child drinking, while perhaps monotonic, is not strong; there are many contradictions to a simple direct relationship. Schuckit, Gunderson, Heckman, and Kalb (1976), for example, have presented evidence indicating that non-alcohol users came from families in which there were severe alcohol-use problems (i.e., alcoholism). On the other hand, there is the evidence, as reviewed by Braucht et al. and in Amark (1951), and Schuckit (1972), for example, that alcohol-abuse occurs within families in which the parents were alcohol abusers.

It is interesting in this respect that factor 4 has near-zero correlations with most of the remaining 20 primary factors of the LSQ. The only factors with which factor 4 has correlations significantly larger than zero are those numbered 3 and 17, both indicating acting-out delinquency (Horn and Wanberg, 1970). The other correlations suggest that the factor has little in common with the other measures. It is particularly noteworthy that the factor correlates near zero with factors indicating extremes of alcohol abuse. For example, the factor correlates only .14 with factor 10, suggesting loss of control over alcohol use; the correlations with other factors (11, 12, 13, 14) indicating excess are even lower. The suggestion is that there is only a very weak link between retrospective reports about parental abuse of alcohol and the extremes of abuse in a group of people all of whom have drinking problems.

Examination of the frequency distributions for the items of scale 4 reinforce this observation. Less than 15 percent of the patients entering the facility (for treatment of alcoholism) report that their parents were problem drinkers; less than 20 percent report that anyone in their immediate family had a drinking problem.

Thus, in a sample of 1130 alcohol-abuse patients, there is very little support for an hypothesis that such patients necessarily, or even frequently, developed in families in which the members had drinking problems; and, indeed, there is little relationship between the familial and self drinking variables. At a practical level this means that one should not (very often) interpret patients' statements that their parent or parents had or did not have a drinking problem as indicative of their own alcohol-abuse problems.

In interpreting this evidence relative to that reviewed by Braucht et al., it should be realized that here the sample is of adults having some degree of drinking problem, whereas in most of the research reviewed by Braucht et al., often the samples were of adolescents just beginning to evince behaviors interpreted as indicative of problems in the use of alcohol. These sampling differences portend a number of measurement and interpretative differences that cannot be considered here, but are nevertheless extremely important.

Subcultural Differences Pertaining to Development

A major problem with developmental theories such as that of Jellinek is that they are too general. They are premature attempts to emulate the grand theories of physics or medicine, in which laws apply to all objects in the universe or to all mammals. No doubt there are all-encompassing laws that apply to the development of

drinking problems, but the knowledge base of understanding of these
phenomena is not yet sufficiently large or firmly established to
support a general theory in which one can, or should, have confidence. Moreover, the existing attempts to state the grand developmental theory have not taken sufficient account of knowledge that
does exist. For example, they have not been variegated enough;
they have not included reference to all the cultural and subcultural
and gender and historical (and other) differences that are known to
exist in problem drinking. They have not sufficiently recognized
that people drink, and have problems with their drinking for quite
different reasons.

   This is not to say, of course, that the people who have stated
developmental theories have not recognized cultural, gender, etc.,
differences. They have. Jellinek (1957), for example, has reported
on the notable differences between problem drinking in Anglo-Saxon
countries as compared with other countries, and of course Jessor
et al. (1968, 1972) have made cultural and subcultural differences a
part of the major theme of their theory. But neither of these
theories, or the theories of which they are illustrative, includes
reference to anything approximating the full range of styles of
drinking revealed by comparisons across cultures, subcultures and
gender. Indeed, existing theories have not yet taken full account
of the individual differences in drinking problems that are found
in samples that are fairly homogeneous with respect to dominant
culture (e.g., the results of the Fort Logan research).

   These points may be illustrated with results as those shown in
Table 4. These data suggest the need to step down the abstraction
ladder from the types and phases discussed in previous sections to
relatively narrow patterns.

   The table contains the means on the AUI primary and second-
order scales for different subcultural samples gathered in the
U.S.A. There are two American Indian groups, one being comprised
of individuals living in the rural setting of a reservation, the
other being a sample of Indians who had entered the Fort Logan
center from Denver, where, however, they tend to reside in close
proximity in a particular section of town and thus retain a sem-
blance of the Indian culture. The individuals of both groups are
males. The significance of difference between the means for these
two groups is indicated in column 3 of the table.

   Columns 4, 5, and 6 contain the means for groups of Spanish
surnamed (Hispano), Black and white males respectively. The week
of admission to treatment at the Fort Logan center was the sampling
unit in gathering these data. From the weeks in which the urban
Indian sample was drawn, a subset of weeks was randomly selected to
provide a sample of Hispanos of about the same size as the urban
Indian sample. All of the Blacks admitted in these weeks were

TABLE 4

SUBCULTURAL DIFFERENCES IN DRINKING STYLES

		1 Rural N=53	2 Urban N=70	3 Two Indian Signif	4 Hispano N=77	5 Black N=48	6 White N=75	7 Main Effect Signif	8 Plan Comp Signif
Daily Average Quantity Alcohol	QUNT	6.83	8.10	.01	7.78	6.10	5.20	.01	.01
Social-Role Maladaptation	SOMA	3.87	5.04	.01	4.47	3.98	3.77	.01	.01
Perceptual Withdrawal Symptoms	DTS	4.66	6.54	.01	6.87	6.67	5.79		.05
Psychophysiological Withdrawal	HANG	3.15	5.36	.01	5.06	5.52	3.97		.01
Drinking Control Loss	CTRL	6.70	8.23	.01	8.55	7.94	7.32		.05
Compulsive Drinking	COMP	2.02	3.39	.05	4.12	4.19	3.65		.05
Daily, Habitual Drinking	HABT	2.19	4.73	.05	5.21	4.92	5.21		
Drinking Worry, Guilt	WOGT	5.72	6.46	.01	7.22	6.48	6.95		.05
Sought External Help	HELP	2.17	3.07	.05	2.78	2.65	2.92		
Drink to Change Mood	MOOD	4.04	4.76	.01	5.21	5.13	5.13		.05
Drink to Improve Sociability	IMSO	4.11	4.96	.01	5.68	5.79	5.04		.05
Drink to Improve Cognition	IMCG	1.43	1.47		2.05	2.29	1.88	.05	.05
Gregarious Drinking	GREG	5.42	5.80	.01	5.30	4.65	4.12	.01	.01
Drink Provokes Marital Problems	DMRP	3.76	3.55	.05	4.51	3.96	4.15		.01
Marital Problems Provoke Drink	MRPD	3.76	3.55	.05	4.51	3.96	4.15		.01
Use Other Drugs	DRUG	0.13	1.36		1.36	1.19	0.76		
Bottom Level Apathetic Deterioration	BOTM	17.04	22.77	.01	22.09	20.54	17.48	.01	.01
Anxious Concerned Deterioration	ACDT	12.19	13.91	.01	15.71	14.54	14.95		.05
Obsessive Ambivalent Deterioration	OBAD	3.68	7.29	.01	9.09	9.02	8.77		.01
Commitment to Drinking	COMT	7.89	8.49	.01	8.64	8.08	7.27		.01

Underlines indicate outliers in the sense that a contrast of this group (or two groups) with others produces a significant difference. A double underline indicates order comparison for three groups, as A < B < C for double underline, single underline and no underline respectively.

selected and the whites were sampled randomly until this group was about the same size as the Hispano sample.

The significance test indicated in column 7 of the tab is for the main effect ANOVA in comparisons of the urban Indian, Hispano, Black and white groups (i.e., omitting the rural Indian group). The test indicated in the last column of the table is for a post hoc planned comparison in which the underline (or underlines) of the means indicates a group that is contrasted with a group comprised of the remaining subjects, the rural Indian group being omitted in all of these comparisons, however. In one test, for GREG, there was an order comparison for three groups, White < Black < the others. These tests were designed primarily to be descriptive of these data rather than to test hypotheses derived from someone's grand theory about drinking styles.

One of the interesting things these results suggest is that if only the broad types discussed in the previous section were used to describe ethnic groups several differences between the groups would be neglected. These differences could be important for under- standing the faces of alcoholism presented in different subcultures. Notice, for example, that the ordering of the ethnic groups with respect to the averages for the BOTM dimension is not a reflection of the same ordering on the primary scales of which BOTM is com- prised. Of the five groups, the urban Indians are best described as reporting the behaviors that characterize bottom level apathetic deterioration; the Hispanos and Blacks report somewhat less of this behavior, and the whites and rural Indians report the least of all. But if this kind of comparison is made with respect to COMP, one of the primary factors of BOTM, the Blacks and Hispanos would be said to report the most behavior that is characteristic of compulsive drinking, the whites and urban Indians being next, the rural In- dians reporting the least amount of such drinking. Similarly, if one were to try to understand the difference between urban and rural Indians using only the broad type, he would conclude, quite simply, that those in the urban group report more of the behavior that is characteristic of alcoholism, including a commitment to drinking (COMT), than do those in the rural group, but if he were to look at the primary scales he would learn that the latter group reports more marriage problems associated with drinking than the former group. Several other differences in the information conveyed through primary factors and that indicated by second-order factors can be found in the table.

Results such as those of Table 4 have led us rather haltingly in the direction of trying to specify separate patterns of problem drinking that at once use the information of subcultural differen- ces, the primary factors of the AUI and correlates from the PAS and LSQ as well. This attempt to specify patterns (or syndromes) il- lustrates the way in which the scale measurements of the AUI, PAS, and LSQ may be used in efforts to provide objective, repeatable definitions of the kinds of people who respond well and not so well to particular kinds of treatment.

In looking over the data of Table 4 one may note that some of the means for different groups tend to be in roughly the same order for different AUI primary scales. For example, the orders of the means for QUNT and SOMA are similar. Similarly, DTS, CTRL, and COMP hang together, as it were, in discriminating between groups. Interpretation of these patterns of means can be enhanced by considering correlates of the AUI factors of a pattern. For example, the DTS, CTRL and COMP scales correlate with anxiety in the PAS and acting-out rebelliousness in the LSQ. By considering relationships of this kind, one can discern the patterns that are roughly sketched below.

1. Alienation pattern. QUNT and SOMA provide a core for this pattern. The means for these factors are low for whites and Blacks relative to the means for urban Indians and Hispanos, the rural Indians being intermediate. The principal correlates of QUNT and SOMA indicate unemployment, low economic and social status, not married, hypersensitivity and maleness.

    The ethnic group that is at one extreme in comparison of the means, the whites, can be viewed as well integrated into the dominant culture. In the mid-range rural Indians might be regarded as well integrated into Indian culture. But urban Indians, at the other pole, are people without a culture, and thus alienated.

    Thus it seems that the face of alcoholism one can see through the QUNT and SOMA factors is that of a person who is alienated from the dominant culture or any culture within which he might be integrated. His means for maintaining a livelihood are very limited. He tends to live alone, unmarried. He drinks a great deal.

2. Rebellion pattern. The primary factors involved here are DTS, CTRL, and COMP. On these scales Hispanos and Blacks tend to score high, whites and rural Indians tend to score low. The principal correlates from the LSQ indicate childhood instability, illegal and antisocial activities, unemployment, lack of sociableness, lack of conventional interests, and resentfulness. In the PAS the correlates indicate strange, eccentric thinking, distrust of others, hypochondriasis and anxiety. As noted before, the COMP factor is associated with youth in males.

    In this pattern, then, through DTS, CTRL, and COMP, there

is view of an angry, untrusting, unsociable, anxious, rebellious, rather youthful person who tends, also, to be associated with groups that are, to some extent, in rebellion, the Hispanos and Blacks being most characteristic of this in Denver.

3. Tranquilizer pattern. Here WOGT, MOOD, and DMRP are involved. Hispanos and whites tend to score high on these factors, rural Indians low, and Blacks and urban Indians intermediate. In the PAS and LSQ the most prominent correlates indicate anxiety, feelings of inferiority and powerlessness, hypersensitive tendermindedness, and femininity. The symptoms of MOOD and WOGT represent acknowledgment of using alcohol to deal with problems of affect, coupled with worry and guilt associated with this use. The symptoms of DMRP suggest a vacant marriage or unrewarding love affair.

    Here the face is of one who is using alcohol as a medicine to deal with emotional problems and doubts about self worth. The face is likely to be that of an Hispano or white female.

4. Function facilitation pattern. On the IMSO and IMCG factors Blacks tend to be extreme, followed by Hispanos, whites, and urban and rural Indians in that order. The correlates of these AUI factors indicate shyness, lack of self-confidence, lack of decisiveness, anxiety, mistrust of others and reports of strange, unusual thoughts.

    It seems, then, that the person reporting the IMCG and IMSO beliefs is regarding alcohol as a means for overcoming inadequacies and expanding awareness. This is the face of one striving to improve his function through use of alcohol. It is likely to be a Black face.

5. Sociability pattern. The gregarious drinking pattern has tended to stand fairly solidly alone in many of our analyses. Here it seems to provide an order of the ethnic groups from whites, on the low end, then Blacks, and the Indian and Hispano groups on the high end. The correlates in the PAS are other variables indicating sociableness. The variables of the LSQ that are most prominently correlated are those of the second-order childhood instability dimension.

    The pattern in this case appears to be one associated with that kind of loneliness that comes from not having had rewarding interactions in the family of origin. Now, in the present, drinking and the conviviality of the bar are being used to make up for the past, to establish a place, as it were, in a social group wherein one can enjoy and be appreciated.

Other of the primary factors, and combinations of them, can be discussed in these ways to suggest different patterns of alcoholism. These examples are probably sufficient to illustrate the major point, however, that concepts based on our multivariate theory can be used in several ways to describe the faces of alcoholism (see also Nerviano, 1976).

It should be noted, incidentally, that rather different treatment plans can be worked out to help alter the faces that have been sketched. This will not be done here, however, in order to use the time and space to mention some very recent work being done at the Fort Logan center.

MEASURES OF USE OF DRUGS OTHER THAN ALCOHOL

Wackwitz, Foster and Deisenhaus have used the basic format of the questions of the AUI to ask also about other drugs in samples in which the patients had given indication that other drugs were in fact used. For example, the question "Do you use alcohol to help you forget?" was phrased as, "Do you use Marijuana to help you forget?" if the respondent reported that he used Marijuana. A set of questions developed to provide flexibility in this kind of inquiry is referred to as the Substance Use Inventory, SUI (Wackwitz and Foster, 1977).

One of the interesting results that has come out of analyses of the SUI is that essentially the same factor structure as has been found for the items when they pertain to alcohol also emerge when the items pertain to some 9 other drugs. This suggests that the basic concepts one needs in a theory about drinking problems are, to a considerable extent, the same as the concepts needed to describe problems associated with the use of other drugs. Moreover, these concepts can be represented operationally in very much the same way (in the AUI and SUI). This is an important finding for recent efforts to merge theory about alcohol use with theory about the use of other drugs.

The results from the Fort Logan research in this area are just now emerging and thus are highly tentative.

These results indicate that although some of the same operationally-based concepts can be used to describe the use of drugs other than alcohol, as noted in mention of the common factorial structure for the SUI, applications of the concepts lead to quite different understandings of the ways individuals who use the substances describe the functions of a drug, problems with it, etc.

The descriptions for alcohol, oddly enough, are most similar, on
the average, to the descriptions for the barbiturates. The profile
of the scale means for alcohol are least similar to the corresponding profile for the amphetamines.

Some of the major features of the data may be briefly
summarized as follows:

1) Alcohol use is described as similar to heroin use in terms
   of social maladaptation (SOMA) and mood change (MOOD).
   Both can be contrasted in this sense with Marijuana. Also, the tranquilizers are low with respect to SOMA, and
   the hallucinogens and amphetamines are not likely to be
   described as used to change mood.

2) The descriptions for alcohol are most similar to those
   for the barbiturates in respect to loss of control (CTRL),
   this being high for both, and use to improve cognition
   (IMCG), this being low for both. The two can be contrasted
   with Marijuana and the tranquilizers in respect to CTRL,
   and with the hallucinogens, cocaine and the amphetamines in
   respect to IMCG. In other words people are most apt to say
   that they use these last-mentioned drugs to improve their
   thinking, and they are least likely to use alcohol and the
   barbiturates for this purpose.

3) On the other hand, alcohol is most frequently described as
   the drug one uses to improve sociability (IMSO). It is
   most dissimilar to the hallucinogens and other opiates in
   this respect.

4) Marijuana and alcohol are similar in that relative to most
   of the other drugs, but particularly heroin, the other
   opiates and the tranquilizers, both are said to be used
   primarily under conditions in which one is socializing
   (GREG) and, relative to the inhalants and hallucinogens,
   they are reported to have relatively weak perceptual withdrawal problems (DT's).

5) In regard to the symptoms of worry and guilt associated
   with use (WOGT), alcohol is a bit on the high side, but
   heroin and amphetamine seem to generate the most of these
   kinds of symptoms, while Marijuana generates the fewest.
   Similarly, alcohol use is intermediate with respect to
   hangover (HANG), and habituation (HABT) symptoms.

6) Rather oddly, alcohol use is reported by users to occasion
   relatively few problems of compulsivity (COMP) and in this
   respect is similar to the hallucinogens and most different
   from heroin and, interestingly, the tranquilizers.

Very dimly, then, a face of alcohol use and abuse can be made out in distinction from faces for other drugs. Not surprisingly, alcohol is seen as the drug, par excellence, for facilitating socializing. It also is the drug that, relative to most other drugs, occasions difficulties in controlling use. Alcohol is not an escape drug when seen in comparison with other drugs, nor is it the drug of preference for improving thinking. In short, the problems for which one is likely to find that alcohol provides a kind of solution are those of establishing satisfying ways to come to know and relate to others.

SUMMARY

1) A considerable amount of research on the nature of problem drinking has come out of the Alcoholism Division of the Fort Logan Mental Health Center in Denver, Colorado, U.S.A. Much of this work has involved multivariate analyses of questionnaires. The sample sizes for these analyses usually have been upwards from 300 to over 2,000.

2) A principal concern in the development of scales to represent factors defined in separate domains of variables-- Background, and Current Situations, Self-Characterization and Attitudes, Alcohol-Use-Symptoms--has been to assure independence: The scale internal consistency reliability is required to be larger than the squared multiple correlation of the scale with all other scales of the domain.

3) Factor-based questionnaires developed on this basis in the Fort Logan research include the

   Alcohol Use Inventory (AUI)

   Follow-up Adjustment Questionnaire (FAQ)

   Life Situation Questionnaire (LSQ)

   Personality Assessment Survey (PAS)

   Substance Use Inventory (SUI)

   These scales, and their precursors, have been used in a number of studies designed to describe, and provide indication of effective treatment of, drinking problems.

4) Problem drinkers and those who diagnose or otherwise describe problem drinkers put somewhat different weight on the symptoms with which they identify alcoholism. In their designation of alcoholism, clinicians are likely to empha-

size social role maladaptation and the patients' indications of desire for help. Drinkers themselves are likely
to emphasize loss of control and dependence on alcohol for
mood change in their self-descriptions of alcoholism.

5) Results of several kinds suggest that abstinence is neither
   a necessary nor a sufficient condition of successful coping
   with alcohol use and its attendant problems.

6) Extreme-score drinking types can be specified by establishing a cut score that separates the upper X percent of
   scores from the 100-X below. Types thus defined are referred to as the Bottom-level Apathetic Deterioration
   (BOTM), Anxious Concerned Deterioration (ACOD), Obsessive
   Ambivalent Deterioration (OBAD), and Commitment to Drinking
   (COMT). Such types are similar, descriptively, to the
   phases discussed in Jellinek's theory and related formulations. The types do not, however, have the relationships
   with age that one can expect for phases. This evidence
   questions the phase theory of Jellinek as well as other
   similar theories.

7) Parental drinking measured by retrospection in adult sample
   is not very highly correlated with problem-drinking factors
   This causions against generalizations of findings indicatin
   that evidence of parental drinking problems provide an accurate tool for predicting offspring drinking problems when
   these offspring become adults.

8) Ethnic group differences in the factors of the AUI suggest
   that description in terms of the broad types alone (BOTM,
   ACOD, etc.) will sacrifice valuable information indicated
   at the primary factor level.

9) The averages for some of the AUI primaries tend to be ordered in the same ways for different ethnic groups. This
   information, together with results indicating relationships
   between the AUI factors and factors of the PAS and LSQ,
   suggests several distinct patterns of problem drinking:

   The Alienation pattern describes individuals whose
   drinking (large quantities) is associated with being
   "out of it" in respect to satisfactory role functions
   within the dominant culture, or, for that matter,
   even within a subculture.

   The Rebellion pattern pertains to individuals whose
   compulsive, low control drinking is associated with
   being "at war," as it were, with the dominant culture.

The <u>Tranquilizer</u> pattern describes people, often women, who use alcohol as one uses medicine to help cure the sickness of doubts about self worth, emotionality, depression, and moodiness.

The <u>Function Facilitation</u> pattern pertains to individuals whose use and abuse of alcohol is linked to their feelings of inadequacy and fears and mistrust of others.

The <u>Sociability</u> pattern indicates convivial drinking to make up for a lack of satisfying relationships with others.

10) The primary factor structure for the AUI is very similar to comparable structures obtained when the AUI questions are reworded (in the SUI) to apply to ther drugs.

11) The concepts represented (operationally) by the comparable factors of the AUI and SUI provide quite different descriptions of the functions and problems associated with use of different drugs. In particular the patient descriptions (on common factor scales) suggest that alcohol is:

Similar to heroin in respect to social role maladaptation and mood change problems.

Similar to the barbiturates in respect to loss of control problems and little use to improve thinking processes.

Similar to Marijuana in its use under socializing conditions and fewness of perceptual withdrawal problems associated with use.

Similar to the hallucinogens in fewness of compulsivity problems associated with use.

Intermediate with respect to the other drugs in worry and guilt, habituation and hangovers associated with use.

By itself as the drug used to improve ability to socialize.

REFERENCES

Amark, C. A study in alcoholism: Clinical social-psychiatric and genetic investigations. <u>Acta-Psychiat. Scand.</u>, 1951, Supp. No. <u>70</u>, 1-283.

Block, J. Advancing the psychology of personality: Paradigmatic shifts for improving the quality of research. In David Magnusson & Norman S. Endler (Eds.), Personality at the cross-roads: Current issues in interactional psychology. Hillsdale, N. J.: Lawrence Erlbaum Assoc., 1977.

Braucht, G. N., Brakarsh, D., Follingstad, D., & Berry, K. L. Deviant drug use in adolescence: A review of psychosocial correlates. Psychological Bulletin, 1973, 79, 93-106.

Cattell, R. B. Personality. New York: McGraw-Hill, 1950.

Cattell, R. B. Personality and motivation structure and measurement. Yonkers, New York: World Book, 1957.

Cattell, R. B. Personality and mood by questionnaire. San Francisco: Jossey-Bass, 1973.

Cronbach, L. J., & Snow, R. E. Aptitudes and instructional methods: Handbook for research on interaction. New York: Halstead Press, 1977.

Dawes, R. M. Fundamentals of attitude measurement. New York: Wiley, 1972.

Eysenck, H. J. Psychology is about people. New York: Penguin, 1977.

Foster, F. M., Horn, J. L., & Wanberg, K. W. Dimensions of treatment outcome. Quarterly Journal of Studies on Alcohol, 1972, 33, 1079-1098.

Glatt, M. M. Alcoholism disease concept and loss of control revisited. British Journal of Addictions, 1976, 71, 135-144.

Hoffman, Paul J. The paramorphic representation of clinical judgment. Psychological Bulletin, 1960, 57, 116-131.

Horn, J. L. The structure of intellect: Primary abilities. In R. H. Dreger (Ed.), Multivariate personality research. Baton Rouge, Louisiana: Claitor, 1972.

Horn, J. L. Human abiliites: A review of research and theory in the early 1970's. Annual Review of Psychology, 1976, 27, 437-485.

Horn, J. L., & Cattell, R. B. Vehicles, ipsatization and the multiple method measurement of motivation. Canadian Journal of Psychology, 1965, 19, 265-279.

Horn, J. L., & Wanberg, K. W. Symptom patterns related to excessive use of alcohol. Quarterly Journal of Studies on Alcohol, 1969, 30, 35-58.

Horn, J. L., & Wanberg, K. W. Dimensions of perception of background and current situation of alcoholic patients. Quarterly Journal of Studies on Alcohol, 1970, 31, 633-658.

Horn, J. L., & Wanberg, K. W. Females are different: On the diagnosis of alcoholism in women. Proceedings of the First Annual Alcoholism Conference of the National Institute on Alcohol Abuse and Alcoholism. M. E. Chafetz (Ed.). Rockville, Maryland: U. S. Department of Health, Education and Welfare, 1973.

Horn, J. L., Wanberg, K. W., & Adams, G. Diagnosis of alcoholism. Quarterly Journal of Studies on Alcohol, 1974, 35, 147-175.

Horn, J. L., Wanberg, J. W., & Appel, M. On the internal structure of the MMPI. Multivariate Behavioral Research, 1973, 8, 131-171.

Humphreys, L. G., & Dachler, H. P. Jensen's theory of intelligence. Journal of Educational Psychology, 1969, 60, 419-426, 432-433.

Jellinek, E. M. Phases of alcohol addiction. Quarterly Journal of Studies on Alcohol, 1952, 13, 673-684.

Jellinek, E. M. The world and its bottle. World Health, 1957, 10, 4-6.

Jellinek, E. M. The disease concept of alcoholism. Highland Park, New Jersey: Hillhouse, 1960.

Jessor, R., Collins, M. I., & Jessor, S. L. On becoming a drinker: Social-psychological aspects of an adolescent transition. Annals of the New York Academy of Sciences, 1972, 197, 199-213.

Jessor, R., Graves, T. D., Hansen, D. C., & Jessor, S. L. Society, personality, and deviant behavior: A study of a tri-ethnic community. New York: Holt, Rinehart & Winston, 1968.

Knox, W. J. Objective psychological measurement and alcoholism: Review of literature, 1971-72. Psychological Reports, 1976, 38, 1023-1050.

MacAndrew, C., & Geertsma, R. H. A critique of alcoholism scales derived from the M.M.P.I. Quarterly Journal of Studies of Alcohol, 1963, 25, 68-76.

Maddox, G. L. (Ed.). The domesticated drug: Drinking among colle-
    gians. New Haven, Connecticut: College and University Press,
    1970.

Miller, W. R., & Caddy, G. R. Abstinence and controlled drinking
    in the treatment of problem drinkers. Journal of Studies on
    Alcohol, 1977, 38, 986-1003.

Mischel, W. Introduction to Personality. New York: Holt, Rine-
    hart & Winston, 1971.

Nerviano, V. J. Common personality patterns among alcoholic males:
    A multivariate study. Journal of Consulting & Clinical Psycho-
    logy, 1976, 44, 104-110.

Overall, J. E., & Patrick, J. H. Unitary alcoholism factor and its
    personality correlates. Journal of Abnormal Psychology, 1972,
    79, 303-309.

Schuckit, M. The alcoholic woman: A literature review. Psychiatri
    Medicine, 1972, 3, 37-44.

Schuckit, M. A., Gunderson, E. K., Heckman, N. A., & Kalb, D. Famil
    history as a predictor of alcoholism in U. S. Navy personnel.
    Journal of Studies on Alcohol, 1976, 37, 1675-1678.

Tarter, R. E. Psychological deficit in chronic alcoholics: A re-
    view. The International Journal of Addictions, 1975, 10, 327-
    368.

Wackwitz, J. H., & Foster, F. M. Patterns of substance use assessed
    by the Substance Use Inventory. Fort Logan Research Report,
    1977.

Wanberg, K. W., & Horn, J. L. Alcoholism symptom patterns of men
    and women. Quarterly Journal of Studies on Alcohol, 1970, 31,
    40-61.

Wanberg, K. W., & Horn, J. L. Alcoholism syndromes related to
    sociological classifications. Journal of Addictions, 1973, 8,
    99-120.

Wanberg, K. W., Horn, J. L., & Fairchild, D. Hospital versus com-
    munity treatment of alcoholism problems. International Journal
    of Mental Health, 1974, 3, 160-176.

Wanberg, K. W., Horn, J. L., & Foster, M. F. A differential model
    for the diagnosis of alcoholism scales of alcohol use question-
    naire (SAUQ). In Proceedings of the 20th Annual Meeting of the
    Alcohol and Drug Problems Association of North America. U. S.
    Government Printing Office, Washington, D. C., 1973.

Wanberg, K. W., Horn, J. L., & Foster, F. M. A differential assessment model for alcoholism: The Scales of the Alcohol Use
Inventory. Journal of Studies on Alcohol, 1977, 38, 512-543.

FOOTNOTES

1) Principal financial support for this research has come from
U.S.P.H.S., N.I.A.A.A. Grant RO1 AA 00221-01-07 and U.S.P.H.S.,
N.I.D.A. Grant 1 H81 DA 1701-01-03.

2) While some aspects of this assertion are controversial, the
basic claim that there are several kinds of problem drinkers
and drinking problems is really not in dispute. Some of the
basic facts in support of this claim are that different people
treated as problem drinkers:

a) manifest quite different symptoms, have quite different
backgrounds of relevance for understanding etiology, and
progress in quite different ways, at different rates, in
developing their drinking problem;

b) respond quite differently to what can be recognized as very
similar therapeutic efforts;

c) make quite different adjustments and adaptations to psychological and social conditions;

d) react quite differently to alcohol, as such, and when alcohol
is ingested with other agents.

Many studies provide the basis of support for these assertions.
Miller and Caddy (1977) provide a concise but comprehensive review of these studies; other reviews focused on this theme have
been given as prologues to some of our reports (Horn and Horn,
1970, 1973; and Wanberg, Horn and Foster, 1973, 1977).

3) Stating it quite simply, Thurstone's metatheory is that in a
study that is well designed to reveal simple structure, a factor,
indicating an influence that is manifested in somewhat different
ways in different variables, will affect only a relatively small
number of the samples variables and, conversely, only a relatively small number of factors will affect any one variable.

4) Josephine Wright of the clinical staff of the Alcoholism Division was very prominently involved in the development of the
early forms of the alcohol-use questionnaires of this research.

5) AUI materials (the questionnaire, preliminary norms, some rele-
    vancy data) are being distributed for public use by a not-for-
    profit organization called the Center for Alcohol-Abuse Research
    and Evaluation (CAARE), P. O. Box 26528, Denver, Colorado,
    U.S.A., 80226.

6) This specifies transivity types, rather than, say, profile types
    or reticular types, because the classification (BOTM alcoholics
    versus not BOTM alcoholics) is made on the basis of only one
    dimension in a manner that retains an order for the type classi-
    fication that is monotonic with the order for scores on the
    scale itself.

ETIOLOGY OF ALCOHOLISM: INTERDISCIPLINARY INTEGRATION

Ralph E. Tarter

Carrier Clinic Foundation, Belle Mead, N. J.

An understanding of etiological factors in alcoholism has remained elusive despite intensive study by researchers in the biological and social sciences over the past several decades. Reviews
of advances in disciplines as disparate as genetics (Cadoret, 1976;
Goodwin, 1976), epidemiology (Cahalan and Cisin, 1976), biochemistry
(Walsh, 1973), cultural anthropology (Bacon, 1973; Stivers, 1976)
and psychology (Hoffman, 1976) have been unable to specify the necessary conditions antecedent to this disorder.

While each discipline has contributed important information on
the factors that predispose a person to drink abusively, the necessary condition for becoming alcoholic has yet to be defined. Part
of the difficulty in delineating specific causes for alcoholism
may be, as pointed out by Tarter and Schneider (1976), due to the
heterogeneous nature of the alcoholic population. Thus, a search
for unidimensional etiological mechanisms is most probably an
oversimplification of what is probably a complex of disorders of
multiple causality. Given the variability of symptom manifestation
in alcoholics (Wanberg, Horn and Foster, 1974) and their sequence
appearance in the addictive process (Jellinek, 1952), it would appear that one heuristic approach would be a multidisciplinary research program aimed at defining homogeneous subgroupings with common etiological and process variables in the drinking population.

It is with this orientation that the author has attempted
during the past several years to differentiate alcoholic subtypes
on the assumption that such information could then lead to both
more accurate diagnosis and improved forms of therapeutic intervention. Beginning with relatively simple notions about personality
typologies as related to age of alcoholism onset (Tarter, 1975), the

research has progressed to more behavioral studies directed at
delineating neuropsychological mechanisms as antecedent factors in
alcoholism. Recent investigations have been directed toward interrelating present drinking style, childhood behavior characteristics
and psychosocial development. Initial findings have led to the hypothesis that childhood hyperactivity or minimal brain dysfunction
is, at the behavioral level of discourse, a putative characteristic
of future adult primary alcoholism. Culling the genetic, longitudinal, psychophysiological and psychological literature provides
encouraging support for this hypothesis which in turn seems to offer
a viable theoretical and empirical basis for defining one subtype of
alcoholism. The remainder of this chapter will review the evidence
as well as speculate on how these findings can both be applied to
future research and help explain certain alcoholic symptomatology
including craving and ethanol analgesia. By working toward an integrative conceptualization across levels of empirical analysis, it
is hoped that a heuristic framework for research can eventually be
developed to identify specific subgroups of alcoholics so that more
rationally based treatment can be implemented.

STATE OF THE ORGANISM

Consuming alcohol far beyond that required for sedation, mood
alteration and social facilitation would tend to suggest that organismic factors of a markedly compelling nature are operative in
the inception and persistence of a drinking bout. While environmental cues and contingency factors play a crucial role in alcohol
usage, a comprehensive analysis of drinking also must take into
consideration the effects of alcohol on organismic variables. The
mode of interaction between a pharmacological agent and ongoing organismic state is of critical importance therapeutically, but its
study has also proved valuable in furthering an understanding of
target physiological systems by the known actions of such psychoactive drugs (Irwin, 1968). A similar approach can be taken in alcoholism research, namely, to determine alcohol's effects on the
organism, taking into consideration ongoing states, so as to elu-
cidate possible mechanisms and desired effects by the individual.
Alcohol, as a chemical, does not affect behavior directly but rather
interacts with other chemicals at a cellular level to produce
changes in tissue, organ and systemic functioning. The resulting
alteration in physiological state then determines the limits and
manner by which the individual copes and interacts with his environment. Thus the mode of action of ethanol at the biochemical and
physiological levels may provide useful information about its possible dissimilar effects in alcoholics and nonalcoholics that may
result in differential behavioral patterns of consumption. The
issue is raised as to whether alcoholics are unique in their reasons
for selection of and response to alcohol and hence consume this
substance in quantities far exceeding social normative amounts in an

effort to modify or manipulate a target system (or systems). Their consumption, therefore, would ultimately have behavioral consequences that alter and possibly increase coping capabilities. Such an interpretation of alcohol usage in alcoholics has an obvious teleological flavor since it presupposes that drinking serves at least a potentially adaptive purpose. As will be elaborated upon more fully later, there is some indication that this may indeed be the case for some alcoholics.

However, before proceeding further, a conceptual framework is proposed so as to allow the reader to establish a context or perspective for the position to be outlined. If, as pointed out earlier, the organismic state is an important determinant of alcohol consumption, then it remains to be elucidated as to what target physiological system or systems the person seeks to modify and toward what altered state the individual is striving. Irwin (1968) summarized some of the more salient organismic conditions from psychopharmacological studies. All of the factors presented in Table I that contribute to organismic state have been suggested as reasons for consumption at one time or another.

A person who voluntarily takes a drug often does so with the intention of altering one or more of the organismic variables described in Table I. For example, a person may ingest amphetamines for a variety of reasons: to produce a euphoric effect of well-being, to stay awake or study for an examination, for increased endurance during a sporting event and so on. The point to be made is that the motivation for and effect sought can vary between individuals and serve diverse functions in a given person in different situations.

This fact is of paramount importance in the study of alcoholism. First, it emphasizes multiple causality in the onset of the addictive process. From a clinical standpoint one often hears that alcohol dependency originally began for such varied purposes as combating insomnia, for alleviating anxiety and depression, for social facilitation and so forth. While the processes and eventual consequences of addiction may bear many similarities between individuals, the above examples serve to illustrate how a consideration of organismic states precludes unidimensional etiological mechanisms as explanations for all alcoholics. A second point is that, in addition to motivational needs and desired altered organismic states that alcohol can induce, another factor for alcohol use may be intrinsic physiological disturbance that is rectified by alcohol. Thus a prealcoholic person who is defective in one aspect of physiological functioning may seek alcohol with essentially medicinal intent, much as a person with adrenocortical insufficiency craves salt (Welkens and Richter, 1940). Thus, the study of excessive alcohol consumption must incorporate into a comprehensive explanatory system an understanding of the state of the organism which under

TABLE I

<u>State of the Organism Variables</u>[1]

Wakefulness

Arousal

Activity

Endurance

Bisocial drives

Set

    attitudes

    expectations

Responsiveness to stimuli

    sensory-motor

    affect

Information processing capacity

Physiological functioning

    autonomic

    neurological

    endocrine

[1]Adapted from Irwin, 1968

appropriate environmental circumstances (cues and contingencies) leads to alcohol use by an individual for its perceived and/or real effects in modifying coping capacity through its mediating action on target physiological systems.

ETIOLOGY OF ALCOHOLISM

From the preceding discussion it is apparent that the motivation for using a pharmacological substance can vary between individuals and that a given chemical may also possess unique effects for certain individuals whose organismic state dispositionally distinguishes them from others.

Of the almost limitless number of organismic variables that can lead to physiological and behavioral disruption, the concept of disturbed arousal has probably received the most attention. Disturbances in arousal mechanisms have been implicated in a variety of psychopathological conditions including hyperactivity (Zentall, 1975), schizophrenia (Venables, 1977), and sociopathy (Quay, 1965). Hyperactivity has been theorized to be an etiological factor in children who, as adults, are at risk for hysteria (Briquet's Syndrome) if female and sociopathy if male (Guze, 1975). The question is whether alcoholism, classified as a personality disturbance like hysteria and anti-social personality in the Diagnostic and Statistical Manual of Mental Disorders (1968) and often shown to be related to sociopathy, may also have a hyperactivity syndrome as an etiological substrate. The tentative hypothesis advanced is that within the population of alcoholics there exists a subgroup for whom symptoms of hyperactivity and minimal brain dysfunction (HK/MBD) were premorbidly extant. Such persons are at risk for primary as opposed to affective disorder alcoholism (Winokur, Rimmer and Reich, 1971). The evidence implicating a premorbid hyperactivity disorder for primary alcoholism is reviewed below.

Genetics

In a comprehensive review of the literature, Goodwin (1976) concluded that there is strong evidence to implicate heritability as a major factor in alcoholism. What exactly is genetically transmitted that is premorbidly manifest is still uncertain but the concept of hyperactivity has been revealed to be one possible variable in several studies. Morrison and Stewart (1973) found an increased incidence of alcoholism and sociopathy in the parents of hyperactive children. Similar results were obtained by Cantwell (1972) who observed that hyperactive children more frequently than normal controls had sociopathic fathers and hysterical mothers. Cadoret, Cunningham, Loftus and Edwards (1975) also found an association between sociopathic male parentage and hyperactivity in the male offspring. Robins, Bates and O'Neal (1962) reported that antisocial fathers were more likely to have children who became alcoholic. These studies suggest that an interrelationship among sociopathy, alcoholism, hysteria and hyperactivity in offspring may exist.

Because the rearing environment was not controlled for in these studies, factors other than genetic may be responsible for the hyperactivity. Several investigations have been conducted to examine the role of environment. Morrison and Stewart (1973) found that the association between hyperactivity and parental alcoholism and sociopathy held only if the parent was biological and not adoptive.

Goodwin et al. (1975), in a study of Danish adoptees, found a significantly higher prevalence of hyperactivity in children who later became alcoholic than nonalcoholics even though they had been reared apart from their biological parents.

While not conclusive, family and adoption studies of alcoholics point to alcoholism's heritability and also illustrate its association with other clinical disorders such as sociopathy and hysteria. Moreover, the rearing environment does not appear to play a vital role, at least as far as the severe alcoholic is concerned (Amark, 1951; Goodwin, 1976). What is inherited may be phenotypically expressed as a hyperactivity disorder.

If hyperactivity is the genetic given as a risk factor in alcoholism, then one might ask: what happens to such children grown up?

Longitudinal Studies

If alcoholism is inherited via hyperactivity, then one would expect that such children, after reaching adulthood, would more likely exhibit drinking problems than children without a hyperactivity disorder. Several studies have been conducted which do suggest that hyperactive children are at greater risk for alcoholism and for sociopathy. Mendelson, Johnson and Stewart (1971) observed that 15% of hyperactive children were abusing alcohol as teenagers. McCord and McCord (1962) found, among other traits, hyperactivity to be more prevalent in children who became alcoholic. Jones (1968) reported that future problem drinkers were rebellious, impulsive, undercontrolled and attempted more testing of authority than their peers, characteristics which have all been identified with HK/MBD.

Retrospective studies have also found hyperactive children to be at greater risk for alcoholism. Robins et al. (1962) compared future alcoholics with problem children who did not become alcoholic and found more "acting out" and more anti-social behavior in the former group.

A link between alcoholism and childhood HK/MBD was observed by Tarter, McBride, Buonpane and Schneider (1977), who requested alcoholics to retrospectively endorse items characteristic of this syndrome that were applicable to them prior to age 12. They found that severe alcoholics reported more than twice as many such symptoms as less severe alcoholics who, in turn, endorsed virtually the same number of symptoms as normals. From this study the authors concluded that, from the general population of alcoholics, there can be identified a subgroup with HK/MBD symptoms in childhood who mature to develop a more severe form of alcoholism than others, as indicated by drinking related characteristics (such as DT's, loss of control and absence of precipitating causes). These results were corroborated by Gomberg (in preparation) who found that young male alcoholics were reported twice as frequently as older alcoholics to have been overactive as children. In a follow-up study, Tarter, Perley, and Sansom (in preparation) found that alcoholics who are classified as essential alcoholics (i.e., lacking clear reasons for drinking onset and exhibiting a history of personal and social inadequacy) reported over twice as many HK/MBD symptoms as reactive alcoholics (persons who become alcohol dependent as a result of a life crisis but otherwise present a well-adjusted premorbid picture). The possibility was raised in this study that HK/MBD may be antecedent to a history of social incompetency, with accompanying psychological and interpersonal ramification, leading in turn to alcohol use for both its physiological properties and to satisfy (albeit maladaptively) social and emotional needs as well. McClelland, Davis, Kalin, and Warner (1972), in an extensive study of drinking motivation, reported that alcohol may satisfy power strivings which, if applicable to the clinical population of severe drinkers, may provide an integrative link by relating neurobehavioral dysfunction of HK/MBD and social incompetency to motivation for alcohol excess.

In support of this position are several documented studies which show that hyperactivity sequelae persist into adulthood. In terms of the basic neuropsychological disturbance, Mann and Greenspan (1976) have identified and proposed treatment for a syndrome they refer to as adult brain dysfunction. They assert that MBD children mature into the adult brain dysfunction syndrome with such characteristics as short attention span, impulsivity, low self-esteem, interpersonal difficulties, anxiety and depression. The disorder is frequently found in conjunction with alcoholism, drug abuse and characterological disorders like sociopathy, explosive personality, hysteria and impulsive personality. In another study, Quitkin and Klein (1969) have reported that a history of MBD in childhood places the child at risk in adulthood for either impulse problems as part of an emotionally unstable character disorder (EUCD) or, alternatively, schizophrenia withdrawal.

Additional evidence for the persistence of hyperactivity symptoms into adulthood has been provided by at least two other studies. Mendelson, Johnson and Stewart (1971), in a 2-5 year follow-up, found that 71% of diagnosed hyperactive children were still overactive, 74% were impulsive, and 77% showed concentration problems. In another study, Wood, Reimherr, Wender and Johnson (1976) found that hyperactive children manifested a variety of problems as adults. All of the adults in their sample reported restlessness; 87% were anxious and emotionally overreactive; 80% were moody and short-tempered; 67% were impulsive; and 50% were immature. It was also observed that 27% exhibited drinking problems and a similar percentage were sociopathic. Weiss, Minde, Werry, Douglas, and

Nemeth (1974) conducted a longitudinal study of hyperactive children and found that, while motor disturbances subsided in adulthood, other symptoms persisted. The children matured to be distractible, restless and rebellious. One quarter of the children began to manifest anti-social behavior; 30% revealed a paucity of friends; and 15% were referred to the courts. Thus, a variety of behavioral and social problems can be found in persons with a history of childhood hyperactivity.

Concomitant clinical and interpersonal problems also persist, supporting the findings of Tarter et al. (in preparation). Stewart, Mendelson and Johnson (1973) interviewed hyperactive children when they were adolescents and observed that over 50% described themselves as lacking ambition, feeling sad and having low self-esteem.

From these studies, it can be safely concluded that HK/MBD is still experienced into adulthood as cognitive and behavioral symptoms although the gross motoric disturbances may be diminished somewhat. The literature also indicates that alcohol abuse is a frequen sequela and that, of the social and psychological problems that evolve into adulthood, the most frequent manifestations are in the form of sociopathy, hysteria, and alcoholism.

Physiological Functioning

If hyperactivity is a common etiological factor in both adult sociopathy and primary alcoholism, then one might expect similarities in physiological functioning between the two. Straightforward comparisons between hyperactive children, sociopaths and alcoholics are, however, not possible because of obvious problems in matching subjects. Furthermore, the gross disturbances in hyperactive children tend to diminish as they mature, even though many of the characteristics persist in type but to a lesser degree into adulthood (Weiss et al., 1974; Wood et al., 1976). In addition, one can never be certain that the physiological characteristics observed in alcoholics preceded their abusive drinking and are not the result of ethanol's effects on the central nervous system. Finally, any comparison between children and adults assumes a certain invariance or stability of functioning through maturation. Until longitudinal studies are conducted, inferences must necessarily be drawn from findings obtained from disparate groups, all the while bearing in mind the tentativeness of conclusions.

The research literature on hyperactivity tends to implicate a deficit in arousal mechanisms. Zentall (1975), in reviewing the field, concluded that hyperactive children are underaroused and consequently self-generate behavior to increase stimulus input to more optimal levels. Psychophysiological studies tend to confirm this general conclusion although some discrepancies in results have been recorded. Cohen and Douglas (1972) observed that hyperactive children were underaroused as measured by resting electrodermal levels. Zahn, Abate, Little and Wender (1975) found no differences in basal arousal but did observe that hyperactive children were autonomically less reactive and, in contrast to normal children, required a higher level of induced arousal for effective performance. Probably the reason hyperactive children bore quickly and exhibit attentional deficits is because of their inability to sustain stable high levels of arousal (Douglas, 1972). Their disturbance in maintaining adequate levels of arousal may also explain the failure of such children to integrate behavior to the ongoing situation in meeting social demands (Sprague, Barnes and Werry, 1970). To function adaptively, the hyperactive child therefore requires a higher state of arousal which can be achieved by an environment with challenging and variable stimulation (O'Malley and Eisenberg, 1973). Otherwise boredom or a state of functional sensory deprivation becomes manifest, to which the child responds by self-stimulating and stimulus-seeking behavior that globally appears as hyperactivity and eventually leads to personal and interpersonal problems in adjustment.

The physiological picture in psychopathy generally conforms to the same pattern. Overall, however, a defect in arousal mechanisms has been theorized to be extant in sociopaths (for a review, see Hare, 1970). Quay (1965) speculated that either basal reactivity is impaired or faster adaptation to stimuli takes place in psychopaths, thereby requiring them to engage in stimulus-seeking behavior. Schacter (1971) reported that the injection of adrenalin, a powerful arousal incrementor, had a relatively greater activating effect on psychopaths than on normals. Not only does it abolish an avoidance learning deficit but it also produced higher pulse rates than in normals. Low anxiety, often thought to be a prime feature of psychopathy (Lykken, 1957; Cleckley, 1964), is often reflected in greater autonomic reactivity during a stressful task (Dykman, Ackerman, Galbrecht and Reese, 1963). Overtly unemotional persons tend to be the most reactive on the GSR (Jones, 1950); and it has also been shown by Valins (1967) that subjects rated high on sociopathy tended to show the greatest cardiac reactivity.

This research suggests that psychopaths may be deficient in modulating arousal level but when injected with the natural, powerful stimulating chemical adrenalin, are able to function adaptively. Furthermore, there is some tentative evidence to suggest that the psychopath is overly reactive, particularly during stress situations, even though outwardly he appears as calm or unemotional (Schacter, 1971). Mawson and Mawson (1977), in a comprehensive theoretical and research review, point out that the essential defect in psychopaths is their variability in both maintaining adequate basal arousal levels and modulating reactivity. This variability is hypothesized to be due to abnormal oscillations in neuro-

transmitter balance. As a result of this defect, there is increased
variability in transition from one arousal state to another (e.g.,
sleep to waking), creating disequilibrium between activating (noradrenergic and dopaminergic) and quiescent (serotonergic and cholin-
ergic) systems. This intrinsic disorder, it is hypothesized, may
be tied to a fundamental disturbance in biological rhythms (e.g.,
circadian) and may explain inconsistency in psychopathic behavior,
inasmuch as level of arousal and reactivity being essentially un-
predictable preclude stable adjustment in meeting social and behavioral demands. Also characteristic of this problem are hyperactive
children; Mawson and Mawson (1977) elaborate upon how this rhythmic
disturbance in neurochemistry may apply to this latter disorder as
well as psychopathy.

Schacter (1971) has provided a psychological explanation for
psychopathic behavior which concurs with the neurochemical position
of Mawson and Mawson (1977) and is also in agreement with the psycho
physiological research. He theorizes that this increased reactivity
and concomitant hypoarousal makes it difficult - if not impossible -
for the psychopath to learn to apply cognitive labels to different
physiological states. As a result there is a failure to learn, experience and differentiate specific emotions. Only after a marked
elevation in arousal is induced so that it is sufficiently discriminable from the usual variable indiscriminate state (e.g., by injection of adrenalin) can the individual respond affectively and
adaptively. Therefore, the essential defect, according to Schacter
(1971), is an inherent incapacity to cognitively distinguish between
excessively reactive physiological states, leading to an absence in
developing appropriate emotional responses for mediating behavior.

Psychophysiological studies of alcoholics have not been as
conclusive, probably because of the heterogeneity of the alcoholic
population studies and also because of a lack of consistency in the
measures employed across investigations. In an early investigation,
Wenger (1948) studied seven alcoholics to derive the autonomic balance score, an index of relative dominance between the sympathetic
and parasympathetic aspects of the nervous system. He found that
baseline or resting levels of this sample were no different from
normals. In what has frequently been viewed as a landmark study in
the area, Kissin, Schenker and Schenker (1959) observed "dried-out"
alcoholics to have increased parasympathetic and decreased sympathetic activity. The alcoholics, on comparison with normals, showed
lower diastolic cold pressor response, greater percentage drop in
the intravenous glucose tolerance curve, and higher sodium concentrations in saliva. They also yielded diminished urinary excretion
of 17-hydroxycorticoids and 17-ketosteroids. These results were
interpreted as reflecting reduced activation levels in the alcoholics.

Another study suggesting disturbances in physiological functioning was reported by Chotlos and Goldstein (1967). They found

that alcoholics had higher heart rate than schizophrenics, hyper-
tensives and normals as well as the highest skin resistance.
Chandler, Parsons and Vega (1975) measured the skin conductance and
heart rate responsivity of alcoholics while at rest and during a
demanding dichotic listening and memory task. No resting level
differences were noted between alcoholics and controls on these
measures, but, during the dichotic task, the alcoholics showed much
more heart rate variability.

   Clues to why alcoholics drink have been obtained by investigating the effects of ethanol on psychophysiological functioning by
alcoholics. In the study by Kissin et al. (1959), in which it was
found that their alcoholic subjects were overactive in PNS and
underactive in SNS responding, the alcoholics were also administered
a dose of alcohol which had a "normalizing" effect on physiological
functioning. Parasympathetic activity declined and less variability
on the Funkenstein mecholyl test was noted. If one can extrapolate
these findings to drinking motivation, it is possible that consumption for the alcoholic may represent an attempt to boost arousal
and simultaneously reduce response variability. Evidence in support of this position has been supplied by Garfield and McBrearty
(1970) who discovered in their alcoholic sample that, upon presenting neutral and stress inducing photographs (mutilated bodies) before and after alcohol ingestion, greater arousal to both types of
stimuli as measured by skin conductance occurred with subjects
under the influence of alcohol but, at the same time, reactivity to
both the neutral and stressful stimuli decreased. These authors
conclude that alcohol acts to increase anxiety (arousal) in alcoholics as inferred from skin conductance changes. This latter finding
is congruent with observations of Nathan, O'Brien and Norton (1971)
who studied alcoholics' drinking in a laboratory setting.

   The finding that alcohol functions dually to increase arousal
and decrease reactivity is intriguing, for within Schacter's hypothesis it illustrates how inebriation creates an organismic state
which is quite discriminable from other (sober) states which, accordingly, allows the individual to apply different cognitive labels. The nature of these changes indicates that they may even be
adaptive, at least in the short run. Alcoholics' failure to discriminate internal cues is theorized to be due to sustained physiological variability and reactivity, preventing the learning of such
cues. That alcoholics cannot discriminate blood alcohol levels as
well as normals may also be tied to this factor of organismic
variability; it will be discussed more fully later.

   In an important study, Coopersmith and Woodrow (1967) obtained
basal skin conductance and galvanic skin response recordings in
alcoholics and nonalcoholics. They reported no differences in basal arousal but that the alcoholics showed greater responsivity to
both positive and negative stimulation. Their data indicated that

the defect in alcoholics was one of incapacity at modulating responses, not simply disturbed basal levels of arousal. In interpreting
their results, the authors present a psychophysiological profile of
the alcoholic that is remarkable in its similarity to Schacter's
(1971) views of the psychopath. In essence, Coopersmith and Woodrow (1967) theorized that the alcoholic is not disturbed in basal
arousal level but, when stressed or stimulated, he becomes both
verbally and physiologically more responsive than normals. To the
extent that he cannot distinguish between neutral and affective
stimuli, he responds maximally to all impinging stimuli - which
leads to even further distressing reactivity. If Garfield and
McBrearty's (1970) results can be applied to these findings, then
one may hypothesize that alcohol consumption for the alcoholic serve
to reduce reactivity, especially under conditions of stress or stimu
lation, thereby allowing for greater discrimination between physiological states and hence potentially better adaptation.

Thus, there does emerge a consistent physiological pattern in
alcoholics, sociopaths and hyperactive children. Accordingly, it is
not surprising that alcohol abuse emerges as a potential problem for
hyperactive children (Mendelson et al., 1971) and that alcohol is
the most frequent drug of abuse in psychopathy (Cleckley, 1964).
While these facts need not necessarily imply a common mechanism in
these disparate disorders, it does conform to previously cited data
indicating a fundamental (possibly genetically determined) interrelationship among them. Moreover, as will be elaborated upon more
fully later, the type and patterning of psychophysiological functioning in alcoholics may have a direct bearing on the mechanisms
of their behavioral impairment in discriminating internal bodily
cues in order to regulate level of intoxication.

Physiological functioning will, as pointed out in the introduction, ultimately affect the scope and style of behavioral adaptation. If, as postulated, hyperactivity is the essential etiological factor in primary alcoholism, then one would expect certain
similarities between hyperactives and alcoholics in processing of
stimulus input. In an experiment designed to test the hypothesis
that alcoholics, like hyperactives, strive to increase stimulus
input, Tarter and Novick (in preparation) administered to a group
of primary and secondary alcoholics the Petrie Perceptual Reactance
Test (Petrie, 1967). This instrument measures the subjective tendencies of a person to either increase or diminish kinesthetic sensory input. It was found, as theorized, that primary alcoholics
(who also retrospectively reported more childhood hyperactivity and
MBD symptoms) augmented sensory input, while secondary alcoholics
were more frequently classified as moderates or reducers insofar
as modulating their sensory experience was concerned. These results
were interpreted to reflect a need by the primary alcoholics to enhance stimulation in an effort to increase arousal level to more
optimal levels. The demand to increase stimulus input may also

serve another coping need, namely, to organize behavior on the basis of external cues inasmuch as the alcoholic cannot do so on the basis of internal cues derived from subtle varying physiological states (Lansky, Nathan and Lawson, in press). When the demand for stimulation is not met, creating a state of stress (Coopersmith and Woodrow, 1967), the alcoholic resorts to alcohol consumption to increase arousal and simultaneously induce a differential physiological state.

The finding that ethanol can increase arousal was previously described in the psychophysiological studies summarized above. Additional results reported by Docter, Naitoh and Smith (1966) also indicate that, in alcoholics, ethanol is activating as measured by heart rate, EMG, rapid eye movement and finger pulse volume. And of great importance was the beneficial value of alcohol in their sample, namely, the increased performance of alcoholics on a vigilance task while intoxicated. This observation suggests that increased arousal and attentiveness may take place in alcoholics while inebriated, thus serving to enhance performance.

In conclusion, psychophysiological studies of alcoholics, while far from definitive, nonetheless implicate a pattern of responsivity different from that of nonalcoholic normals. The pattern of functioning, to a large extent, resembles that seen in hyperactive children and sociopathic adults. Although methodological limitations and scarcity of studies mitigate against a definitive conclusion, there is an accumulating body of evidence which supports the hypothesis of defective arousal mechanisms in alcoholics (or at least one subtype of alcoholic) similar to that seen in hyperactives and sociopaths. The disturbance seems to be one of excessive physiological reactivity which, as posited by Mawson and Mawson (1977), is tied to a defect in neurotransmitter regulation of biological rhythms. Thus, while basal arousal is not necessarily disturbed, the failure to maintain arousal stability creates a pattern of excessive physiological reactivity which, in turn, is responsible for the alcoholic's failure to distinguish interoceptive cues. Difficulties in applying differential cognitive labels to subtle variations in physiological states can result ultimately in impairment in discriminating intoxication levels and in the need for more alcohol than normals to induce a discriminable state. Alcohol functions to bolster autonomic arousal and, at the same time, reduce reactivity. In the absence of extrinsic stimulation, it can serve, therefore, to augment arousal and, as demonstrated in at least one experiment, enhance vigilance performance where sustained attention is required. Thus, in the short-term at least, alcohol may have some beneficial and adaptive value.

Response to Drugs

One method by which organismic disturbances can be investigated
is to determine the mode of drug action on neurochemical processes
which, in turn, can help in the understanding of physiological and
psychological processes. From such pharmacological research it may
then be possible to deduce the source of organismic disturbance
(see variables cited in Table 1) which can then be rectified by appropriate pharmacotherapy.

With respect to the specific problem of hyperactivity, it has
long been known that stimulant type drugs exert a "paradoxical"
calming effect in many such children (Bradley and Bowers, 1941).
Clinical trials of methylphenidate (Ritalin) are routinely introduced to diagnose hyperactivity and, where positive responses occur,
dramatic behavioral improvements at home and school are frequently
observed (Gittleman-Klein, Klein, Katz, Saraf and Pollack, 1976).
Ritalin has been shown to improve self-control and behavioral integration as well as to reduce hyperactive behavior (Conners, 1972).

The therapeutic action of amphetamine-like drugs, while not
entirely known, has been hypothesized as due to the arousal-augment-
ing effects of such drugs (Zentall and Zentall, 1976). Prichep,
Sutton and Hakarem (1976) reported also that methylphenidate had a
normalizing effect in hyperactive children by increasing arousal
and simultaneously reducing deficits in attention.

Of the other drugs that have been utilized for this purpose,
only imipramine (a tricyclic anti-depressant) has met with comparable success. Several studies have shown that, for certain hyperactive children, imipramine is even more effective than stimulant
drugs (Gross, 1973). Although not as extensively used as the stim-
ulant-type drugs, there is substantial evidence to indicate a positive therapeutic effect of imipramine in ameliorating hyperactivity
symptomatology (Gualtieri, 1977).

The question is raised as to the effects of these two distinctl
different drugs, one a stimulant and the other an anti-depressant, c
the central nervous system. Undoubtedly the neurochemical mode of
action is complex and multifaceted, but one factor which stands out
is their common action on brain catecholamines; specifically, both
drugs enhance noradrenergic transmission by increasing norepinephrir
availability in the synaptic cleft. These findings conform to the
catecholamine hypothesis of hyperactivity and minimal brain dysfunction proposed by Wender (1971). Considering the neuroanatomical
representation of this system - the medial forebrain bundle tract
reaching anterior to the forebrain and posterior to the hypothala-
mus and brainstem - it is not surprising that its ultimate organismi
influence would be varied and critically affect such diverse aspects
of physiological and behavioral functioning as attention, goal-di-

rected behavior, arousal, primary drives, reward mechanisms, aggression and biological rhythms, to mention but a few processes.

Thus, there is some indication to implicate a catecholamine disturbance theory of hyperactivity. As pointed out above, such a neurochemical disorder would have a far-reaching effect on the physiological and behavioral functioning of the organism. The focus on noradrenergic mechanisms, specifically norepinephrine activity, as the source of disorder in hyperactivity is probably an oversimplification of complex dynamic relationships existing among all neurochemical systems. Nonetheless, it emphasizes how chemistry can be integrated with physiology and behavior and, in the specific case of hyperactivity, it reveals how increasing the availability of this neurotransmitter can have dramatic and positive effects on organismic functioning.

With this pharmacological and neurochemical background, we can proceed to inquire if similar mechanisms might be extant in psychopaths and alcoholics. Such would be expected if, as hypothesized earlier, hyperactivity is the phenotypic expression of the genetic disturbance. As discussed earlier, Schacter (1971) reported hypersensitivity to adrenalin in his sample of psychopaths and found that an avoidance learning deficit could be completely ameliorated after injection of this chemical. Hill (1944) studied a group of psychopaths (defined as hostile, alcoholic and antisocial) before and after amphetamine administration and reported temporary improvement. While there is some indication that short-lived therapeutic gains may be observed, there is also the possibility of addiction to such stimulant-like drugs. Indeed, even the antidepressant monoamine-oxidase inhibitor (MAOI) tranylcypromine (Parnate) has been shown to lead to tolerance and progressive build-up of self-administered dosage to the level of abuse. In three such reported cases, Shopsin and Kline (1976) found this to be evident in persons who had sociopathic tendencies as well as other stimulant drug and/or alcohol abuse. The reasons for their dependence and increased tolerance to this drug were speculated to be its action on biogenic amines, namely the increased amount of norepinephrine available for receptor stimulation. Thus, it can be seen that there is, in the psychopath, a heightened sensitivity to norepinephrine and at least a short-lived positive response upon its administration. However, there is also an insurmountable problem which mitigates against the therapeutic use of these drugs. Drugs with stimulant-like properties, that is, those that enhance transmission in the noradrenergic system by augmenting norepinephrine availability at the synapse, for example, also pose a risk for dependency, especially so in sociopathic persons with a history of stimulant drug and alcohol abuse.

The question then logically arises: What process might underlie alcohol abuse? Ethanol has been reported to increase the activity of central norepinephrine (Fuxe, Hokfelt, and Ungerstedt, 1970) and also to be an MAO inhibitor (Schenker, Kissin, and Maynar 1967). Ingestion of alcohol might be a mechanism for increasing norepinephrine availability and so used by alcoholics to increase arousal. It might also explain the increased autonomic arousal in alcoholics previously described and therefore its beneficial effect on performance. These findings illustrate that ethanol may have stimulant-like effects via its action as an MAO inhibitor and, considering the fact that the monoaminergic system is under largely genetic control, the possibility is raised, although unproven, that this might be the source of the inherited defect. Persons with thi defect consume excessive quantities of alcohol to correct this aspect of organismic disturbance.

Support for the therapeutic benefits of increased norepinephri levels in the brain is scarce, possibly because of the justified hesitancy on the part of clinicians to administer such drugs to alcoholics for fear of cross-addiction. However, Kissin and Gross (1968) report some success with the combined administration of chlordiazepoxide and imipramine. Nialomide (Niamid), a MAO inhibitor, did not alter symptom ratings when compared to placebo, but did result in substantial improvement on cognitive-attentional task such as Raven's Progressive Matrices and Trail Making Test. Lithiu has also been tested in alcoholics with encouraging preliminary results. Among its other CNS actions, lithium carbonate also increas norepinephrine availability but its exact mode of action is unknown Kline, Wren, Cooper, Varga and Canal (1974) found that patients receiving this drug were more likely to remain abstinent from alcohol and manifest less disabling drinking episodes. A seventy-five percent reduction in drinking episodes was observed after one year. While these results are provocative and merit further investigation this study was based on too small a sample to allow for more than a tentative conclusion and groundwork for further study.

In conclusion, the response to drugs by detoxified alcoholics suggests in several studies a short-term positive effect from those agents that increase CNS norepinephrine availability. However, a caveat is in order. First, such drugs, particularly the stimulants and possibly MAO inhibitors as well, may be a source of abuse becau of increasing tolerance and dependency; second, such drugs have not yet received sufficient clinical trial to recommend therapeutic adoption. In a review of the psychopharmacology literature, Cole and Ryback (1976) present a pessimistic view of drug treatment of detoxified alcoholics with a generally negative conclusion about the usefulness of anti-anxiety agents such as chlordiazepoxide and the benzodiazepenes. In one study they reviewed, patients receivin chlordiazepoxides were more inclined toward illicit drinking and showed more cognitive impairment despite that a psychiatric interviewer rated them as less anxious.

In summary, the pharmacological and neurochemical mechanisms in hyperactivity, sociopathy, and alcoholism provide preliminary indication for disturbed catecholamine metabolism. Particular emphasis has focussed on noradrenergic mechanisms, especially the positive benefits obtained from increasing CNS norepinephrine availability. Alcohol has been demonstrated to be an MAO inhibitor, and this fact may explain its use. Other drugs such as the tricyclic antidepressants (e.g., imipramine) may have the same end effect and thus should be considered for further clinical research inasmuch as early research by Kissin and Gross (1968) and Kissin, Platz and Su (1970) suggest that it may work in combination with chlordiazepoxide. Considering the neurophysiological, psychological and behavioral processes subserved by the noradrenergic system in motivation, reward, arousal and so forth, it is obvious that overall organismic functioning is strongly influenced by alterations of this system. How this system interacts with other neurochemical systems eventuating in disordered physiology and ultimately placing the person at risk for alcoholism is still unknown, but perhaps is the crucial issue in alcoholism research.

Personality

Studies of the personality of alcoholics have consistently led to a rejection of an "alcoholic personality." At best, only rough typologies have been reported, with the most objective and replicable findings achieved with the MMPI. In terms of single scale performance, Hoffman (1976) reports that the psychopathic deviate scale is most often the peak score. The most frequent combination of scales is 4-2 or 2-4, in which the psychopathic score is elevated concurrently with depression. Goldstein and Linden (1969) factor analyzed MMPI performance from an alcoholic sample and derived four subtypes: 1) psychopathic and emotional instability; 2) psychoneurotic; 3) mixed psychopathic; and 4) alcoholism with paranoia and drug addiction. Hill (1962) also factor analyzed the MMPI in a group of alcoholics and obtained three factors: 1) undifferentiated psychopath; 2) primary psychopath; and 3) neurotic psychopath. Parthington and Johnson (1969) report five alcoholic subtypes on the MMPI, of which the most prominent was the young, unstable anti-social type. Thus, from these studies, it is apparent that psychopathic features enter very frequently into the personality description of alcoholics.

Given the prevalence of the psychopathic designation of alcoholic MMPI's, it is important to determine precisely what this scale measures. Factor analysis of the psychopathic deviate scale (Astin, 1959) has indicated that it assesses at least five diverse aspects of personality and behavior: self-esteem, hypersensitivity, social maladaptation, impulse control and emotional deprivation. Accordingly, while this scale is most frequently elevated in alcoholism, it embodies clinical features beyond simply anti-social behavior. In fact, it includes traits often observed in HK/MBD. As a single predictor of autonomic reactivity, the psychopathic scale has been found to correlate +.63 with palmar skin potential on a variety of stress tests such as cold pressor and electric shock (Learmonth, Ackerly and Kaplan, 1959), thereby adding credence to the presumption of increased lability in alcoholics.

The MMPI has also differentiated the prealcoholic samples from normal controls. Essentially, the prealcoholic presents as more impulsive, nonconforming and extroverted than his peers (Loper, Lammeier and Hoffman, 1973). While these traits may presumably simply reflect personality features of a neurotic prealcoholic disposition, they are equally likely behavioral expressions of HK/MBD.

Thus, personality studies of alcoholics reveal certain behavioral inclinations both in the morbid and premorbid states that are also observed in psychopathy and hyperactivity. While such observations cannot by themselves conclusively implicate a common underlying mechanism, they are at least in corroboration with the research reviewed in prior sections of this chapter.

EMPIRICAL INTEGRATION

The preceding discussion has focussed on the state of the organism as an important determining factor in alcoholism. A line of reasoning was advanced to implicate a fundamental disturbance in physiological arousal and reactivity in alcoholics in a fashion similar to that seen in psychopathy and childhood hyperactivity. As stated at the outset, the characteristic disorder is not hypothesized to be extant in all alcoholics, but rather in the primary alcoholic for whom drinking is a pervasive disorder beginning early in life without precipitating extrinsic cause and in the absence of other psychiatric disturbance. A person possessing these characteristics is theorized to be the "genetic alcoholic."

Evidence culled from research conducted in genetics, maturation, psychophysiology, psychopharmacology, personality and behavio can be organized into a heuristic and conceptually integrative fram work across levels of analysis. While there is a dearth of researc in several key areas, it is hoped that the present formulation will offer impetus and perspective for subsequent investigation.

In searching for organismic conditions that place a person at risk for future alcoholism, the question is posed as to "what is given?". What is genetically transmitted that renders the person vulnerable? In the language of behavior, _it_ is hypothesized to be hyperactivity. In the language of physiology it is arousal and reactivity and, for neurochemistry, neurotransmitter disequilibrium

Whatever the level of discourse, however, the presumption is that alcohol serves adaptive value even though short-lived until the afflicted individual is eventually overwhelmed by the negative consequences of addiction.

As pointed out earlier, hyperactivity may be the behavioral substrate for what later emerges in a significant proportion of persons as sociopathy and alcoholism. Hyperactivity is the phenotypic expression of what is transmitted and evidence accrued to date suggests that this disorder may be a heritable phenomenon (Cantwell, 1975; Wender, 1971). Children manifesting this disorder are more likely to have a parent psychiatrically disturbed, perhaps in the form of a sociopathic father or hysterical mother. The psychosocial development of hyperactive children is frequently marked by academic problems, failure to relate to peers, poor self-control and low self-esteem. They are also more likely to be problem drinkers as adolescents. Perhaps at this point, as the gross motoric and attentional disturbances of childhood subside (but not disappear), alcohol becomes utilized for adaptive pharmacological effect in boosting arousal and simultaneously diminishing excessive reactivity to stimulus input. Attentional capacity may be enhanced (Docter, et al., 1966); well-learned behavioral skills undergo minimal deterioration (Carpenter, 1962) during a drinking episode. Furthermore, psychological needs are satisfied; low self-esteem, history of failure, and peer rejection are overcome by the subjective experience of power and social ascendance. The excessive reactivity, now tempered by alcohol, permits more accurate cognitive labelling of physiological states to define emotional experience. The individual feels almost "normal"; aroused, self-confident, feels emotion.

This composite profile of the addictive process is based on the research evidence previously described. As should be apparent by now, the study of alcoholism involves investigating a number of different disorders united by the common feature of abusive drinking. Different etiological mechanisms may exist for these disorders and so it is important to emphasize that the information presented in this chapter applies to the primary alcoholic. This person is male, without psychiatric disturbance (other than, perhaps, some sociopathic elements), for whom drinking begins early in life without specific precipitating cause. This conceptualization of alcoholism etiology and process is not argued to be extant in other forms of alcoholism such as female alcoholism, reactive alcoholism (drinking in response to a life stress), or affective disorder. Nor is it implied to be operative for persons who are heavy abusive drinkers but not evidencing signs of the addictive process.

If the above formulations of primary alcoholism are heuristic and offer an integrative perspective, then they should also possess explanatory power beyond the data themselves. Can the information be extrapolated to account for other phenomena of alcoholism symptomatology?

The next section will briefly consider two such aspects of alcohol usage: alcohol analgesia and craving.

EXTRAPOLATION OF THE THEORY: TWO EXAMPLES

Alcohol Analgesia

The use of ethanol for relief from pain can be traced to ancient times. Pain itself is a complex process consisting of cognitive and sensory components and, while alcohol has been utilized as an analgesic, its effectiveness is still in doubt.

In a series of experiments comparing problem drinkers, alcohol- ics and normals, there have emerged group differences in the effect: of alcohol on pain stimulation. Recording pain thresholds of pressure and temperature, Cutter and his associates (Brown and Cutter, 1977; Cutter, Maloof, Kurtz and Jones, 1976; Maloof, 1975) observed that ethanol was an analgesic only for alcoholics and not for non-alcoholics. In a study of college drinkers, it was found that solitary barroom drinkers experienced pain relief after alcohol ingestion but such was not the case for nonproblem social drinkers. In fact, some indication was obtained to suggest that pain may be increased in some nonproblem drinkers during intoxication while the converse effect is seen in alcoholic and problem drinkers. And, finally, it was shown that experienced pain was greater in sober alcoholics and problem drinkers than nonproblem drinkers. These studies are intriguing in that they illustrate a differential response to ethanol in alcoholics and nonalcoholics and also suggest that pain experience in the sober state may also distinguish the groups.

At first glance, one might theorize that alcohol's pain-reliev- ing qualities are due to its powerful expectancy effects or perhaps attributable to enhanced well-being. Psychological factors do play an important role in pain experience but a more parsimonious explana tion can be advanced on the basis of the previously described formu- lation to explain the alcoholic's unique response to pain in the sober and intoxicated states.

It is theorized that the reason alcoholics, while sober, exper- ience more pain than normals is a combined effect of their tendency to augment and maximally respond to stimulus input and their increased physiological reactivity, a state which is itself stressful. Additional input, therefore, is likely to be aversive, especially if the intensity and perception of the stimulus is cognitively associated with pain. This interpretation of lower pain thresholds in alcoholics conforms to the observations by this author of the tendency by severe alcoholics to subjectively enhance stimulus input on the Petrie Perceptual Reactance Test and, by inference, is in agreement with the findings of Cutter and colleagues in alcoholics who are solitary (and probably more severe) drinkers. By consuming alcohol, they shift from augmenting to reducing stimulus input (Petrie, 1967), physiological reactivity diminishes, and more effective labelling of their state can be achieved, thereby allowing for better interpretation of the emotion of pain. Thus, it can be seen that integrating stimulus seeking and information processing characteristics in alcoholics with physiological patterns of functioning can help explain and perhaps predict their heightened sensitivity to pain and its reduction after alcohol consumption.

Craving

Many believe that craving is the cardinal feature of an addictive process. Ludwig and colleagues (Ludwig, Wikler and Stark, 1974) have theorized that craving is a conditioned cognitive labelling process utilized by the individual to explain a state of physiological arousal. Craving in this framework is hypothesized to be an emotion, determined by the person's past experience in identifying internal interoceptive cues. Often the craving for alcohol is associated with other affective states such as anxiety and depression (Hore, 1974; Tarter and Sugerman, in press) but whatever, craving has been variously invoked to explain drinking onset after a period of sobriety, loss of control after drinking onset, and as a physiological need state during withdrawal (Isbell, 1955).

The question is raised as to what properties ethanol has for the alcoholic that could lead to craving. As pointed out earlier, ethanol seems to have a normalizing effect which, psychophysiologically, is reflected as an increment in autonomic arousal and diminution of reactivity. Affectively, this may be experienced as increased anxiety, depression or other features of psychopathology (Nathan, O'Brien and Norton, 1971). Alcohol thus becomes a vehicle for inducing an altered state sufficiently discriminable from the sober condition of excessive reactivity. This discriminable physiological state becomes cognitively labelled, leading to the experience or emotion of craving. Since the alcoholic seeks and responds maximally to exteroceptive input of both a neutral and affective quality (Coopersmith and Woodrow, 1967), expectancy of the effects derived from the discriminable intoxicated state could result in anticipatory craving from extrinsic cues such as sight, smell or social setting and hence precipitate the first drink. While the craving can be identified as another affect (e.g., anxiety), it is important to point out that the subjective experience may be different from the objective physiological change induced by alcohol. For example, Steffen, Nathan and Taylor (1974) reported

that their sample of four alcoholics experienced heightened anxiety
although EMG from the <u>frontalis</u> muscle showed decreased activity
during intoxication. Moreover, it is speculated that the alcoholic's tendency to augment stimulus input (Tarter and Novick, in preparation) would predispose him to discover stimulus cues more rea-
dily and at perhaps a lower threshold than the nonalcoholic which,
by virtue of set and expectancy, may result in anticipatory craving

From this discussion, in which theory and observation are
interwoven, one can see how craving as a cognitive construct utilize
to explain a physiological state can evolve in the alcoholic. While
speculative in nature, the information pertaining to organismic,
physiological, and stimulus-seeking characteristics can be applied
to explain craving in a manner that is congruent with existing facts
Furthermore, a consideration of physiological factors may also assis
in elucidating the behavioral aspects of alcohol consumption. For
example, the inability of alcoholics to discriminate blood alcohol
levels of intoxication based on internal cues (Lansky, Nathan and
Lawson, in press) but to do so with external cues follows from the
information presented above.

Because alcoholics exhibit greater physiological reactivity
than normals, thereby resulting in a failure to learn to discriminate internal cues, it stands to reason that they would drink more
and of a sufficient amount to achieve a specific organismic state,
speculated to be in the range of their alcohol tolerance. Drinking
therefore, in the absence of conflicting drives, is likely to proceed more rapidly than with normals as, for example, by gulping
drinks to the tolerance range at which point a stable level of intoxication is maintained. Studies of alcoholics allowed noncontingent access to alcohol reveal that they do not always drink to unconsciousness, but rather consume quantities that maintain stable
blood alcohol levels (Mello and Mendelson, 1971). However, this
itself has been shown by the above authors not to be invariant but
rather manipulatable by the introduction of contingencies and programming of alcohol availability. Pattern of drinking is modifia-
ble, but of note is the finding that higher alcohol tolerance and
blood alcohol levels are attained when unrestricted access to alcohol is allowed.

In conclusion, the psychophysiological literature can be interpreted with the cognitive behavioral research in understanding
the craving phenomenon. While it has been shown that cravers are
more obsessive-compulsive in their drinking pattern (Tarter and
Sugerman, in press), it also follows that this holds only up to a
point, namely drinking to an optimal physiological state speculated
to be at around the tolerance level. This position specifies drinking to a subjective discriminable state at which time, in the absence of competing drives or stimuli, stable levels of intoxication
are maintained. It does not follow that drinking would continue

until unconsciousness in what has traditionally been thought to reflect loss of control. Recent behavioral research strongly argues
against the loss of control phenomenon and this is congruent with
the present theory. The present formulation additionally indicates
how expectancy set can play a prime role in predisposing craving
for alcohol prior to the first drink through exteroceptive stimuli,
namely by conditioned anticipation of a physiologically discriminable state which the organism may have found beneficial or adaptive
in the past, despite other adverse consequences from excessive consumption.

REFERENCES

Amark, C. A study in alcoholism: Clinical, social-psychiatric and
    genetic investigations. Acta Psychiatrica Neurologica Scan-
    dinavica, Supplement #70, 1951.

Astin, A. A factor study of the MMPI psychopathic deviate scale.
    Journal of Consulting and Clinical Psychology, 1959, 23, 550-
    554.

Bacon, M. Cross cultural studies of drinking. In P. Bourne and R.
    Fox (Eds.), Alcoholism: Progress in research and treatment.
    New York: Academic Press, 1973.

Bradley, C. & Bowers, M. Amphetamine (Benzadine) therapy of chil-
    dren's behavior disorders. American Journal of Orthopsychiatry,
    1941, 11, 92-103.

Brown, R. & Cutter, H. Alcohol, customary drinking behavior and
    pain. Journal of Abnormal Psychology, 1977, 86, 179-188.

Cadoret, R. Genetic determinants of alcoholism. In R. Tarter and
    A. Sugerman (Eds.), Alcoholism: Interdisciplinary approaches
    to an enduring problem. Reading, Mass.: Addison-Wesley Pub-
    lishing Co., 1976.

Cadoret, R., Cunningham, L., Loftus, R. and Edwards, J. Studies of
    adoptees from psychiatrically disturbed biologic parents.
    Behavioral Pediatrics, 1975, 87, 301-306.

Cahalan, D. & Cisin, I. Epidemiological and social factors asso-
    ciated with drinking problems. In R. Tarter and A. Sugerman
    (Eds.), Alcoholism: Interdisciplinary approaches to an endur-
    ing problem. Reading, Mass.: Addison-Wesley Publishing Co.,
    1976.

Cantwell, D. P. Psychiatric illness in the families of hyperactive
    children. Archives of General Psychiatry, 1972, 27, 414-417.

Cantwell, D. Familial-genetic research with hyperactive children.
    In D. Cantwell (Ed.), The hyperactive child: Diagnosis, man-
    agement, current research. New York: Spectrum Publications,
    1975.

Carpenter, J. Effects of alcohol on some psychological processes:
    A critical review with special reference to automobile driving
    skill. Quarterly Journal of Studies on Alcohol, 1962, 23, 274-
    314.

Chandler, B., Parsons, O. & Vega, A. Autonomic functioning in al-
    coholics: A study of heart rate and skin conductance. Quar-
    terly Journal of Studies on Alcohol, 1975, 36, 566-577.

Chotlos, J. & Goldstein, G. Psychophysiological responses to mean-
    ingful sounds. Journal of Nervous and Mental Diseases, 1967,
    145, 314-325.

Cleckley, H. The Mask of Sanity. St. Louis, Mo.: Mosby, 1964.

Cohen, N. & Douglas, V. Characteristics of the orienting response
    in hyperactive and normal children. Psychophysiology, 1972,
    9, 238-245.

Cole, J. & Ryback, R. Pharmacological therapy. In R. Tarter and A.
    Sugerman (Eds.), Alcoholism: Interdisciplinary approaches to
    an enduring problem. Reading, Mass.: Addison-Wesley Publish-
    ing Co., 1976.

Conners, C. Stimulant drugs and cortical evoked responses in learn-
    ing and behavior disorders in children. In W. Smith (Ed.),
    Drugs, development and cerebral function. Springfield, IL:
    C. C. Thomas, 1972.

Coopersmith, S. & Woodrow, K. Basal conductance levels of normals
    and alcoholics. Quarterly Journal of Studies on Alcohol, 1967,
    28, 27-32.

Cutter, H., Maloof, B., Kurtz, N., & Jones, W. Feeling pain: The
    differential response of alcoholics and nonalcoholics before
    and after drinking. Journal of Studies on Alcohol, 1976, 37,
    273-277.

Diagnostic and Statistical Manual of Mental Disorders. American
    Psychiatric Association, Washington, D. C., 1968.

Docter, R., Naitoh, P. & Smith, J. Electroencephalographic changes
    and vigilance behavior during experimentally induced intoxi-
    cation with alcoholic subjects. Psychosomatic Medicine, 1966,
    28, 605-615.

Douglas, V. Stop, look and listen: The problem of sustained attention and impulse control in hyperactive and normal children. Canadian Journal of Behavioral Science, 1972, 4, 259-282.

Dykman, R., Ackerman, P., Galbrecht, C. & Reese, W. Physiological reactivity to different stressors and methods of evaluation. Psychosomatic Medicine, 1963, 25, 37-39.

Fuxe, K., Hokfelt, T. & Ungerstedt, N. Central monoaminergic tracts. In W. Clark and J. del Guidice (Eds.), Principles of Psychopharmacology. New York: Academic Press, 1970.

Garfield, Z. & McBrearty, J. Arousal level and stimulus response in alcoholics after drinking. Quarterly Journal of Studies on Alcohol, 1970, 31, 832-838.

Gittleman-Klein, R., Klein, D., Katz, S., Saraf, K. & Bollack, E. Comparative effects of methylphenidate and thioridazine in hyperkinetic children. Archives of General Psychiatry, 1976, 33, 1217-1230.

Goldstein, S. & Linden, J. Multivariate classification of alcoholics by means of MMPI. Journal of Abnormal Psychology, 1969, 74, 661-669.

Gomberg, E. The young male alcoholic--A pilot study. (In preparation)

Goodwin, D., Schulsinger, F., Hermanssen, L., Guze, S. & Winokur, G. Alcoholism and the hyperactive child syndrome. Journal of Nervous and Mental Disease, 1975, 160, 349-353.

Gross, M. Imipramine in the treatment of minimal brain dysfunction in children. Psychosomatics, 1973, 14, 183-285.

Gualtieri, C. Imipramine and children: A review and some speculations about the mechanism of drug action. Diseases of the Nervous System, 1977, 38, 368-375.

Guze, S. The validity and significance of the clinical diagnosis hysteria (Briquet's syndrome). American Journal of Psychiatry, 1975, 32, 138-141.

Hare, R. Psychopathy: Theory and research. New York: Wiley, 1970.

Hill, D. Amphetamine in psychopathic states. British Journal of Addictions, 1944, 44, 50-54.

Hill, E., Haertzen, C. & Davis, H. An MMPI factor analysis study of alcoholics, neurotic addicts and animals. Quarterly Journal of Studies on Alcohol, 1962, 23, 411-431.

Hoffman, H. Personality measurement for the evaluation and prediction of alcoholism. In R. Tarter and A. Sugerman (Eds.), Alcoholism: Interdisciplinary approaches to an enduring problem. Reading, Mass.: Addison-Wesley Publishing Co., 1976.

Hore, B. Craving for alcohol. British Journal of Addictions, 1974, 69, 137-140.

Irwin, S. A rational framework for the development, evaluation and use of psychoactive drugs. American Journal of Psychiatry, 1968, 124, February Supplement, 1-19.

Isbell, H. Craving for alcohol. Quarterly Journal of Studies on Alcohol, 1955, 16, 38-42.

Jellinek, E. Phases of alcohol addiction. Quarterly Journal of Studies on Alcohol, 1952, 13, 673-684.

Jones, H. The study of patterns of emotional expression. In M. Reymert (Ed.), Feelings and emotions. New York: McGraw-Hill, 1950.

Jones, M. Personality correlates and antecedents of drinking patterns in adult males. Journal of Consulting and Clinical Psychology, 1968, 32, 2-22.

Kissin, B. & Gross, M. Drug therapy in alcoholism. American Journal of Psychiatry, 1968, 125, 69-79.

Kissin, B., Platz, A. & Su, W. Social and psychological factors in the treatment of chronic alcoholism. Journal of Psychiatric Research, 1970, 8, 13-27.

Kissin, B., Schenker, V. & Schenker, A. The acute effects of ethyl alcohol and chlorpromazine on certain physiological functions in alcoholics. Quarterly Journal of Studies on Alcohol, 1959, 20, 480-492.

Kline, N., Wren, J., Cooper, T., Varga, E. & Canal, G. Evaluation of lithium therapy in chronic and periodic alcoholism. American Journal of the Medical Sciences, 1974, 268, 15-22.

Lansky, D., Nathan, P. E. & Lawson, D. Blood alcohol level discrimination by alcoholics: The role of internal and external cues. Journal of Consulting and Clinical Psychology, in press.

Learmonth, G., Ackerly, W. & Kaplan, M. Relationship between palmar skin potential during stress and personality variables. Psychosomatic Medicine, 1959, 21, 156-157.

Loper, R., Kammeier, M. & Hoffman, H. MMPI characteristics of college freshman males who later become alcoholics. Journal of Abnormal Psychology, 1973, 82, 159-162.

Ludwig, A. & Stark, L. Alcohol craving: Subjective and situational aspects. Quarterly Journal of Studies on Alcohol, 1974, 35, 899-905.

Ludwig, A., Wikler, A. & Stark, L. The first drink: Psychological aspects of craving. Archives of General Psychiatry, 1974, 30, 539-547.

Lykken, D. A study of anxiety in the sociopathic personality. Journal of Abnormal and Social Psychology, 1957, 55, 6-10.

Maloof, B. Alcohol stress and coping: An examination of the differential impact of alcohol on the ability of alcoholics and nonalcoholics to cope with a stressful situation. Doctoral Dissertation, Brandeis University, 1974. Cited in Dissertation Abstracts International, 1975, 36, 1099.

Mann, H. & Greenspan, S. The identification and treatment of adult brain dysfunction. American Journal of Psychiatry, 1976, 133, 1013-1017.

Mawson, A. & Mawson, C. Psychopathy and arousal: A new interpretation of the psychophysiological literature. Biological Psychiatry, 1977, 12, 49-74.

McClelland, D., Davis, W., Kalin, R. & Warner, E. The drinking man. New York: The Free Press, 1972.

McCord, W. & McCord, J. A longitudinal study of the personality of alcoholics. In D. Peltman and C. Snyder (Eds.), Society, culture and drinking patterns. New York: Wiley, 1962.

Mello, N. & Mendelson, J. Drinking patterns during work contingent and noncontingent alcohol acquisition. In N. Mello and J. Mendelson (Eds.), Recent advances in studies of alcoholism: A interdisciplinary symposium. Superintendant of Documents, U. S. Government Printing Office, Washington, D. C., 1971.

Mendelson, W., Johnson, N. & Stewart, M. Hyperactive children as adolescents: A follow-up study. Journal of Nervous and Mental Disease, 1971, 153, 273-279.

Morrison, J. & Stewart, M. The psychiatric status of the legal
     families of adopted hyperactive children. Archives of General
     Psychiatry, 1973, 28, 888-891.

Nathan, P. E., O'Brien, J. & Norton, D. Comparative studies of the
     interpersonal and affective behavior of alcoholics and nonal-
     coholics during prolonged experimental drinking. In N. Mello
     and J. Mendelson (Eds.), Recent advances in studies of alcohol-
     ism: An interdisciplinary symposium. Superintendant of Docu-
     ments, U. S. Government Printing Office, Washington, D. C.,
     1971.

O'Malley, J. & Eisenberg, L. The hyperkinetic syndrome. Seminars
     in Psychiatry, 1973, 5, 95-103.

Parthington, J. & Johnson, F. Personality types among alcoholics.
     Quarterly Journal of Studies on Alcohol, 1969, 30, 21-33.

Petrie, A. Individuality in Pain and Suffering. Chicago: Univer-
     sity of Chicago Press, 1967.

Prichep, L., Sutton, S. & Hakerem, G. Evoked potentials in hyper-
     kinetic and normal children under certainty and uncertainty:
     A placebo and methylphenidate study. Psychophysiology, 1976,
     13, 419-428.

Quay, H. Psychopathic personality as pathological stimulation seek-
     ing. American Journal of Psychiatry, 1965, 122, 180-183.

Quitkin, F. & Klein, D. Two behavioral syndromes in young adults
     related to possible minimal brain dysfunction. Journal of
     Psychiatric Research, 1969, 7, 131-142.

Robins, L., Bates, W. & O'Neal, P. Adult drinking patterns of
     former problem children. In D. Pittman and C. Snyder (Eds.),
     Drinking patterns. New York: Wiley, 1962.

Schacter, S. Emotion, obesity and crime. New York: Academic
     Press, 1971.

Schenker, V., Kissin, B. & Maynard, L. The effects of ethanol on
     amine metabolism in alcoholism. In R. Maickel (Ed.), Biochem-
     ical factors in alcoholism. London: Oxford Press, 1967.

Shopsin, B. & Kline, N. Monoamine oxidase inhibitors: Potential
     for drug abuse. Biological Psychiatry, 1976, 11, 451-456.

Sprague, R., Barnes, K. & Werry, R. Methylphenidate and thiorida-
     zine: Learning, reaction time, activity and classroom behavior
     in disturbed children. American Journal of Orthopsychiatry,
     1976, 40, 615-628.

Steffen, J., Nathan, P. E. & Taylor, H. Tension reducing effects of alcohol: Further evidence and some methodological considerations. Journal of Abnormal Psychology, 1974, 83, 542-547.

Stewart, M., Mendelson, W. & Johnson, N. Hyperactive children as adolescents: How they describe themselves. Child Psychiatry and Human Development, 1973, 4, 3-11.

Stivers, R. Culture and alcoholism. In R. Tarter and A. Sugerman (Eds.), Alcoholism: Interdisciplinary approaches to an enduring problem. Reading, Mass.: Addison-Wesley Publishing Co., 1976.

Tarter, R. Personality characteristics of male alcoholics. Psychological Reports, 1975, 37, 91-96.

Tarter, R. Psychological deficit in chronic alcoholics: A review. International Journal of the Addictions, 1975, 10, 327-368.

Tarter, R., McBride, H., Buonpane, N. & Schneider, D. Differentiation of alcoholics according to childhood history of minimal brain dysfunction, family history and drinking pattern. Archives of General Psychiatry, 1977, 34, 761-768.

Tarter, R. & Novick, L. Perceptual reactance in alcoholics and its relationship to hyperactivity symptomatology in childhood (in preparation).

Tarter, R., Perley, R. & Sansom, C. Psychological history, minimal brain dysfunction and differential drinking patterns of male alcoholics (in preparation).

Tarter, R. & Schneider, D. Blackouts: Relationship with memory capacity and alcoholism history. Archives of General Psychiatry, 1976, 33, 1492-1496.

Tarter, R. & Sugerman, A. Craving for alcohol: Role of drinking pattern, psychosocial history, cognitive style, motor control and personality variables. In M. Gross (Ed.), Alcohol intoxication and withdrawal. New York: Plenum Press, 1977.

Venables, P. The electrodermal physiology of schizophrenics and children at risk for schizophrenia: Controversies and developments. Schizophrenia Bulletin, 1977, 3, 28-38.

Walsh, M. The biochemical aspects of alcoholism. In P. Bourne and R. Fox (Eds.), Alcoholism: Progress in research and treatment. New York: Academic Press, 1973.

Wanberg, K., Horn, J. & Foster, M. A differential model for the diagnosis of alcoholism. Scales for the Alcohol Use Questionnaire (unpublished manual). Fort Logan Mental Health Center, Alcoholism Division, Denver, Colorado, 1974.

Weiss, G., Minde, K., Werry, J., Douglas, V. & Nemeth, E. Follow-up studies of children who present with symptoms of hyperactivity. In C. Conners (Ed.), Clinical use of stimulant drugs in children. New York: American Elsevier, 1974.

Welkens, L. & Richter, C. A great craving for salt by a child with corticoadrenal insufficiency. Journal of the American Medical Association, 1940, 114, 866-868.

Wender, P. Minimal brain dysfunction in children. New York: Wiley, 1971.

Wenger, M. Studies of autonomic balance in Army Air Force Personnel. Comparative Psychology Monograph, 1948, 19, (Serial # 101).

Winokur, G., Rimmer, J. & Reich, T. Alcoholism IV: Is there more than one type of alcoholism? British Journal of Psychiatry, 1971, 118, 525-531.

Wood, D., Reimherr, F., Wender, P. & Johnson, G. Diagnosis and treatment of minimal brain dysfunction in adults. Archives of General Psychiatry, 1976, 33, 1435-1460.

Zahn, T., Abate, F., Little, B. & Wender, P. Minimal brain dysfunction, stimulant drugs and autonomic nervous system. Archives of General Psychiatry, 1975, 32, 381-387.

Zentall, S. Optimal stimulation as a theoretical basis of hyperactivity. American Journal of Orthopsychiatry, 1975, 45, 549-549-563.

Zentall, S. & Zentall, T. Amphetamine's paradoxical effects may be predictable. Journal of Learning Disabilities, 1976, 9, 67-68.

TOWARD A MULTIVARIATE ANALYSIS OF ALCOHOL ABUSE

Glenn R. Caddy

Old Dominion University

"Excessive drinking," "problem drinking," and "alcoholism" are societal designations of individuals' relationships with alcohol. Although the labels differ in connotation and denotation, all involve conceptions of alcohol abuse which have been instrumental in shaping popular, professional, and political opinion (Linsky, 1972).

Many Americans have definite opinions about alcoholism (Albrecht, 1973) yet current conceptions regarding alcohol abuse and alcoholism are neither clear nor consistent (Linsky, 1972; Marconi, 1967). Siegler, Osmond, and Newell (1968), for example, describe eight separate models for alcoholism which Caddy, Goldman, and Huebner (1967a; 1967b) cluster into the disease model, the symptomatic model, and the behavioral model. From the viewpoint of current knowledge and general influence, symptomatic and disease concepts may be seen as combining to form the traditional approach to alcoholism - what Pattison, Sobell, and Sobell (1977) have aptly called the "fold-science" model of alcoholism. This position contrasts sharply with the "academic science" approach which appears to be moving toward a behaviorally-oriented multivariate account of alcoholism.

THE TRADITIONAL APPROACH

The beliefs, values and ideologies which comprise the traditional approach to alcohol dependence view it as an identifiable unitary disease process. Various disease conceptualizations of alcoholism have appeared in the literature over the past forty years (Alcoholics Anonymous, 1939, 1957; American Medical Associa-

tion, 1968; Ausubel, 1961; Gitlow, 1973; Jellinek, 1952; 1960;
Keller, 1962; Mann, 1968; among many others). While these authorities offer differing explanations of the nature of disease, it is
possible to draw together the themes they espouse to outline the
elements of a "traditional" approach to alcoholism. These themes
variously indicate that: alcoholics are different from non-alcoholics; this "difference" either leads to or includes psychologi-
cal/sociological and/or biochemical/physiological changes; these
changes become part of a progressive and irreversible disease process; the disease is characterized by "an inability to abstain"
and/or a "loss of control" over alcohol. It has been hypothesized
that the supposed "difference" between alcoholics and non-alcoholics
is based on a psychological predisposition (Rado, 1958; Shae, 1954;
Wall, 1953), an allergic alcohol reaction (Alcoholics Anonymous,
1939, 1955, 1957; Randolph, 1956; Silkworth, 1937), or some nutri-
tional deficit which may or may not be genetically influenced (Madsen, 1974; Mardones, 1951; Sirnes, 1953; Williams, 1954). The
traditional approach dictates that treatment must emphasize the
permanent nature of the alcoholic's "difference" and, in so doing,
stresses that the disease can be arrested only by abstinence which
must be life long.

Just as the appropriateness of disease models for describing
behavioral problems has been questioned recently (Szasz, 1961, 1970;
Ullmann and Krasner, 1969), so too have the models and postulates
of the traditional view of alcoholism received increasing criticism.
The problem is that, despite continued widespread acceptance in both
professional and lay circles, especially within the fellowship of
Alcoholics Anonymous, the traditional approach has failed to win
empirical support. Keller (1972a), for example, has summarized the
many studies examining "differences" between alcoholics and non-
alcoholics and has noted that "alcoholics are different in so many
ways that it makes no difference" (p. 1147). In similar vein, research examining the construct of "craving," the "loss of control"
hypothesis, and the "irreversibility" aspects inherent in most of
the disease concepts of alcoholism has shown the traditional explanations severely lacking (see for example, Caddy, Note 1; Cohen,
Liebson, and Faillace, 1971; Cutter, Schwaab, and Nathan, 1970;
Engle and Williams, 1972; Gottheil, Crawford, and Cornelison,
1973; Keller, 1972b; Marlatt, Demming, and Reid, 1973; McNamee,
Mellow, and Mendelson, 1968; Mello, 1972; Merry, 1966; Pattison,
Sobell, and Sobell, 1977; Robinson, 1972; Sobell, Sobell, and
Christelman, 1972; and Wilson, Leaf, and Nathan, 1975).

Further, there now exists provocative evidence indicating
that, for at least some alcoholics, abstinence does not represent
the only possible treatment alternative (see for example, Armor,
Polich, and Stambul, 1976; Caddy and Lovibond, 1976; Lovibond and
Caddy, 1970; Miller and Caddy, 1977; Sobell and Sobell, 1972, 1973,
1975, 1976; and the reviews by Hamburg, 1975; Lloyd and Salzberg,

1975; and Sobell and Sobell, 1975) and that abstinence does not
necessarily indicate improvement in other areas of "life health"
(Flaherty, McGuire, and Gatski, 1955; Gerard, Saenger, and Wile,
1962; Pattison, 1966, 1968). Such evidence has led to the serious
questioning of the validity of the traditional models of alcoholism
by an increasing number of scientists working in the field. Despite what may even be the continuing growth of the influence of the
disease view generally, the essentially unidimensional perspective
inherent in the current disease conceptualizations has proved incapable of adequately accounting for the complex behavioral and
other phenomena generally associated with alcohol abuse and alcohol
dependence. Some theorists, in fact, have asserted that the traditional approach may be hindering rather than helping our understanding of alcohol dependence (Maisto and Schefft, in press).

THE SOCIAL LEARNING/BEHAVIORAL MODELS

These models involve an elaboration of the learning theory-
based drive/tension reduction account (Conger, 1956; Dollard and
Miller, 1950) to include the socio-cultural factors which have
also been indicated as important in the development of alcoholism
(Chafez and Demone, 1962; McCord and McCord, 1960; Schmidt, Smart,
and Moss, 1968). According to the social learning/behavioral approach, alcoholism is fundamentally a manner of drinking alcohol
(Sobell, Note 2). Drinking by the alcoholic, like drinking by the
non-alcoholic, is initiated and maintained by its antecedent cues
and consequent reinforcers (Bandura, 1969; Hunt and Azrin, 1973;
Ullmann and Krasner, 1965, 1969). Drinking is learned within a
social-cultural context with the term "alcoholic" being both a
label applied to some aspects of that drinking (Goffman, 1963a,
1963b; Szasz, 1970), and a socially ascribed role taken on by some
drinkers (Roman and Trice, 1968, 1970; Steiner, 1971). Social-
learning models typically support the goal of abstinence as the
treatment goal of choice for alcohol dependent persons. However,
acceptance of abstinence as the only treatment goal is not a necessary requirement for acceptance of the social-learning approach
as is the case with the traditional models.

Advocates for traditional approaches to alcoholism have typically accepted elements of the social learning-behavioral models.
For example, Alcoholics Anonymous (1957) has agreed that psychological and situational factors are important in the initiation of
drinking by an alcoholic following long periods of abstinence and
Jellinek (1960) considered that Conger's (1956) learning approach
at least partly complemented his disease model.

While the social learning/behavioral approach has facilitated
an impressive array of empirical research advancing our knowledge
of alcoholism, it nevertheless has developed as a unitary trait

approach and so, again, is somewhat limited. This limitation is especially apparent when one examines the capacity of the approach to account for the importance of the cognitive features which recent research is indicating to be of major significance in alcoholism (see Marlatt, 1977).

THE MULTIVARIATE APPROACH

The observations that the most significant element common to persons diagnosed "alcoholic" is that they drink too much and that the range of physiological, psychological, and socio-cultural correlates of alcoholism is vast have led many to reject global etiological theorizing and univariate linear conceptualizations of alcoholism in favor of the development of a multivariate approach (see, for example, Edwards, 1974; Goldstein and Linden, 1969; Horn and Wanberg, 1969, 1970; Partington and Johnson, 1969; Pattison, 1974a, 1974b; Wanberg and Knapp, 1970). The multivariate approach views alcohol dependence not as an entity represented by symptoms but as an array of behaviors and cognitions that collectively produce different types of problems which subsequently are labeled. To underline this perspective Horn and Wanberg (1969) have recommended that terms like "alcoholism" and "alcoholic" not be used, for they argue that these terms denote that a specific attribute "alcoholism" exists in the unitary fashion implied by the terms (see, also, Cahalan, 1970).

The disease models of alcoholism began to be undermined by the early reports of successful social drinking in a small number of alcoholic patients (Bailey and Stewart, 1967; Davies, 1962, 1963; Kendall, 1965). Within several years of Davies' now famous article, some clinicians and theorists like Chafetz (1966), Pattison (1966, 1968) and Scott (1968) were laying the foundations for a multivariate view of alcoholism. Chafetz (1966), for example, provocatively asserted that "...we...must conclude that alcoholic excesses, alcoholic problems, alcoholism or any label you care to affix is produced by complex, multidimensional factors, and that, in fact, there is no such thing as an alcoholic" (p. 810)

The multivariate approach owes much of its conceptual development to the empirically-based criticisms of the disease models which have been noted previously, as well as other data addressing the epidemiology of alcohol misuse (see Knupfer, 1967 and, especially, Cahalan, 1970; Cahalan, Cisin and Crossley, 1969; and Cahalan and Room, 1974). The growing appreciation of the multivariate nature of alcohol problems has also been facilitated by the recent application of factor analytic techniques to clinical data in the field. Horn and Wanberg (1969) undertook the factor analysis of drinking history data from 2300 alcoholic patients and identified thirteen independent primary factors of etiologic sig-

nificance. These same investigators (Horn and Wanberg, 1970) have
also identified a set of seven background factors (like youthful
rebellion, parental drinking problems) and eight current status
factors (like work status, social stress, and introversion) from
the analyses of social history data also drawn from large patient
populations.

Perhaps most significantly, however, the multivariate approach
to alcoholism has been facilitated by the entry of a small number
of behaviorally-oriented clinical researchers into the alcoholism
field (see, for example, Lovibond and Caddy, 1970; Sobell and
Sobell, 1972, 1976). These and other clinical pragmatists have
been less concerned with models of alcoholism and more concerned
with broad spectrum approaches (Lazarus, 1965, 1971) to the treatment of individuals for whom alcohol use has become a serious problem. The approach of these investigators to the treatment of alcohol dependence has been idiographic in character.

Up to the present time the development of a multivariate approach to alcohol dependence may be characterized most accurately
as a social systems approach (see Holder and Stratas, 1972; Nathan,
Lipson, Vettraino and Solomon, 1968; Steinglass, Weiner and
Mendelson, 1971a, 1971b; and Ward and Faillace, 1970). One such
systems approach, for example, that of Pattison (1974a, 1974b),
suggests that there are several alcoholic populations that may be
treated by several different methods leading to different patterns
of outcome.

> "It may be possible to match a certain type of patient with
> a certain type of facility and treatment method, to yield the
> most effective outcome...treatment programs can maximize ef-
> fectiveness by clearly specifying what population they propose
> to serve, what goals are feasible with that population, and
> what methods can be expected to best achieve those goals"
> (Pattison, 1974b, p. 59)

This and other systems approaches represent major advances over the
still widely-held view that there exists essentially one population
of alcoholics to be treated by one best method (often through the
fellowship of Alcoholics Anonymous), with only one therapeutic outcome in mind. However, at the present time, the multivariate systems approaches are limited because the technology by which patients could be matched to treatment techniques and outcomes is not
yet available.

Pattison, Sobell, and Sobell (1977) have provided an excellent
integration of the current clinical and laboratory research evidence in the alcoholism field. The most significant conclusions
drawn by these investigators are as follows: 1) Alcohol dependence
summarizes a variety of syndromes defined by drinking patterns and

the adverse physical, psychological and/or social consequences of
such drinking. These syndromes are best considered as a serious
health problem; 2) Alcohol dependence syndromes can be considered
as lying on a continuum from non-pathological to severely pathological; 3) A variety of factors may contribute to differential sus-
ceptibility to alcohol problems. These factors per se do not produce alcohol dependence. Any person who uses alcohol can develop a
syndrome of alcohol dependence; 4) The development of alcohol
problems follows variable patterns over time and does not necessarily proceed inexorably to severe or fatal stages. A given set of
alcohol problems may progress or be reversed through either naturalistic or treatment processes; 5) Alcohol problems are typically
interrelated with other life problems, especially when alcohol dependence is long established. While these investigators cautiously
have avoided the hazards of model building, they do stress that
the dynamic complexity of alcoholism connot be assessed unless the
disorder is conceptualized multidimensionally.

    In this paper, I hope to develop a multivariate, ideographic
approach to alcohol use and dependence by examining what I consider
to be the basic factors which bring about and maintain drinking
practices, whether moderate or alcoholismic. I will propose that
alcoholism and other alcohol-related problems can be best understood as behavioral disorders which may be established and maintained by individuals as a result of the unique interaction (both
direct and reciprocal) of social, incentive and discriminative
elements, all of which function with varying degrees of general
cognitive mediation. Within the framework of this multivariate
approach, it is assumed that each of these elements or dimensions
is interactive and yet each is sufficiently discrete to preserve
its own, albeit cognitively mediated, locus of control. Such a
multivariate approach permits an assessment of both the nature and
extent of the involvement of each of the modalities (behavioral,
discriminative, incentive, and social), which are variously integrated within the overall cognitive functioning of the individual
and which account for that person's alcoholism or lack thereof.
From this type of assessment, it is possible to establish operational hypotheses regarding a drinker's unique interactions with
alcohol. Such an assessment also facilitates the development of
treatment planning which takes into consideration the many elements
that support problem drinking. I will begin this multivariate approach to alcoholism with a review of the behavioral specification
of alcohol use and dependence.

                    GENERAL BEHAVIORAL SPECIFICATION

    In most cases the diagnosis of alcoholism is not made follow-
ing an examination of a person's drinking practices. Rather, it is
made following an evaluation of the consequences of alcohol use.

Until quite recently, there appeared little reason to analyze the drinking behavior of an alcoholic patient. Jellinek (1960) had indicated that the disease concept of alcoholism did not apply to excessive drinking but solely to "loss of control" drinking. The alcoholic state was generally considered to be discontinuous with a person's antecedent drinking behavior and, in accordance with this widely held view, the specific parameters of an alcoholic's drinking were considered to be of little significance in the treatment process. Treatment was directed toward the alcoholism (ideally in the absence of alcohol) and not to the act of drinking.

At the present time, however, there is an emerging belief held by an increasing number of academic alcohologists that alcoholism is essentially a manner of drinking alcohol. While some data are available on American drinking practices (Cahalan, 1970; Cahalan, Cisin, and Crossley, 1969; Cahalan and Room, 1974), Vogel-Sprott (1974) has made the point that there exists no established criteria specifying normal drinking in our society. Hayman (1967) has also made this point. There would be considerable objection, he argued, "if we defined social drinkers who have not been formally diagnosed as alcoholics, but this is hardly more imprecise than most attempts at defining social drinking (p. 585). Indeed, Pattison (1976) has stated that much drinking which passes as "social" is actually pathological and that many novitiate drinkers are introduced to drinking practices that are potentially alcoholismic while de Lint and Schmidt (1971) have concluded that what we call alcoholism appears to differ only in degree from the normal state of health and conduct.

The observation that many alcoholics and many other drinkers not so diagnosed consume large quantities of alcohol in relatively short periods of time is hardly revolutionary. The deliberate administration of alcohol to alcoholic volunteers in laboratory settings for the purpose of studying the topography of their drinking behavior as well as their physiological, metabolic, cognitive, and psychomotor functioning was, however, controversial (see Docter and Bernal, 1964; Talland, Mendelson and Ryack, 1964; Talland and Kasschau, 1965; and the reviews by Mello, 1972, and Nathan and Lisman, 1976). These researchers were concerned predominantly with the concomitants and determinants of alcoholismic drinking. They explored acute drinking behavior in terms of a macro (daily or weekly) assessment rather than a micro (moment-to-moment) assessment.

The first micro assessment study comparing the parameters of the drinking practices of alcoholics and non-alcoholics was conducted by Schaefer, Sobell, and Mills (1971). These investigators used an experimental bar setting in which alcohol was freely available up to generous limits for the purpose of comparing the drinking parameters of alcoholic and non-alcoholic subjects. Simultaneously recording several different components of their subjects' drinking behavior, these investigators reported that the alcoholic drinkers they studied differed from the non-alcoholic drinkers in at least the following ways: (1) Alcoholics chose more often to drink "straight" drinks rather than mixed drinks, wine, or beer; (2) Alcoholics usually consumed their drinks by gulping them (taking larger sips); (3) Alcoholics exhibited a longer inter-sip-interval; and (4) Alcoholics finished their drinks more rapidly. Subsequent research by the same authors (Sobell, Schaefer, and Mills, 1972) and by Williams and Brown (1974) have since confirmed these observations.

The other recent approach to the examination of behavioral dimensions in alcoholism has involved sending trained observers into actual drinking establishments to record parametric data on the drinking practices of target subjects (Kessler and Gomberg, 1974; Reid, Note 3; Sommer, 1965). Kessler and Gomberg's study, for example, was particularly detailed and included data on the number of drinks consumed, time to consume each drink, total time in bar, drinking alone or with others, and height and weight estimates of their unsuspecting subjects. While the extent of alcohol-related problems in these subjects was not known, it may be possible to categorize them as likely "moderate" or "high risk" drinkers from an estimation of their peak blood alcohol concentrations (derived from an estimate of their weight and the quantity of alcohol consumed). Such a categorization would be in keeping with Caddy's attempt (1972) to operationally discriminate between "responsible" drinkers who rarely consumed more than the equivalent of 60 or 70 milliliter of absolute alcohol at a single sitting and "problem drinkers" who frequently consumed in excess of 130 milliliters of absolute alcohol during one drinking session. Interestingly, the data recorded for Kessler and Gomberg's subjects differed from those recorded for social drinkers in the experimental bar setting by Sobell, Schaefer, and Mills (1972). It may be that factors such as age, socioeconomic status, setting and geography must be taken into account before meaningful conclusions about the drinking practices of different categories of drinkers can be made.

The behavioral dimension represents the most basic element of the ideographic multivariate approach to alcohol dependence. Drinking is easy to define, identify, and quantify and there are numerous social and other antecedents and consequences associated with it. Thus, drinking and the problems associated with it are particularly appropriate for behavioral analysis (see Sobell, Sobell, and Sheahan, 1976). The best evidence available in the alcoholism field today (see the reviews by Mello, 1972; and Pattison, Sobell, and Sobell, 1977) indicates that alcohol dependence is regulated not by an insidious physiological process but involves predictable stimulus response sequences operating within the life-space of the drinker. As Pattison (1976) asserts, drinking behavior construed as indicating "loss of control" is neither random, indiscriminate nor out of control. The alcoholic, in fact, does exert considerable control over his/her drinking behavior but he/she does so within the context of a different view of what constitutes an "acceptable" alcohol use pattern.

The treatment of individuals experiencing alcohol problems, of course, involves more than a concentration on the behavior of drinking, whether the ultimate therapeutic goal is the complete avoidance of alcohol or its "controlled" use. The multivariate approach would indicate that a behavioral and cognitive restructuring is required in order to achieve either of these goals. Such a restructuring probably also would require adjustments in the discriminative, incentive, and social domains of the drinker's life-space, for all of these may influence one's drinking behavior. The initial emphasis on the behavioral dimension, however, provides both the alcoholic patient and his/her therapist with a clearer understanding of the various contingencies which influence that patient's drinking. A behavioral analysis, in fact, provides the only data base from which the treatment of the alcoholic may proceed.

The increasing concentration on behavioral parameters which has emerged with the multivariate conceptualizations of alcohol abuse has also contributed to changes in the methodology employed to perform treatment outcome evaluation studies (see Caddy, Note 4; Sobell, in press; and the review by Crawford and Chalupsky, 1977). Until the work of Sobell and Sobell (1972, 1973, 1976), virtually all alcoholism treatment outcome studies presented outcome data in the form of descriptive summaries of groups. There were virtually no studies that reported objective pre- and post-treatment drinking data on individual patients. Further, virtually all treatment outcome studies evaluated the alcoholic patient dichotomously (abstinent or drunk) on the behavioral dimension. Such a categorization essentially precluded outcome measures based on multiple drinking parameters and other life health measures. While the multiple and objective measurements of the behavioral dimension have just begun to appear in the alcoholism treatment evaluation literature (Caddy, Note 5; Caddy, Addington, and Perkins, Note 6; Vogler, Compton, and Weissbach, 1975; Sobell and Sobell, 1972, 1973, 1976) it is becoming increasingly obvious that multivariate indices of drinking behavior and related areas of life health will see increased service as outcome evaluation research taps the multivariate nature of alcoholism.

DISCRIMINATION VARIABLES

It is widely recognized that drinking is a conditioned response subject to stimulus control (Fitzsimmons, 1972; Wayner and Carey, 1973; Weissman, 1972). It is also a high probability re-

sponse in certain clearly defined social situations. For some occasional drinkers, the social setting represents a necessary precursor for the elicitation of the behavior. In cases where this
initially rather specific drinking response becomes generalized to
a wide variety of social and other contexts, the frequency of alcohol use is likely to increase. Further, with increasing use, there
is likely to be an increase in the quantity of alcohol ingested on
any one drinking occasion, for the subjective effects of intoxication following the consumption of a set amount of alcohol are lessened by regular alcohol use in accordance with the phenomenon of
acquired tolerance (see Kalant and LeBlanc, 1971). The often validated assumption here is that the discrimination of cues indicating
certain levels of intoxication modifies or halts the drinking behavior of most individuals in most drinking situations. The act of
restricting one's level of alcohol intoxication involves a discriminative learning task. Okulitch and Marlatt (1972) have noted that
both the act of drinking and the physiological changes which accompany it are, in fact, discriminative stimuli which set the stage
for further drinking. Of course there are many unique reinforcers
and motivational variables (i.e., cognitively mediated components
of the social and incentive dimensions) which necessarily interact
with the discrimination variables to determine the ultimate behavioral outcome (see for example, the studies by Mello and Mendelson,
1965, 1970; Mendelson and Mello, 1966; Nathan, Titler, Lowenstein,
Solomon, and Rossi, 1970; and Nathan and O'Brien, 1971).

It may be argued, however, that some alcohol dependent people
are less victims of their own limited motivation to restrict their
drinking than they are their limited capacity to discriminate their
internal alcohol related state. There is evidence, for example,
suggesting that, by comparison with moderate drinkers, alcoholics
poorly discriminate the changes which occur as their blood alcohol
concentration (BAC) varies. Caddy (in press), Bois and Vogel-Sprott
(1974) and Huber, Karlin, and Nathan (1976) have all noted that,
following BAC discrimination training (the accurate feedback of
BAC's to drinking subjects who are required to attend to internal
and external drinking cues and estimate their BAS; see Lovibond &
Caddy, 1970), non-alcoholic drinkers can achieve considerable accuracy in estimating their own BAC's. In fact, Huber et al.
found that social drinkers can discriminate BAC equally well when
trained to attend either to internal cues (feelings and sensations)
or to external cues (BAC-dose relationships).

The training of alcoholic subjects by Silverstein, Nathan and
Taylor (1974), however, did not show these subjects to be particularly accurate on this task (though Silverstein's subjects were required to reach significantly higher BAC's for much longer periods
of time than subjects in the other studies). Caddy (in press) has
suggested that if, in fact, alcoholics are less able to accurately
discriminate their BAC's from internal cues than are non-alcoholics,

this inability may be a function of the increase in alcohol tolerance which parallels the development of alcohlismic drinking. According to this hypothesis, alcohol-tolerant drinkers are minimally
affected by low BAC's and, thus, they are unable to formulate accurate BAC discriminations, especially at low BAC's. Addressing
this same phenomenon, Huber, Karlin, and Nathan (1976) have suggested that if alcoholics prove to be poor monitors of their own
BAC's, this lack of sensitivity to internal cues might well stem
from the shifting levels of tolerance experienced by virtually all
alcoholics during their lengthy drinking histories. As a result of
these shifts in tolerance levels, the discrete sets of internal
cues which may be associated with specific BAC's, in the case of
moderate drinkers, become associated with many BAC's in the case of
alcoholismic drinkers. The most recent study of BAC discrimination
to date (Lansky, Nathan and Lawson, in press) compared the ability
of four alcoholic subjects who were given BAC feedback emphasizing
internal cues (involving the presentation of a relaxation-facili-
tated focus on internal sensations of intoxication) with four essentially matched alcoholics who experienced BAC discrimination
training emphasizing external cues (involving the presentation of an
individually tailored programmed learning booklet explaining BAC-
dose relationships and the rate of alcohol metabolism). Data from
these subjects were then compared with the data from the non-alcoholic subjects of Huber's study. These comparisons indicated that,
before BAC discrimination training began, the alcoholic subjects
were less able to monitor BAC changes than were the non-alcoholics.
After the feedback training was halted, the "internally" and "externally" trained non-alcoholics in Huber's study were equally well
able to monitor changes in their BAC's and to estimate BAC accurately. However, only the alcoholics who were "externally" trained
showed levels of BAC estimation accuracy comparable to the accuracy
of non-alcoholic subjects during this post-feedback phase. Lansky
and his colleagues concluded that there was little doubt that external cue training in BAC discrimination was more effective than
internal cue training for their alcoholic subjects and that, unlike
non-alcoholic subjects, alcoholics appear unable to acquire BAC
discrimination accuracy by using internal cues.

    There also exist other data bearing on the discrimination
capacity of alcoholics. According to Schaefer (Note 7), alcoholics
are less capable of discriminating among the tastes of various distilled spirits than are social drinkers. While this observation
may provide simply one more indication of the generally poor discrimination capacity of alcohol dependent individuals, Sobell,
Sobell, and Schaefer (1971) have suggested that, in this instance,
the poorer discrimination capacity of alcoholic subjects may be due
to a labeling deficit since these investigators have noted that alcoholics are poorer at recalling the names of spirits in popular
mixed drinks (see also Williams and Brown, 1974). Thus, Schaefer's
observation may be explained in terms of a discrimination difficulty

on the part of his alcoholic subjects or a cognitively mediated labeling difficulty or, perhaps, a combination of the two.

There is also evidence suggesting that alcoholics may not respond to internal fluid regulatory stimuli in the same manner as
non-alcoholics. Marlatt, Demming, and Reid (1973) found that when
alcoholics and non-alcoholics under-took a taste rating task and
were presented with an ad lib supply of either an alcoholic or non-
alcoholic beverage, the alcoholics consumed more of both beverages
than did the control subjects. In a similar vein, Gowardman,
Brown, and Williams (Note 8) have indicated that their hospitalized
alcoholics exhibited a higher daily fluid intake than did non-alcoholic hospitalized controls (see also Brown and Williams, 1975).
These fluid intake differences may be related to the observation
that excessive alcohol ingestion can lead to overhydration which,
in turn, may cause a drinker to experience symptoms of dehydration.
Paradoxically, these symptoms are not relieved by continued drinking
(Lolli, Rubin, and Greenberg, 1944). It may be hypothesized that
if these dehydration stimuli are frequently experienced by an alcoholic, he/she may learn to ignore them as signals for drinking and
therefore other drinking relevant cues may become predominant. An
alternate explanation, of course, is that alcoholics have adapted
to consuming greater fluid volumes than non-alcoholics. The work
of Devereaux and McCorkick (1972) and Holmes and Montgomery (1951)
provides some indirect support for this adaptation hypothesis.
Investigating the effects of the regular ingestion of increased
fluid volumes in humans and animals respectively, these researchers
found that such drinking led to a significant daily increase in the
volume of the fluid ingested by their subjects.

If, as the evidence suggests, alcohol dependent people have
difficulty discriminating with reasonable accuracy internally-based
regulatory stimuli, there is reason to believe that the drinking
behavior of alcoholics is especially influenced by external stimuli.
Cues such as the visibility of alcohol (including its advertising),
the time of day, and social context may be discriminative stimuli
of particular potency for alcohol dependent people. Brown (1974)
conducted a series of studies which provide support for such a hypothesis. Brown found that, by comparison with non-alcoholics,
alcoholics were less responsive to internal cues which accompanied
fluid deprivation or preloading, were more responsive to the external cues involved in personal drink preference and were especially responsive to the drinking environment. Further, alcoholics
drank significantly more (non-alcoholic beverages) than control
subjects when they drank in a bar setting, yet when drinking was
conducted in a "standard room" condition, the consumption of the
alcoholics did not differ from that of the non-alcoholics. Brown
concluded that many alcoholics may be environmentally vulnerable
drinkers just as many obese people may be environmentally vulnerable
eaters (see also Eaglen, Note 9; and Shettleworth, 1972).

There is also evidence suggesting that alcohol dependent individuals generally are substantially more field-dependent than non-
alcoholics (Chess, Neuringer, and Goldstein, 1971; Jacobson, 1968;
Witkin, Karp, and Goodenough, 1959). In terms of Witkin's (1965)
theory of psychological differentiation, highly field-dependent individuals may be characterized by a cognitive style which includes
excessive reliance upon external cues for the definition and identification of internal feelings and sensations. That alcoholics
seem to have particular difficulty discriminating their BAC's on
the basis of internal sensations may reflect the same processes
which underlie the field-dependence phenomenon. Alcoholics may not
be able to formulate these discriminations because they rely so
heavily upon environmental cues to define their internal states.

Data from the study of the eating behavior of obese people offers an intriguing analogue for alcoholism of particular relevance
to the discrimination dimension. We know that both the obese and
the alcoholic, in contrast to moderate eaters and drinkers, emit
their behavior at a high rate in a wide variety of circumstances
and both have difficulty limiting the rate and quantity of ingestion of their respective substances of abuse. There is also evidence that some obese people use food to such excess that it produces sedation with symptoms of psychological dependence emerging
when their food intake is restricted (Swanson and Dinello, 1970).
Other investigators have also reported these symptoms of dependence
together with "binge" eating in obese patients (de Lint, 1971;
Saltzman, 1972; Stunkard, 1959). The most intriguing possible
similarity between obese and alcohol dependent individuals, however,
involves data bearing directly on the discrimination dimension.
Obese individuals, in contrast to their normally weighted peers,
have been shown to be especially responsive to external stimuli
such as time of day (Schachter and Gross, 1968), the taste of food
(Nisbett, 1968a), and the sheer presence of food (Nisbett, 1968b).
Schachter (1971), in fact, has asserted that "eating by the obese
seems unrelated to any internal visceral state (gastric contrac-
tions, hypoglycemia), but is determined by external food relevant
cues such as sight, smell, and taste" (p. 130). More recent evidence, however, has indicated that it is only when external food
relevant cues are highly salient that obese individuals are disproportionately influenced by them (see Johnson, 1974; Mahoney, 1975a,
1975b; Ross, 1974; Wooley, 1972; and the review by Wooley and
Wooley, 1975). There is also recent evidence that obese individuals
are generally more sensitive to salient external (non-food-related)
cues). Rodin, Herman, and Schachter (1974) measured performance on
a disjunctive reaction time task and on a task involving recall of
briefly presented visual material, both designed to tap external
cue responsiveness. On both tasks, obese subjects outperformed
normals, lending empirical support to the notion of a generalized
responsiveness to external cues on the part of the obese. Further,
a series of studies by Pliner (1973a, 1973b, 1976), examining the

responsiveness of obese individuals in a number of different situations across a variety of response modalities, provides strong evidence that salient environmental stimuli consistently have a greater
impact on the behavior of obese individuals than on that of normally
weighted subjects.

From the recent evidence appearing in the alcoholism literature, especially the research involving BAC discrimination training,
there is considerable support for the hypothesis that, like the
comparison between the obese and normally weighted individuals, the
alcoholismic drinker is predominantly influenced by external (alcohol-related) cues whereas the moderate drinker experiences a significant ongoing interaction between internal and external alcohol-
related sensations. While alcoholismic drinking may be based in
the behavioral dimension (i.e., a drinker may rapidly consume such
large quantities of alcohol that intoxication becomes a fait
accompli), it may be that, for at least some alcoholismic drinkers,
intoxication is, at least in part, the consequence of an inability
to accurately discriminate the internal cues indicating increasing
BAC. Certainly one might expect that the subtle and changing cues
noticed during moderate drinking would be reduced for the person
drinking large quantities quite rapidly and that this person would,
therefore, face a more difficult BAC discrimination task than that
faced by the moderate drinker. Another possibility, of course, is
that when alcoholics restrict their drinking to match moderate
drinker ingestion rates, they do receive internal alcohol-relevant
sensations in much the same way as moderate drinkers (especially
in those instances where their tolerance has been reduced). However, while the moderate drinker is labeling these cues as indices
of intoxication, the alcoholic is not debilitated by them, does
not consider himself/herself "intoxicated," and so de-emphasizes
them.

Lovibond and Caddy (1970) developed a treatment program for
alcoholics which addressed the discrimination dimension directly
by including BAC discrimination training as a significant component. This procedure, when incorporated into a multivariate individualized treatment program, has been shown to be quite successful
in facilitating the non-problem use of alcohol by at least some
previously alcohol dependent persons (see Caddy and Lovibond, 1976;
Sobell and Sobell, 1972, 1973, 1976; and Vogler, Compton, and
Weissbach, 1975). Given the current controversy which exists
about the whole issue of "controlled drinking," it may be that
research examining the variables which govern or otherwise influence the discrimination capacity of alcohol dependent persons may
be especially relevant to future efforts to determine those patients for whom abstinence would prove the most suitable drinking
related treatment goal and those who might attempt to drink in a
restricted manner.

INCENTIVE VARIABLES

According to the multivariate approach to alcohol dependence, there may be as many incentives to drink alcoholismically as there are drinkers to enact such behavior, for the incentives which influence the behavior of each drinker are mediated by that individual's idiosyncratic cognitions regarding use of alcohol.

Nevertheless, the reinforcement contingencies for alcohol use and dependence may be summarized generally in terms of the following operant elements:

1) positive reinforcement associated with the psychopharmacological properties of alcohol (e.g., euphoria and relaxation).

2) positive reinforcement associated with the social aspects of alcohol use (e.g., acceptance into a drinking group).

3) negative reinforcement associated with aversive environmental aspects (e.g., relief of boredom or temporary escape from unpleasant living conditions).

4) negative reinforcement related to non-drug-induced aversive physical states (e.g., relief of chronic or acute pain due to injury or illness) or to drug-induced aversive physical states (e.g., relief from the physical discomfort of the withdrawal state).

5) negative reinforcement related to attempts by a drinker to alter his/her psychological state (e.g., removing anxiety or inducing a temporary positive change in self-concept).

The view that alcohol serves as a negative reinforcer by enhancing one's psychological state via tension reduction has been widely cited as an explanation for alcoholismic drinking (see Freed, 1967, 1968; Hughes, Forney, and Gates, 1963; Masserman, Jacques, and Nicholson, 1945; Smart, 1965; Vogel-Sprott, 1967). The classic statement of the negative reinforcement view was presented by Conger (1956) as the tension-reduction hypothesis. This hypothesis stated simply that the existence of a tension state energized the drinking response and the relief of tension provided by alcohol reinforced this response. In a recent review, Cappell (1975) has integrated much of the animal and human literature in this area and concluded that "while the tension reduction hypothesis may be quite plausible intuitively, it has not been convincingly supported empirically" (p. 59). Subsequent reviews by Nathan and Lisman (1976) and Marlatt (1976) have also reached the same conclusion. Like so much of the research seeking to explore the basis of "alcoholism," however, the evidence bearing on the tension-reduction hypothesis generally misses the point that alcohol problems are

multivariate in nature and, therefore, more often than not, defy
research efforts to account for alcoholismic drinking via parsimon-
eous general explanations. Such explanations, in fact, often have
little to do with the problems of the individual case.

Nathan and O'Brien (1971) have conducted a clinical study which
was perhaps more relevant to the general tension-reduction issue in
alcoholism than most of the experimental literature reviewed by
Cappell. These researchers compared the behavior of alcoholics and
non-alcoholics under conditions of experimental intoxication and
reported that, during an initial 12 to 24 hour drinking period, a
modest decrease in anxiety was noted, but that subsequent drinking
led to an increase in anxiety and depression. Another important
study examining the progression of mood states during a period of
experimental intoxication was conducted by Tamerin and Mendelson
(1969). During this study, four male alcoholics participated in
two weeks of baseline observations, three weeks of programmed
drinking, ten days of alcohol withdrawal, three weeks of free access drinking and, finally, a ten-day withdrawal period. When
asked before onset of drinking to anticipate the effects of alcohol
consumption on their mood, all four subjects looked forward to becoming more relaxed, more comfortable, and less depressed. Contrasted with these anticipated pleasurable effects, descriptions of
self while sober were characterized by chronic anxiety, depression,
self-depreciation, feelings of hopelessness and inability to cope
with problems. Early drinking stages were uniformly experienced as
pleasurable, with subjects reporting feelings of relaxation, ela-
tion, and reduced inhibition. As drinking continued, however,
these positive states were superseded by feelings of remorse, self-
depreciation, guilt, anxiety, aggression, and hostility. Finally,
following the period of intoxication, subjects had little recollec-
tion of their alcohol induced unpleasant mood state (see also
McNamee, Mello, and Mendelson, 1968; Mendelson, LaDou, and Solomon,
1964; Mendelson and Mello, 1966; Vanderpool, 1969).

Studies such as these suggest that, even for those individuals
who have experienced extremely negative consequences of their alcoholismic drinking, the expectations of positive effects following
the use of alcohol provide a significant part of the initial incentive to drink again. These expectations of positive effects no
doubt play an important role in supporting the continued alcoholismic drinking of many alcoholics and others not so labeled. Such
individuals, seeking to maintain or recapture the initial positive
effects of alcohol, drink themselves into "states of consciousness"
which, in some instances, prove to be anything but that which they
were striving to achieve.

The continuous "relief-avoidance" drinking of the physically
dependent alcoholic, however, is probably motivated more by avoidance of the negative effects of withdrawal than by the positive

expectations of any "altered state of consciousness." Certainly
the evidence suggests that much of what has been referred to as
"loss-of-control" drinking is, in fact, an attempt to control and
ameliorate the withdrawal symptoms produced by falling blood alcohol levels (see Hershon, 1973, 1977). Unfortunately, the alcoholic
who drinks to avoid or relieve the withdrawal symptoms becomes involved in a vicious circle for his action also perpetuates these
withdrawal symptoms.

While it may be possible to explain the initiation and continuation of drinking in terms of the operation of either the positive
or negative reinforcement paradigms, it is probably more likely, as
Keehn (1970) has suggested, that the act of drinking, alcoholismically or otherwise, depends upon multiple schedule control by both
positive and negative reinforcements. Certainly, however, the combined effects of the immediacy and the potency of the reinforcements that follow the act of drinking are more than adequate on
most occasions to enable most alcoholismic drinkers to overcome the
disincentives which may derive from the awareness of the possibility of the negative consequences that follow such drinking.

There are a variety of rather global incentive-based theoretical accounts of alcoholismic drinking which may be relevant to the
initiation or continuation of drinking in an individual case. Some
investigators (Bacon, Barry and Child, 1965; Blane, 1968; McCord
and McCord, 1960; Tahaka, 1966) have implicated unfulfilled dependency needs as the psychological antecedents of alcoholismic drinking. Other researchers (Kilpatrick, Sutker, and Smith, 1976;
Zuckerman, Weary, and Brustman, 1970) have suggested that the
search for novelty and experiential alteration may be the principal
incentive for much alcoholismic drinking. Perhaps the most influential recent motivational theory of alcoholismic drinking, however, is that proposed by McClelland and his co-workers (McClelland,
David, Kalin, and Wanner, 1972). Following a series of investigations ranging from the use of projective techniques to reviews of
cross-cultural studies, McClelland concluded:

> Men drink primarily to feel strong. Those for whom personal-
> ized power is a particular concern drink more heavily. Alcohol
> in small amounts in restrained social settings, and in re-
> strained people tends to increase thoughts of social power.
> In larger amounts, in supportive settings and in impulsive
> people, it leads to an increase in thoughts of personalized
> power (p. 334).

While McClelland's hypothesis would appear to account for at least
some of the incentive to drink moderate quantities of alcohol, it
is rather hard to imagine that many individuals incapacitated by
alcohol to the point of stupor see themselves as becoming more
powerful as they become more intoxicated. Certainly, alcohol de-

pendent individuals who undergo treatment experiences involving
videotaped self-confrontation of their drunken comportment do not
ultimately experience an increase in their sense of power; quite
the contrary, in fact (see Schaefer, Sobell and Mills, 1971; Vogler
Compton, and Weissbach, 1975).

Over the past four decades, etiological theorizing and treatment endeavor in the alcoholism field has been concentrated quite
heavily on a variety of incentive elements. These efforts, however, have been limited by the very conceptualizations of the disorder on which they were based. For example, numerous aversion
therapy programs perhaps rather simplistically have sought to reduce the alcoholic's incentive to drink by employing escape, avoid·
ance, and punishment paradigms extracted directly from the learning
laboratory (see Rachman and Teasdale, 1969), without adequate consideration of the complexity of either the precise incentive elements operating in an individual drinker or the interaction of
these elements with other drinking variables. Employing a different strategy, Alcoholics Anonymous has attempted to develop the al·
coholic's motivation to overcome the recurring desire to drink by
avoiding alcohol entirely and by drawing the motivation to accomplish this feat from an external power such as religious conviction
Both the aversion therapy and the A.A. approaches, however, have
developed within the framework of the traditional approach with it∶
emphasis on the "loss of control" hypothesis.

Future treatment strategies may not view the alcoholic as a
person lacking control over his/her encounters with alcohol, but a∶
an individual whose incentive to drink in either a normal or alcoholismic fashion depends upon the unique interplay among the behavioral, discriminative, social, and cognitive variables that are
relevant to the determination of alcohol use and dependence. Such
a theoretical perspective, requiring assessment of the incentive
motivation provoking drinking in the individual case, has significant implications for treatment. For example, if an individual's
incentive to drink was found to be highly related to seeking personalized power, then treatment strategies may be directed to re-
versing or re-directing that person's desire for power (perhaps by
joining organizations, running for office, or modifying aspects of
the power structure within his family), so that his existing power
needs are met in non-damaging, socially acceptable ways. Clearly,
a fine grained analysis of the incertive forces provoking and per-
petuating alcoholismic drinking is required in order that specific
treatment elements can be directed to the specific difficulties
found in the individual case.

SOCIAL VARIABLES

Studies of socio-cultural variables have typically concluded that these elements correlate with the existence of alcoholism (Cahalan, 1970; Chafetz and Demone, 1962; Field, 1962; Heath, 1975; McClelland, Davis, Kalin, and Wanner, 1972; McCord and McCord, 1960; Pittman and Snyder, 1962; among many others). According to Encel and Kotowicz (1970), relationships between heavy drinking and alcoholism in a given group "seems to lie in the social psychology of conformity and deviance" (p. 608). If heavy drinking is normative behavior, more people who conform to the norm are exposed to the risk of alcoholism. If heavy drinking is proscribed, only those drinkers who are willing to disobey social sanctions are at risk. This point is also implied in Jellinek's (1960) "vulnerability-acceptance" hypothesis. In somewhat similar vein, MacAndrew and Edgerton (1969) propose that:

> "Persons learn about drunkenness what their societies impart to them, and comporting themselves in consonance with these understandings they become living confirmations of their societies' teachings...Because our society's teachings are neither clear nor consistent, we lack unanimity of understanding; and where unanimity of understanding is lacking, we would argue that uniformity of practice is out of the question. In such a situation,...what people actually do when they are drunk will vary enormously; and this is precisely what we find when we look around us." (p. 172)

Research in other segments of the social dimension also indicates the extent to which sociocultural and social class factors are related to the seeking of treatment by alcoholics, the treatments provided, and the outcomes of the actual interventions (see, for example, Blane, Overton, and Chafetz, 1963; Chafetz, Blane, and Hill, 1970; Edwards, Kyle, and Nicholls, 1974; and Schmidt, Smart, and Moss, 1968). In this context, one is reminded of Cahn's (1970) recommendation that treatment services for alcoholism patients be instituted on a social class basis so that such services might better meet the needs of those seeking treatment. In a similar vein, Pittman and Gordon (1958) expressed serious concerns that the special social characteristics of the skid row alcoholic require that he (or she) receive special residential care involving provision of an alternate total social milieu (see also Pattison, Coe, and Doerr, 1973; Pattison, Coe, and Rhodes, 1969).

Unfortunately, the views which I have reviewed thus far do not contribute greatly to our current knowledge of the development of alcoholism in the individual or to our clearer understanding of its management. Obviously, the drinking of alcohol is learned and conducted in a social context. For some novitiate drinkers this context both condones and supports a drinking pattern which is deviant from the outset (see Jessor, Collins, and Jessor, 1972; Jessor and Jessor, 1973). Yet the assessment of social influences on drinking, moderate or otherwise, is difficult for it is the individual drinker's unique perception of his/her social context that, in many cases, provides the effective social stimulus components for the act of drinking. Williams and Brown (1974) investigated the effects of socialization on the drinking behavior of alcoholics and non-alcoholics in a structured social drinking experiment and noted that socializing did not influence the alcohol consumption of alcoholics (perhaps because of a ceiling effect) but that it led "normal" drinkers to increase their consumption so that, in a number of cases, their consumption approximated that of the alcoholics. Brown (1974) reported similar results and suggested that drinking does not seem to be correlated with social interaction in the same way for alcoholics and non-alcoholics.

Such findings, of course, do not imply any lack of social support for the act of drinking alcoholismically. Mello (1972), for example, notes that intoxication often permits the taking on of intergroup roles, sexual and otherwise, which are not normally available to a person during sober periods. Further, Marlatt (Note 10) reports that over fifty percent of his alcoholic patients who relapse cite the social pressure exerted by friends and former drinking associates as major factors in their relapse. Even in the case of the solitary drinker, the lack of social interaction does not mean that social reinforcement is not maintaining that person's behavior. Possible intermittent social reinforcements for the solitary drinker may even include the occasions on which he/she is taken to hospital or some social support agency and describes his/her condition to a sympathetic listener.

Bacon (1973) has presented a socio-dynamic account of the development of alcoholism which, though he does not describe them as such, includes behavioral, cognitive, and incentive components. According to Bacon, the essential process in the development of alcoholism (in males) is dissocialization. This process involves, first, a reduction in the number and variety of the drinker's social activities and, then, a movement into social groups more tolerant of his drinking. According to this formulation, alcohol is used repeatedly to ease difficulties so that drinking becomes individually rather than socially motivated.

The interaction between the behavioral dimension and certain social and incentive elements is particularly implicated in alcoholismic drinking in those instances where social modeling causes a drinker to drink greater quantities when exposed to a heavy drinking model rather than a light drinking model or no model (see Caudill and Marlatt, 1975). Also, just as alcohol functions to facilitate social intercourse, so too some social sub-systems facilitate and provide support for the maintenance of alcoholismic drinking. The extremes of these sub-systems may be seen in the matriarchal alcoholic marriage (Jackson, 1954; Paredes, 1973), the revolving-door chronic drunkenness offender (Pittman and Gordon,

1958), and on skid row (Blumberg, Shipley, and Moor, 1971; Blumberg, Shipley, and Barsky, 1977). In each of these systems secondary gains appear through the incentive dimension which aid in the support of the social status quo of the alcoholic.

Recent developments in social-learning theories of alcoholism have concentrated less on social-incentive parameters and more on interactions between the social and cognitive elements of the drinking problems. In processes such as dissocialization there is a two-phase sequential development. First, the drinker becomes labeled and, after a period which Clancy (1960) considers to reflect the procrastination of the alcoholic, this labeling leads to the incorporation of a "sick role" (Roman and Trice, 1968). To the extent that the label influences cognitive elements such as the drinker's self-image, subsequent behavior is also likely to be influenced and, so, the acceptance of the label and the taking on of the role "alcoholic" aid in the development, maintenance, and legitimization of alcoholism. As Roman and Trice (1968) point out: "Deviant drinking behavior is legitimized through the disease label in the sense that the individual is no longer held responsible for this behavior and this behavior is very rewarding" (pp. 246-267).

In our society the label "alcoholic" tends to be withheld until repeated occurrences of alcoholismic drinking, with attendant negative consequences, force acceptance of the diagnosis. An individual in such a society can possess a long history of abusive drinking and yet not suffer the aversive consequences which follow the imposition of sanctions which appear with the label "alcoholic" (Canter, 1968; Negrete, 1973). Further, as noted previously, with relatively few exceptions our society accepts abstinence as the only goal of alcoholism therapy and the price for alcohol abuse. In a "drinking" society such a therapy requirement necessarily places considerable strain on the alcoholic whose treatment involves abstinence and whose problem involves a failure to refrain from drinking.

The increasing appreciation of the relevance of the alcoholic's social system to the establishment and maintenance of abusive drinking has led to the development of treatment strategies directed toward this system. Thus, Foy, Miller, Eisler, and O'Toole (1976) treated two alcoholics who showed particular difficulty resisting the social pressure to drink by providing specific social skills training, directed toward helping them refuse more effectively (e.g., modeling specific resistance strategies and recommending, with focused instruction, the use of alternate behavior). A three month follow-up indicated significant improvement in the ability of these subjects to resist the pressure to drink.

Hunt and Azrin (1973) provide the best current model of the therapeutic manipulation of an alcoholic's social reinforcement

system. These authors undertook the treatment of 16 inpatient alco
holics (half of whom were assigned to a matched control condition)
in a program designed to re-arrange the vocational, familial and
social reinforcers of their patients so that time-out from these
reinforcers would occur if the patient began to drink. For example
marital counseling attempted to structure reinforcement to make the
drinking of alcohol incompatible with the marital relationship.
Social counseling attempted to restore and improve the alcoholics'
social relationships and to make continuation of these improved
relationships dependent upon sobriety rather than drinking. These
strategies showed marked and consistent improvement of the experimental over the control groups on all measures of sobriety, employment, home attendance and extent of institutionalization over a
six-month period.

While procedures directed toward halting abusive drinking by
manipulating the social elements or the interaction between the social and incentive dimensions associated with such drinking have a
long history, study of the effective components of such procedures
has not yet been undertaken. Further investigation of the complex
interaction between the social/cognitive components of labeling and
the social/incentive elements inherent in community reinforcement
approaches to the treatment of alcoholism are likely to contribute
greatly to the understanding of alcoholism in the individual case.

COGNITIVE VARIABLES

In recent years, an increasing awareness that human behavior
cannot be accounted for in simple stimulus-response terms and that
cognitive processes play a central role in accounting for most huma
actions has developed (Bem, 1970; Brewer, 1974). With this awareness has come an increasing clinical interest in the mediating role
of cognitive processes (see, for example, Bandura, 1974; Brady,
1967; Davison, 1968; Ellis, 1962; Lazarus, 1968, 1977; Mahoney,
1974).

A number of subtle cognitive elements correlate with alcohol
dependence. Witkin, Karp, and Goodenough (1959), for example,
have shown that alcoholics are more field dependent than non-alcoholic controls (see also the review by Sugerman and Schneider, 1976
Of course, whether field dependence in some way contributes to the
development of alcoholism or is a general consequence of prolonged
drinking and/or other aspects of the life style of alcoholismic
drinkers remains to be determined. The general philosophy underlying the multivariate approach espoused in this paper is that if
a cognitive element (like field dependence) is relevant to an individual's alcoholismic drinking, it likely represents but one variable among many others contributing to the behavior of that individual.

There also may be a relationship between the generally more field dependent cognitive style of alcoholics and the difficulty they have discriminating internal cues. Following a study of the effects of sensory deprivation on the field dependence scores of alcoholics, Jacobson (1971) hypothesized that, to the extent that alcohol serves as a chemical means of suppressing interoceptive stimuli that conflict with exteroceptive stimuli, it acts as a psychophysiological reinforcement for field dependent behavior. This hypothesized perceptual reliance on external sources of stimulation may be also related to the observation that alcoholics tend to achieve higher "external" scores than non-alcoholics on measures of locus of control; that is, they indicate that external forces rather than internal forces control their actions (see Sugerman, Reilly, and Albahary, 1965). Such evidence is consonant with Shapiro's (1965) view of the "addictive character style" which he considers to be characterized by "an absence of a sense of having chosen to act in a particular way."

Evidence bearing on more general "personality" aspects of the cognitive dimension also suggests differences between alcoholismic drinkers and normal drinkers. For example, Kalin (1972) writes that excessive drinking by college students has a "wild, narcissistic social show-off" quality about it. Similarly, Blane (Note 11) points out that while alcoholics are frequently engaging people who enjoy the company of others, their sociability often co-exists with an inability to maintain long-term relationships based on mutual give and take (see also Force, 1958; Machover and Puzzo, 1959).

While these observations of "cognitive style" and "personality" may describe some of the cognitive elements impinging on an individual drinker, they do not constitute evidence of an "alcoholic" cognitive style or an "alcoholic" personality. Certain self-perceptions, however, are central to all multivariate behavior disorders. In the case of alcoholism, these variables include the way a drinker sees himself/herself, the way he/she sees drinking, and the conceptualizations which he/she and his/her significant others hold regarding alcohol dependence.

Linsky (1972) used survey methodology to explore relationships between beliefs about the etiology of alcoholism and preferred methods of control. He noted that the different models of alcoholism held by different individuals brought with them different treatment implications (see, also, Aubert and Messinger, 1958; Caddy, Goldman, and Huebner, 1976a, 1976b; Glock, 1964). Stoll (1968) has summarized these general findings as follows:

1) To the extent that individuals believe non-conformity to be conscious defiance of rules...they will prefer to restrict and castigate deviants, and

2) To the extent that individuals believe non-conformity to be
   the result of external forces...they will prefer to treat o
   cure deviations without accompanying opprobium (p. 121).

The extent to which alcoholics are able to exert control over
their drinking behavior appears to be the single most important
controversy in the current alcoholism literature. The traditional
approach to alcoholism, of course, has yielded a number of formulations regarding the matter (Jellinek, 1952, 1960; Keller, 1972b;
Marconi, 1959; and Marconi, Fink, and Moya, 1967), all of whom propose that most or all alcoholics are incapable of regulating their use
of alcohol. Jellinek (1960), for example, hypothesized that gamma
alcoholics are affected by alcohol consumption in such a way that:

The ingestion of one alcoholic drink sets up a chain reaction
so that they are unable to adhere to their intention to have
one or two drinks only but continue to ingest more and more--
often with quite some difficulty and disgust--contrary to
their own volition (p. 41).

The evidence bearing on the whole issue of control in alcoholism, however, does not support the view that the alcoholic is
powerless with respect to his/her use of alcohol (see Pattison,
Sobell, and Sobell, 1977; and the reviews by Maisto and Schefft, in
press; and Marlatt, 1977).

The multivariate approach proposed in this paper addresses the
question of control (or lack thereof) in very different terms from
those of the traditional orientation. From the multivariate perspective drinking, like all other behaviors, is "controlled" in
that when it occurs it is the consequence of the unique interaction
of various cognitive, behavioral, discriminative, incentive, and
social antecedents. Moreover, drinking continues because of reinforcements coming from these same dimensions. Such behavior is not
random or indiscriminant and may involve very stringent controls.
From this perspective, the repeated apparent failure of an individual to regulate his/her use of alcohol is seen as somewhat akin to
learned helplessness (Seligman, 1972). Such an individual has
either never learned or has unlearned the principal components of
certain self-regulatory behaviors which permit the control of the
use of alcohol. The alcohol dependent individual is not, however,
the "victim of an insidious, progressive disease" over which he can
exert no influence.

Given the overwhelming evidence recommending the abandonment
of the notion that alcoholics cannot regulate their drinking behavior, it is ironic that this view continues to be a central tenet
of public and professional opinion regarding alcoholism. Since
this cognition is widely accepted, it is extremely influential in
shaping the self-perceptions of alcoholismic drinkers as well as

as attitudes to alcoholism by non-alcoholics. If, for example, an alcoholismic drinker believes that the inability to control the use of alcohol is the <u>sine qua non</u> of alcoholism and if he/she then interprets this inability in terms of a <u>complete</u> lack of control, then such a cognitive distortion is likely to contribute to the failure of that drinker to appreciate the extent of his/her own alcohol problems. There are certainly many problem drinkers who maintain that, as long as they possess the capacity to regulate their drinking (even if they rarely exercise this option), then they are not alcoholic. Further, many of these individuals expand still further their cognitive distortion of the nature of alcoholism, for the more they continue to drink alcoholismically, the more they tend to categorize "the alcoholic" in terms of "the skid row derelict."

Acceptance of one's "alcoholism" may be seen best as a cognitive surrender to the application of a socially-based label. At the time of this surrender, alcoholism is judged not by the observation of alcoholismic drinking, but by an overriding concern with abstinence and acceptance of the relevance of the "loss of control" construct to the individual drinker. With the eventual acceptance of the diagnosis, many individuals reorient their stance on the issue of control and view their demonstrated failure (now inability) to regulate their alcohol intake as the index of their alcoholism. Many also employ the loss of control construct as the explanation (and typically also the excuse) for previously unacceptable alcohol related behaviors and/or continued drinking.

The extent to which alcoholics believe that they cannot control their drinking affects their acceptance of the treatment goal of abstinence. Orford (1973) compared alcoholics who believed that their drinking was "mainly controlled" with others who believed their drinking was "totally uncontrolled." He found that the "totally uncontrolled drinkers" considered themselves alcoholics and preferred abstinence as a treatment goal, whereas the "mainly controlled drinkers" did not consider themselves alcoholics, even though they were in treatment for alcohol abuse. Further, the "mainly controlled" drinkers were found to be less willing to accept abstinence-oriented treatment even though they were ready to admit the disadvantages of their drinking behavior.

Other components of the cognitive dimension also affect facets of alcoholismic drinking. In reviewing the principal incentive elements associated with alcohol use earlier in this paper, expectancies about the effects of alcohol were indicated to be important in the initiation and continuation of such behavior. These expectancies would appear to be particularly relevant to the construct of craving (a subjectively experienced intense desire for alcohol) which has been employed in attempts to explain the initiation of drinking and relapse after abstinence (Isbell, 1955; Jellinek, 1952, 1960; Mardones, 1955). Despite recent attempts empirically to legitimize the construct of craving (Ludwig and Stark, 1974; Ludwig and Wikler, 1974; Ludwig, Wikler, and Stark, 1974), this construct, like loss of control, seems to have questionable scientific utility (see the reviews by Maisto and Schefft, in press; and Marlatt, 1977).

The significance of cognitive expectancies about alcohol and its effects has been established in a series of recent studies addressing issues surrounding loss of control and craving. For example, Marlatt, Demming, and Reid (1973) told half of a group of non-abstinent male alcoholics and half of a group of matched social drinkers they were to rate alcohol beverages (vodka and tonic) and the other half of the groups that they were to rate tonic water only. In fact, half of the subjects in each instructional conditio received alcohol beverages. The results of the study were clear-cu The only significant determinant of overall beverage consumption - and subjects' later estimates of the alcohol content of their respective drinks - was the expectancy factor. Regardless of the actual alcohol content of the drinks, both alcoholic and social drinker subjects consumed significantly more beverage when they believed they were sampling drinks containing vodka. Similarly, subjects assigned to the told tonic/received alcohol condition consumed relatively little beverage, whether or not the drinks actually contained alcohol (see also Briddell, Rimm, Caddy, Krawitz Scholis, and Wunderland, Note 12; Engle and Williams, 1972; Lang, Goecker, Adesso, and Marlatt, 1975; Maisto, Lauderman, and Adesso, 1977; Williams, 1970; Wilson and Lawson, 1976). All these studies indicate that prior expectations may exert a stronger influence than the physiplogical effects of alcohol on a variety of behaviors by both alcoholic and social drinkers.

Recently, Marlatt (1977) has offered a compelling analysis of relapse across different consummatory behaviors; it massively implicates the cognitive dimension. Marlatt proposes the existence of an Abstinence Volition Effect (AVE) which he postulates to occur when: (a) an individual is personally committed to an extended or indefinite period of abstinence from a specific behavior; and (b) the behavior occurred during this period of voluntary abstinence. The AVE is characterized by two key cognitive elements:

(1) A cognitive dissonance effect (Festinger, 1957, 1964), wherein the occurrence of the previously restricted behavior is dissonant with the cognitive definition of oneself as abstinent. Cognitive dissonance is experienced as a conflict state and underlies what most people would define as guilt for having given in to temptation.

(2) A personal attribution effect (Jones, Kanouse, Kelley, Nisbett, Valins, and Weiner, 1972), wherein the individua attributes the occurrence of the taboo behavior to internal

weakness or personal failure (e.g., a "lack of will power" or "insufficient personal control" over one's behavior) rather than to external situational or environmental factors (Marlatt, 1977, p. 53).

According to Marlatt's analysis, a full-blown alcoholic relapse occurs under the following circumstances: (a) The abstinent alcoholic feels "in control" until he encounters a high-risk situation which challenges his perception of control; (b) The individual lacks an appropriate method of coping with the high-risk situation or fails to engage in a coping response; (c) He has positive expectancies about the effects of alcohol and alcohol is available; (d) He takes the first drink; (e) He experiences one or both components of the AVE; and (f) The probability of continued drinking markedly increases. Clearly, this analysis of the events antecedent to taking the first drink is multivariate in nature. So, a multivariate analysis, involving elements from the behavioral, discriminative, incentive, social, and cognitive dimensions, is required to account for drinking beyond the first drink.

A number of treatment and prevention implications have their roots in the cognitive dimension. In fact, a detailed analysis of the cognitive variables that set the stage for the use of alcohol as well as those operating during the act of drinking is critical to an understanding of alcoholismic drinking. From such a base, specific treatment components could be directed to an alcoholismic drinker or alcoholic in order to correct his/her cognitively-based distortions and to develop adaptive attitudes toward himself/herself, and the nature of his/her cognitively-based difficulties.

CONCLUSIONS

This paper attempts a systematic analysis of the response patterns and cognitive mediational variables which combine in the development of alcoholism. The multivariate approach espoused here views alcoholism not as a unitary disease but as a diagnosis made when the use of alcohol results in pervasive problems that are no longer avoidable. Recognition of the multidimensional nature of alcohol dependence virtually forces attention to be directed to the uniqueness of each individual case. An analysis of the numerous elements which may be implicated in the disorder, based on examination of the behavioral, discriminative, incentive, social, and cognitive variables associated with alcoholism, permits the drinker, his/her family, and the therapist to have a far clearer view of the complex called alcoholism. From such a vantage point more specific treatment elements could be directed at specific difficulties and specific research questions could be developed and tested.

Until quite recently alcoholism treatment methods have been

based on the supposition that essentially one population of alcohol
ics have only one major problem, alcoholism, and the correlary tha
alcoholics require one form of therapy which ultimately leads to on
treatment outcome, abstinence. While some characteristics are shar
by many people with alcohol problems and alcoholic sub-populations
may be clustered according to attitudes, role ascriptions, degree o
dysfunction, etc. (Hurwitz and Lebos, 1968), it is clear that different alcoholic populations have distinctly different characteristics (see English and Curtin, 1975; Pattison, Coe, and Doerr, 1973;
Pattison, Coe, and Rhodes, 1969) and that the population of people
with alcohol problems is a multivariate one.

The treatment implications of viewing alcoholism from the multivariate perspective are considerable. By integrating data from
the behavioral, discriminative, incentive, social, and cognitive
dimensions, the clinician can better appreciate the role of the
various components of an individual's drinking matrix and how these
components relate to many other aspects of his/her life health.
From such a perspective, future research may permit the clinician
to reliably match patients to specific treatment components, thereb,
maximizing outcome.

ACKNOWLEDGMENTS

The author would like to thank Dr. Steve Maisto and Dr. Arthur
Alterman for their valuable criticisms of an earlier draft of this
paper.

REFERENCE NOTES

1. Caddy, G. R. Abstinence or cure: A re-evaluation of the natur
of alcoholism and implications for treatment. Paper presented
at the 29th International Congress on Alcoholism and Drug Dependence. Sydney, Australia, 1970.

2. Sobell, M. B. Alcohol abuse as an operant; treatment implications. Paper presented at the Fourth Annual Southern California Conference on Behavior Modification, Los Angeles, October,
1972.

3. Reid, J. B. Workshop at Conference on Behavioral Approaches to
Alcoholism and Drug Dependencies. University of Washington,
Seattle, 1975.

4. Caddy, G. R. Problems in the Field of Alcoholism Treatment
Outcome Evaluation: A review with special reference to blind
and independent research programs. Manuscript submitted for
publication, 1977.

5. Caddy, G. R. <u>Individualized behavior therapy for alcoholics,</u> <u>third year followup: Advantages and disadvantages of conducting</u> <u>double blind treatment outcome evaluation studies.</u> Paper presented at the Southeastern Psychological Association Convention, Hollywood, Florida, May 4, 1977.

6. Caddy, G. R., Addington, J. H., & Perkins, D. <u>Individualized</u> <u>behavior therapy for alcoholics: A third year independent</u> <u>double-blind followup.</u> Manuscript submitted for publication, 1977.

7. Schaefer, H. H. Aversive conditioning with alcoholics. Progress report, April 1970, Patton State Hospital, California. (NIMH Research Grant No. 16547)

8. Gowardman, M., Brown, R. A., & Williams, R. J. <u>Fluid intake in</u> <u>hospitalized alcoholic and psychiatric patients.</u> Unpublished manuscript, Kingseat Hospital, New Zealand, 1974.

9. Eaglen, A. <u>The effect of experimentally manipulated failure</u> <u>and success on the drinking behavior of alcoholic and nonalco-</u> <u>holic subjects.</u> Unpublished masters thesis, University of Auckland, 1974.

10. Marlatt, G. A. <u>A comparison of aversive conditioning proce-</u> <u>dures in the treatment of alcoholism.</u> Paper presented at the Western Psychological Association Annual Convention, Anaheim, California, April, 1973.

11. Blane, H. T. The personality of the alcoholic. The role of the nurse in the care of the alcoholic patient in a general hospital. Paper presented at the Conference at Chatham, Massachusetts, 1960. In D. C. McClelland, W. N. Davis, R. Kalin, and E. Wanner (Eds.), <u>The drinking man.</u> New York: The Free Press, 1972.

12. Briddell, D. W., Rimm, D. C., Caddy, G. R., Kravitz, G., Scholis, D., and Wunderland, R. <u>Alcohol, cognitive set, and</u> <u>sexual arousal to deviant stimuli.</u> Manuscript submitted for publication, 1977.

REFERENCES

Albrecht, G. L. The alcoholism process: A social learning viewpoint. In P. G. Bourne and R. Fox (Eds.), <u>Alcoholism: Pro-</u> <u>gress in research and treatment.</u> New York: Academic Press, 1973.

Alcoholics Anonymous. The story of how more than 100 men have re-
    covered from alcoholism. New York: Works Publishing Company,
    1939.

Alcoholics Anonymous. The story of how many thousands of men and
    women have recovered from alcoholism. New York: Alcoholics
    Anonymous Publishing, 1955.

Alcoholics Anonymous comes of age. New York: Harper & Brothers,
    1957.

American Medical Association. Manual on Alcoholism. American
    Medical Association, 1968, 9801, 468-32E-25 M.O.P. 185.

Armor, D. J., Polich, J. M., & Stambul, H. B. Alcoholism and
    treatment. (R-1739-NIAAA). Santa Monica, California: Rand
    Corporation, 1976.

Aubert, U., & Messinger, F. The criminal and the sick. Inquiry,
    1958, 1, 137-160.

Ausubel, D. P. Personality disorder is disease. American Psycho-
    logist, 1961, 61, 69-74.

Bacon, S. D. The process of addiction to alcohol: Social aspects.
    Quarterly Journal of Studies on Alcohol, 1973, 34, 1-27.

Bacon, M. K., Barry, H., III, & Child, L. Quarterly Journal of
    Studies on Alcohol Supplement, 1965, 3, 29-48.

Bailey, M. B., & Stewart, J. Normal drinking by persons reporting
    previous problem drinking. Quarterly Journal of Studies on
    Alcohol, 1967, 28, 305-315.

Bandura, A. Principles of behavior modification. New York: Holt,
    Rinehart, & Winston, 1969.

Bandura, A. Behavior theory and the models of man. American Psy-
    chologist, 1974, 29, 859-869.

Bem, D. Beliefs, attitudes and human affairs. Belmont, California
    Wadsworth Publishing Company, 1970.

Blane, H. T. The personality of the alcoholic: Guises of depen-
    dence. New York: Harper & Row, 1968.

Blane, H. T., Overton, W. F., & Chafetz, M. E. Social factors in
    the diagnosis of alcoholism. Quarterly Journal of Studies on
    Alcohol, 1963, 24, 640-643.

Blumberg, L. U., Shipley, T. E., Jr., & Moor, J. O., Jr. The skid row status community, with perspectives on their future. Quarterly Journal of Studies on Alcohol, 1971, 32, 909-941.

Blumberg, L. U., Shipley, T. E., Jr., & Barsky, S. F. Liquor and Poverty: Skid row as a human condition. New Brunswick, New Jersey: Rutgers Center of Alcohol Studies (Monograph No. 13), 1977.

Bois, C., & Vogel-Sprott, M. Discrimination of low blood alcohol levels and self-titration skills in social drinkers. Quarterly Journal of Studies on Alcohol, 1974, 35, 86-97.

Brady, J. P. Psychotherapy, learning theory, and insight. Archives of General Psychiatry, 1967, 16, 304-311.

Brewer, W. F. There is no convincing evidence for operant or classical conditioning in adult humans. In W. B. Weimer & D. S. Palermo (Eds.), Cognition and the symbolic processes. New York: Halsted Press, 1974.

Brown, R. A. A comparison of the control of alcoholic and non alcoholic drinking. Unpublished Doctoral dissertation, University of Auckland, 1974.

Brown, R. A., & Williams, R. J. Internal and external cues relating to fluid intake in obese and alcoholic subjects. Journal of Abnormal Psychology, 1975, 84, 660-665.

Caddy, G. R. Behaviour modification in the management of alcoholism. Unpublished Doctoral dissertation, University of New South Wales, 1972.

Caddy, G. R. Blood alcohol concentration discrimination training: Development and current status. In G. A. Marlatt & P. E. Nathan (Eds.), Behavioral approaches to the assessment and treatment of alcoholism. New Brunswick, New Jersey: Center for Alcohol Studies, 1978.

Caddy, G. R., Goldman, R. D., & Huebner, R. Group differences in attitudes towards alcoholism. Addictive Behaviors, 1976, 1, 281-286 (a).

Caddy, G. R., Goldman, R. D., & Huebner, R. Relationships among different domains of attitudes towards alcoholism: Model, cost and treatment. Addictive Behaviors, 1976, 1, 159-167 (b).

Caddy, G. R., & Lovibond, S. H.   Self-regulation and discriminated aversive conditioning in the modification of alcoholics drinking behavior. Behavior Therapy, 1976, 7, 223-230.

Cahalan, D. Problem drinkers; a national survey. San Francisco: Jossey-Bass, 1970.

Cahalan, D., Cisin, I. H., & Crossley, H. American drinking practices: A national survey of behavior and attitudes. New Brunswick, New Jersey: Rutgers Center of Alcohol Studies (Monograph No. 6), 1969.

Cahalan, D., & Room, R. Problem drinking among American men. New Brunswick, New Jersey: Rutgers Center of Alcohol Studies (Monograph No. 7), 1974.

Cahn, S. The treatment of alcoholics: An evaluative study. New York: Oxford University Press, 1970.

Canter, F. M. The requirements of abstinence as a problem in institutional treatment of alcoholics. Psychiatric Quarterly, 1968, 42, 217-231.

Cappell, H. An evaluation of tension models of alcohol consumption. In R. J. Gibbons, Y. Israel, H. Kalant, R. E. Pophan, W. Schmidt, & R. G. Smart (Eds.), Research Advances in Alcohol and Drug Problems (Vol. 2). New York: Wiley, 1975.

Caudill, B. D., & Marlatt, G. A. Modelling influence in social drinking: An experimental analogue. Journal of Consulting and Clinical Psychology, 1975, 43, 405-415.

Chafetz, M. E. Alcohol excess. Annals of the New York Academy of Sciences, 1966, 133, 808-813.

Chafetz, M. E., Blane, H. T., & Hill, M. J. (Eds.). Frontiers of alcoholism. New York: Science House, 1970.

Chafetz, M. E., & Demone, H. W. Alcoholism and society. New York: Oxford University Press, 1962.

Chess, S. B., Neuringer, C., & Goldstein, G. Arousal and field dependency in alcoholics. Journal of General Psychology, 1971, 85, 93-102.

Clancy, J. Procrastination: A defense against sobriety. Quarterly Journal of Studies on Alcohol, 1960, 21, 269-276.

Cohen, M., Liebson, I. A., & Faillace, L. A. The modification of drinking in chronic alcoholics. In N. K. Mello and J. H. Mendelson (Eds.), Recent advances in studies of alcoholism. Washington, D. C.: U. S. Government Printing Office, 1971.

Conger, J. J. Reinforcement theory and the dynamics of alcoholism. Quarterly Journal of Studies on Alcohol, 1956, 17, 296-305.

Crawford, J. J., & Chalupsky, A. B. The reported evaluation of alcoholism treatments, 1968-1971: A methodological review. Addictive behaviors, 1977, 2, 63-74.

Cutter, H. S. G., Schwaab, E. L., Jr., & Nathan, P. E. Effects of alcohol on its utility for alcoholics and non-alcoholics. Quarterly Journal of Studies on Alcohol, 1970, 31, 369-378.

Davies, D. L. Normal drinking in recovered alcohol addicts. Quarterly Journal of Studies on Alcohol, 1962, 23, 94-104.

Davies, D. L. Normal drinking in recovered alcohol addicts. (Comment by various correspondents). Quarterly Journal of Studies on Alcohol, 1963, 24, 109-121, 321-332.

Davison, G. C. Systematic desensitization as a counter-conditioning process. Journal of Abnormal Psychology, 1968, 73, 91-99.

de Lint, J. The status of alcoholism as a disease. British Journal of Addiction, 1971, 66, 108-109.

de Lint, J., & Schmidt, W. The epidemiology of alcoholism. In Y. Israel & J. Mardones (Eds.), The biological basis of alcoholism. New York: Wiley, 1971.

Deveraux, M. W., & McCorkick, R. A. Psychogenic water intoxication: A case report. American Journal of Psychiatry, 1972, 129, 628-630.

Docter, R. F., & Bernal, M. E. Immediate and prolonged psychophysiological effects of sustained alcohol intake in alcoholics. Quarterly Journal of Studies on Alcohol, 1964, 25, 438-450.

Dollard, J., & Miller, N. E. Personality and psychotherapy. New York: McGraw-Hill, 1950.

Edwards, G. Drugs: Drug dependence and the concept of plasticity. Quarterly Journal of Studies on Alcohol, 1974, 35, 176-195.

Edwards, G., Kyle, E., & Nicholls, P. Alcoholics admitted to four hospitals in England. Quarterly Journal of Studies on Alcohol 1974, 35, 499-522.

Ellis, A. Reason and emotion in psychotherapy. New York: Lyle Stuart, 1962.

Encel, S., & Kotowicz, K. Heavy drinking and alcoholism: Preliminary report. Medical Journal of Australia, 1970, 1, 607-612.

Engle, K. B., & Williams, T. K. Effects of an ounce of vodka on alcoholics' desire for alcohol. Quarterly Journal of Studies on Alcohol, 1972, 33, 1099-1105.

English, G. E., & Curtin, M. E. Personality differences in patients at three alcoholism treatment agencies. Journal of Studies on Alcohol, 1975, 36, 52-61.

Festinger, L. A theory of cognitive dissonance. Stanford: Stanford University Press, 1957.

Festinger, L. Conflict, decision and dissonance. Stanford: Stanford University Press, 1964.

Field, P. B. A new cross-cultural study of drunkenness. In D. J. Pittman & C. R. Snyder (Eds.), Society, culture and drinking patterns. New York: Wiley, 1962.

Fitzsimmons, J. T. Thirst. Physiological Review, 1972, 52, 468-561.

Flaherty, J. A., McGuire, H. T., & Gatski, R. L. The psychodynamics of the "dry drunk". American Journal of Psychiatry, 1955, 112, 460-464.

Force, R. C. Development of a court test for the detection of alcoholism by a keying of the Kuder Preference Record. Quarterly Journal of Studies on Alcohol, 1958, 19, 72-78.

Foy, D. W., Miller, P. M., Eisler, R. M., & O'Toole, D. H. Social skills training to teach alcoholics to refuse drinks effectively. Journal of Alcohol Studies, 1976, 37, 1340-1345.

Freed, E. X. The effect of alcohol upon approach-avoidance conflict in the white rat. Quarterly Journal of Studies on Alcohol, 1967, 28, 236-254.

Freed, E. X. Effect of self-intoxication upon approach-avoidance conflict in the rat. Quarterly Journal of Studies on Alcohol, 1968, 29, 323-329.

Gerard, D. L., Saenger, G., & Wile, R. The abstinent alcoholic. Archives of General Psychiatry, 1962, 16, 83-95.

Gitlow, S. E. Alcoholism: A disease. In P. G. Bourne, & R. Fox (Eds.), Alcoholism: Progress in research and treatment. New York: Academic Press, 1973.

Glock, G. Y. Image of man and public opinion. Public Opinion Quarterly, 1964, 28, 539-546.

Goffman, E. Behavior in public places. New York: Free Press, 1963 (a).

Goffman, E. Stigma. Englewood Cliffs: Prentice-Hall, 1963 (b).

Goldstein, S. G., & Linden, J. D. Multivariate classification of alcoholics by means of the MMPI. Journal of Abnormal Psychology, 1969, 74, 661-669.

Gottheil, E., Crawford, H. D., & Cornelison, F. S. The alcoholics' ability to resist available alcohol. Disorders of the Nervous System, 1973, 34, 80-82.

Hamburg, S. Behavior therapy in alcoholism: A critical review of broad-spectrum approaches. Journal of Studies on Alcohol, 1975, 36, 69-87.

Hayman, M. The myth of social drinking. American Journal of Psychiatry, 1967, 125, 585-594.

Heath, D. B. A critical review of ethnographic studies of alcohol use. In R. J. Gibbins, Y. Israel, H. Kalant, R. E. Popham, W. Schmidt, & R. G. Smart (Eds.), Research Advances in Alcohol and Drug Problems. New York: Wiley, Vol. 2, 1975.

Hershon, H. I. Alcohol withdrawal symptoms: Phenomenology and implications. British Journal of Addiction, 1973, 68, 295-302.

Hershon, H. I. Alcohol withdrawal symptoms and drinking behavior. Journal of Studies on Alcohol, 1977, 38, 953-971.

Holder, H. D., & Stratas, N. E. A systems approach to alcoholism programming. American Journal of Psychiatry, 1972, 129, 32-37.

Holmes, J. H., & Montgomery, A. V. Physiological changes induced by high fluid intakes. Federation Proceedings, 1951, 10, 66.

Horn, J. L., & Wanberg, K. W. Symptom patterns related to the excessive use of alcohol. Quarterly Journal of Studies on Alcohol, 1969, 30, 35-58.

Horn, J. L., & Wanberg, K. W. Dimensions of perception of back-
    ground and current situation of alcoholic patients. Quarterly
    Journal of Studies on Alcohol, 1970, 31, 633-658.

Huber, H., Karlin, R., & Nathan, P. E. Blood alcohol level discrim-
    ination in non-alcoholics: The role of internal and external
    cues. Journal of Studies on Alcohol, 1976, 37, 27-39.

Hughes, F. W., Forney, R. B., & Gates, P. W. Performance in human
    subjects under delayed auditory feedback after alcohol, a
    tranquilizer (Benquinamide) or Benquinamide-Alcohol combina-
    tion. Journal of Psychology, 1963, 55, 25-32.

Hunt, G. M., & Azrin, N. H. A community-reinforcement approach to
    alcoholism. Behaviour Research and Therapy, 1973, 11, 91-101.

Hurwitz, J. I., & Lelos, D. A multilevel interpersonal profile of
    employed alcoholics. Quarterly Journal of Studies on Alcohol,
    1968, 29, 64-76.

Isbell, H. Craving for alcohol. Quarterly Journal of Studies on
    Alcohol, 1955, 16, 38-42.

Jackson, J. K. The adjustment of the family to the crisis of al-
    coholism. Quarterly Journal of Studies on Alcohol, 1954, 15,
    562-586.

Jacobson, G. R. Reduction of field dependence in chronic alcoholic
    patients. Journal of Abnormal Psychology, 1968, 73, 547-549.

Jacobson, G. R. Sensory deprivation and field dependence in alco-
    holics. (University Microfilms, No. 72-22, 839). Unpublished
    Doctoral dissertation, Illinois Institute of Technology, 1971.

Jellinek, E. M. The phases of alcohol addiction, Quarterly
    Journal of Studies on Alcohol, 1952, 13, 673-684.

Jellinek, E. M. The disease concept of alcoholism. Highland Park,
    New Jersey: Hillhouse Press, 1960.

Jessor, R., Collins, M. I., & Jessor, S. L. On becoming a drinker:
    Social aspects of an adolescent transition. Annals of the New
    York Academy of Science, 1972, 197, 199-213.

Jessor, R., & Jessor, S. L. Problem drinking in youth: Personality,
    social and behavioral antecedents and correlates. Proceedings
    of the Second Annual Alcoholism Conference. Washington, D. C.:
    NIAAA, 1973.

Johnson, W. The effect of cue prominence and subject weight on human food-directed performance. In S. Schachter & J. Rodin (Eds.), Obese humans and rats. Potomac, Maryland: Erlbaum, 1974.

Jones, E. E., Kanouse, D. E., Kelley, H. H., Nisbett, R. E., Valins, S., & Weiner, B. (Eds.), Attribution: Perceiving the causes of behavior. Morristown, N. J.: General Learning Press, 1972.

Kalant, H., & LeBlanc, A. E. Tolerance to, and dependence on, ethanol. In Y. Israel & J. Mardones (Eds.), Biological Basis of Alcoholism. NewYork: Wiley, 1971.

Kalin, R. Self-descriptions of college problem drinkers. In D. C. McClelland, W. N. Davis, R. Kalin, & E. Wanner (Eds.), The drinking man. New York: The Free Press, 1972.

Keehn, J. D. Reinforcement of alcoholism: Schedule control of solitary drinking. Quarterly Journal of Studies on Alcohol, 1970, 31, 28-39.

Keller, M. The definition of alcoholism and the estimation of its prevalence. In D. J. Pittman & C. R. Snyder (Eds.), Society, culture and drinking patterns. New York: Wiley, 1962.

Keller, M. The oddities of alcoholics. Quarterly Journal of Studies on Alcohol, 1972, 33, 1147-1148 (a).

Keller, M. On the loss-of-control phenomenon in alcoholism. British Journal of Addiction, 1972, 67, 153-166 (b).

Kendall, R. E. Normal drinking by former alcoholic addicts. Quarterly Journal of Studies on Alcohol, 1965, 26, 247-257.

Kessler, M., & Gomberg, C. Observations of barroom drinking: methodology and preliminary results. Quarterly Journal of Studies on Alcohol, 1974, 35, 1392-1396.

Kilpatrick, D. G., Sutker, P. B., & Smith, A. D. Deviant drug and alcohol use: The role of anxiety, sensation seeking, and other personality variables. In M. Zuckerman & C. D. Spielberger (Eds.), Emotions and anxiety: New concepts, methods and applications. New York: Wiley, 1976.

Knupfer, G. The epidemiology of problem drinking. American Journal of Public Health, 1967, 57, 974-986.

Lang, A. R., Goeckner, D. J., Adesso, V. J., & Marlatt, G. A. The effects of alcohol on aggression in male social drinkers. Journal of Abnormal Psychology, 1975, 84, 508-518.

Lansky, D., Nathan, P. E., & Lawson, D. M. Blood alcohol level dis-
    crimination by alcoholics: The role of internal and external
    cues. Journal of Consulting and Clinical Psychology, in press.

Lazarus, A. A. Toward the understanding and effective treatment of
    alcoholism. South African Medical Journal, 1965, 39, 736-741.

Lazarus, A. A. Variations in desensitization therapy. Psychother-
    apy: Theory, Research and Practice, 1968, 5, 50-52.

Lazarus, A. A. Behavior therapy and beyond. New York: McGraw-Hill,
    1971.

Linsky, A. S. Theories of behavior and social control of alcoholism.
    Social Psychiatry, 1972, 7, 47-52.

Lloyd, R. W., Jr., & Salzberg, H. C. Controlled social drinking:
    An alternative to abstinence as a treatment goal for some
    alcohol abusers. Psychological Bulletin, 1975, 82, 815-842.

Lolli, G., Rubin, M., & Greenberg, L. A. The effect of ethyl alco-
    hol on the volume of extra cellular water. Quarterly Journal
    of Studies on Alcohol, 1944, 5, 1-4.

Lovibond, S. H., & Caddy, G. R. Discriminated aversive control in
    the moderation of alcoholics' drinking behavior. Behavior
    Therapy, 1970, 1, 437-444.

Ludwig, A. M., & Stark, L. H. Alcohol craving: Subjective and
    situational aspects. Quarterly Journal of Studies on Alcohol,
    1974, 35, 899-905.

Ludwig, A. M., & Wikler, A. "Craving" and relapse to drink.
    Quarterly Journal of Studies on Alcohol, 1974, 35, 108-130.

Ludwig, A. M., Wikler, A., & Stark, L. H. The first drink: Psycho-
    biological aspects of craving. Archives of General Psychiatry,
    1974, 30, 539-547.

McAndrew, C., & Edgerton, R. B. Drunken Comportment: A social ex-
    planation. Chicago: Aldine, 1969.

Machover, S., & Puzzo, F. S. Clinical and objective studies of
    personality variables in alcoholism. Quarterly Journal of
    Studies on Alcohol, 1959, 20, 505-519.

Madsen, W. The American alcoholic. Springfield, Illinois: Charles
    C. Thomas, 1974.

Mahoney, M. J. Cognition and behavior modification. Cambridge, Mass.: Ballinger, 1974.

Mahoney, M. J. Fat fiction. Behavior Therapy, 1975, 6, 416-418 (a).

Mahoney, M. J. The obese eating style: Bites, beliefs, and behavior modification. Addictive Behaviors, 1975, 1, 47-53 (b).

Maisto, S. A., Lauderman, R., & Adesso, V. J. A comparison of two experimental studies of the role of cognitive factors in alcoholics' drinking. Journal of Studies on Alcohol, 1977, 38, 145-149.

Maisto, S. A., & Schefft, B. K. The constructs of craving for alcohol and loss of control drinking: Help or hindrance to research. Addictive Behaviors, in press.

Mann, M. New primer on alcoholism. New York: Holt, Rinehart, & Winston, 1968.

Marconi, J. The concept of alcoholism. Quarterly Journal of Studies on Alcohol, 1959, 20, 216-235.

Marconi, J. Scientific theory and operational definitions in psychotherapy with special references to alcoholism. Quarterly Journal of Studies on Alcohol, 1967, 28, 631-640.

Marconi, J., Fink, K., & Moya, L. Experimental study on alcoholics with "inability to stop". British Journal of Psychiatry, 1967, 113, 543-545.

Mardones, R. J. On the relationship between deficiency of B vitamins and alcohol intake in rats. Quarterly Journal of Studies on Alcohol, 1951, 12, 563-575.

Mardones, R. J. "Craving" for alcohol. Quarterly Journal of Studies on Alcohol, 1955, 16, 51-53.

Marlatt, G. A. Alcohol, stress and cognitive control. In S. D. Spielberger & I. G. Sarason (Eds.), Stress and anxiety (Vol. 3). New York: Hemisphere Publishing Co., 1976.

Marlatt, G. A. Craving for alcohol, loss of control, and relapse: A cognitive-behavioral analysis. (Technical Report No. 77-05) Seattle: University of Washington, Alcoholism and Drug Abuse Institute, 1977.

Marlatt, G. A., Demming, B., & Reid, J. B. Loss of control drinking in alcoholics: An experimental analogue. Journal of Abnormal Psychology, 1973, 81, 233-241.

Masserman, J. H., Jacques, M. B., & Nicholson, M. R. Alcohol as preventive of experimental neuroses. Quarterly Journal of Studies on Alcohol, 1945, 6, 281-299.

McClelland, D. C., Davis, W. N., Kalin, R., & Wanner, E. (Eds.), The drinking man. New York: The Free Press, 1972.

McCord, W., & McCord, J. Origins of alcoholism. Stanford: Stanford University Press, 1960.

McNamee, H. B., Mello, N. K., & Mendelson, J. H. Experimental analysis of drinking patterns of alcoholics: Concurrent psychiatric observations. American Journal of Psychiatry, 1968, 124, 1063-1069.

Mello, N. K. Behavioral studies of alcoholism. In B. Kissin & H. Begleiter (Eds.), The biology of alcoholism (Vol. 2, Physiology and Behavior). New York: Plenum Press, 1972.

Mello, N. K., & Mendelson, J. H. Operant analysis of drinking patterns of chronic alcoholics. Nature, 1965, 206, 43-46.

Mello, N. K., & Mendelson, J. H. Experimentally induced intoxication in alcoholics: A comparison between programmed and spontaneous drinking. Journal of Pharmacology and Experimental Therapeutics, 1970, 173, 101-116.

Mendelson, J. H., La Dou, J., & Solomon, P. Experimentally induced chronic intoxication and withdrawal in alcoholics: Psychiatric findings. Quarterly Journal of Studies on Alcohol, Supplement No. 2, 1964, 40-52.

Mendelson, J. H., & Mello, N. K. Experimental analysis of drinking behavior of chronic alcoholics. Annals of the New York Academy of Sciences, 1966, 133, 828-845.

Merry, J. The "loss of control" myth. Lancet, 1966, 1, 1257-1258.

Miller, W. R., & Caddy, G. R. Abstinence and controlled drinking in the treatment of problem drinkers. Journal of Studies on Alcohol, 1977, 38, 986-1003.

Nathan, P. E., Lipson, A. G., Vettraino, A. P., & Solomon, P. The social ecology of an urban clinic for alcoholism. International Journal of Addictions, 1968, 3, 55-64.

Nathan, P. E., & Lisman, S. A. Behavioral and motivational patterns of chronic alcoholics. In R. E. Tarter & A. A. Sugerman (Eds.) Alcoholism: Interdisciplinary Approaches to an Enduring Problem Reading, Massachusetts: Addison-Wesley, 1976.

Nathan, P. E., & O'Brien, J. S. An experimental analysis of the behavior of alcoholics and non-alcoholics during prolonged experimental drinking. Behavior Therapy, 1971, 2, 455-476.

Nathan, P. E., Titler, N. A., Lowenstein, L. M., Solomon, P., & Rossi, A. M. Behavioral analysis of chronic alcoholism: Interaction of alcohol and human contact. Archives of General Psychiatry, 1970, 22, 419-430.

Negrete, J. C. Cultural influences on social performance of alcoholics. Quarterly Journal of Studies on Alcohol, 1973, 34, 905-916.

Nisbett, R. E. Determinants of food intake in human obesity. Science, 1968, 159, 1254-1255 (a).

Nisbett, R. E. Taste, deprivation and weight determinants of eating behavior. Journal of Personality and Social Psychology, 1968, 10, 107-117 (b).

Okulitch, P. V., & Marlatt, G. A. Effects of varied extinction conditions with alcoholics and social drinkers. Journal of Abnormal Psychology, 1972, 79, 205-211.

Orford, J. A comparison of alcoholics whose drinking is totally uncontrolled and those whose drinking is mainly controlled. Behaviour Research and Therapy, 1973, 11, 565-576.

Paredes, A. Marital-sexual factors in alcoholism. Medical Aspects of Human Sexuality, 1973, 7, 98-114.

Partington, J. T., & Johnson, F. G. Personality types among alcoholics. Quarterly Journal of Studies on Alcohol, 1969, 30, 21-34.

Pattison, E. M. A critique of alcoholism treatment concepts; with special reference to abstinence. Quarterly Journal of Studies on Alcohol, 1966, 27, 49-71.

Pattison, E. M. Abstinence criteria: A critique of abstinence criteria in the treatment of alcoholism. International Journal of Social Psychiatry, 1968, 14, 268-276.

Pattison, E. M. Drinking outcomes of alcoholism treatment. Abstinence, social, modified, controlled, and normal drinking. In N. Kessel, A. Hawker, & H. Chalke (Eds.), Alcoholism: A medical profile. London: B. Edsall and Co., 1974 (a).

Pattison, E. M. The rehabilitation of the chronic alcoholic. In B. Kissin & H. Begleiter (Eds.), The Biology of Alcoholism. (Vol. 3). New York: Plenum Press, 1974 (b).

Pattison, E. M. Non-abstinent drinking goals in the treatment of alcoholism. In R. J. Gibbins. Y. Israel, H. Kalant, R. E. Popham, W. Schmidt, & R. G. Smart (Eds.), Research advances in alcohol and drug problems (Vol. 3). New York: Wiley, 1976.

Pattison, E. M., Coe, R., & Doerr, H. D. Population variation between alcoholism treatment facilities. International Journal of the Addictions, 1973, 8, 199-229.

Pattison, E. M., Coe, R., & Rhodes, R. A. Evaluation of alcoholism treatment: Comparison of three facilities. Archives of General Psychiatry, 1969, 20, 478-488.

Pattison, E. M., Sobell, M. B., & Sobell, L. C. (Eds.). Emerging concepts of alcohol dependence. New York: Springer, 1977.

Pittman, D. J., & Gordon, C. W. Revolving door: A study of the chronic police case inebriate. New Brunswick, New Jersey: Rutgers Center of Alcohol Studies (Monograph No. 3), 1958.

Pittman, D. J., & Snyder, C. R. (Eds.). Society, culture and drinking patterns. New York: Wiley, 1962.

Pliner, P. Effects of cue salience on the behavior of obese and normal subjects. Journal of Abnormal Psychology, 1973, 82, 226-232 (a).

Pliner, P. Effects of external cues on the thinking behavior of obese and normal subjects. Journal of Abnormal Psychology, 1973, 82, 233-238 (b).

Pliner, P. External responsiveness in the obese. Addictive Behaviors, 1976, 1, 169-175.

Rachman, S., & Teasdale, J. Aversion therapy and behaviour disorders: An analysis. London: Routledge and Kegan Paul, 1969.

Rado, S. Narcotic bondage: A general theory of the dependence on narcotic drugs. In P. H. Hock & J. Zubin (Eds.), Problems of addiction and habituation. New York: Grune and Stratton, 1958.

Randolph, T. G. The descriptive features of food addiction. Addictive eating and drinking. Quarterly Journal of Studies on Alcohol, 1956, 17, 198-224.

Robinson, D. The alcohologist addiction: Some implications of having lost control over the disease concept of alcoholism. Quarterly Journal of Studies on Alcohol, 1972, 33, 1028-1042.

Rodin, J., Herman, C. P., & Schachter, S. External sensitivity and obesity. In S. Schachter & J. Rodin (Eds.), Obese humans and rats. Potomac, Maryland: Earlbaum, 1974.

Roman, P. M., & Trice, H. M. The sick role, labelling theory and the deviant drinker. The International Journal of Social Psychiatry, 1968, 14, 245-251.

Roman, P. M., & Trice, H. M. The development of deviant drinking behavior. Archives of Environmental Health, 1970, 20, 424-435.

Ross, L. Cue and cognition controlled eating among obese and normal subjects. In S. Schachter & J. Rodin (Eds.), Obese humans and rats. Potomac, Maryland: Earlbaum, 1974.

Saltzman, L. Obsessive-compulsive aspects of obesity. Psychiatry in Medicine, 1972, 3, 29-36.

Schachter, S. Some extraordinary facts about obese humans and rats. American Psychologist, 1971, 26, 129-144.

Schachter, S., & Gross, L. Manipulated time and eating behavior. Journal of Personality and Social Psychology, 1968, 10, 98-106.

Schaefer, H. H., Sobell, M. B., & Mills, K. C. Baseline drinking behaviors in alcoholics and social drinkers;kinds of drinking and sip magnitude. Behavior Research and Therapy, 1971, 9, 23-27.

Schmidt, W., Smart, R., & Moss, M. Social class and the treatment of alcoholism. Toronto: Addiction Research Foundation (Brookside Monograph No. 7), 1968.

Scott, P. D. Offenders, drunkenness and murder. The British Journal of Addiction, 1968, 63, 221-226.

Seligman, M. E. P. Learned helplessness. Annual Review of Medicine, 1972, 23, 407-412.

Shae, J. E. Psychoanalytic therapy and alcoholism. Quarterly Journal of Studies on Alcohol, 1954, 15, 595-605.

Shapiro, D. Neurotic styles. New York: Basic Books, 1965.

Shettleworth, S. J. Stimulus relevance in the control of drinking
        and conditioned fear responses in domestic chicks (Gallus
        Gallus). Journal of Comparative and Physiological Psychology,
        1972, 80, 175-198.

Siegler, M., Osmond, H., & Newell, S. Models of alcoholism. Quar-
        terly Journal of Studies on Alcohol, 1968, 29, 571-591.

Silkworth, W. D. Alcoholism as manifestation of allergy. Medical
        Records, 1937, 145, 249-251.

Silverstein, S. J., Nathan, P. E., & Taylor, H. A. Blood alcohol
        level estimation and controlled drinking by chronic alcoholics.
        Behavior Therapy, 1974, 5, 1-15.

Sirnes, T. B. Voluntary consumption of alcohol in rats with cir-
        rhosis of the liver. A preliminary report. Quarterly Journal
        of Studies on Alcohol, 1953, 14, 3-18.

Smart, R. G. Effects of alcohol on conflict and avoidance behavior.
        Quarterly Journal of Studies on Alcohol, 1965, 26, 187-205.

Sobell, L. C. Empirical assessment of alcoholism treatment outcome
        evaluation: Past, present and future. In G. A. Marlatt and
        P. E. Nathan (Eds.), Behavioral Assessment and Treatment of
        Alcoholism. New Brunswick, N. J.: Rutgers Center of Alcohol
        Studies, 1978.

Sobell, L. C., Sobell, M. B., & Christelman, W. C. The myth of
        "one drink." Behaviour Research and Therapy, 1972, 10, 119-
        123.

Sobell, L. C., Sobell, M. B., & Schaefer, H. H. Alcoholics name
        fewer mixed drinks than social drinkers. Psychological Re-
        ports, 1971, 28, 493-494.

Sobell, M. B., & Sobell, L. C. Individualized behavior therapy for
        alcoholics. Sacramento: California Department of Mental
        Health (Research Monograph No. 13). 1972.

Sobell, M. B., & Sobell, L. C. Individualized behavior therapy for
        alcoholics. Behaviour Research and Therapy, 1973, 4, 49-72.

Sobell, M. B., & Sobell, L. C. The need for realism, relevance and
        operational assumptions in the study of substance dependence.
        In H. D. Cappell & A. E. LeBlanc (Eds.), Biological and Be-
        havioral Approaches to Drug Dependence. Toronto: Addiction
        Research Foundation, 1975.

Sobell, M. B., & Sobell, L. C.   Second year treatment outcome of alcoholics treated by individualized behavior therapy: Results. Behaviour Research and Therapy, 1976, 14, 195-215.

Sobell, M. B., Schaefer, H. H., & Mills, K. C.   Differences in baseline drinking behavior between alcoholics and normal drinkers.   Behaviour Research and Therapy, 1972, 10, 257-267.

Sobell, M. B., Sobell, L. C., & Sheahan, D. B.   Functional analysis of drinking problems as an aid in developing individual treatment strategies.   Addictive Behaviors, 1976, 1, 127-132.

Sommer, R.   The isolated drinker in the Edmonton beer parlor. Quarterly Journal of Studies on Alcohol, 1965, 26, 95-110.

Steiner, C. N.   Games alcoholics play: The analysis of self-scripts. New York: Grove Press, 1971.

Steinglass, P., Weiner, S., & Mendelson, J. H.   Interactional issues as determinants of alcoholism.   American Journal of Psychiatry, 1971, 128, 275-280 (a).

Steinglass, P., Weiner, S., & Mendelson, J. H.   A systems approach to alcoholism.   Archives of General Psychiatry, 1971, 24, 401-408 (b).

Stoll, C. S.   Images of man and social control.   Social Forces, 1968, 47, 119-127.

Stunkard, A. J.   Eating patterns and society.   Psychiatric Quarterly, 1959, 33, 284-295.

Sugerman, A. A., Reilly, D., & Albahary, R. S.   Social competence and the essential-reactive distinction in alcoholism.   Archives of General Psychiatry, 1965, 12, 552-556.

Sugerman, A. A., & Schneider, D. U.   Cognitive styles in alcoholism. In R. E. Tarter & A. A. Sugerman (Eds.), Alcoholism: Interdisciplinary approaches to an enduring problem.   Reading, Mass.:   Addison-Wesley, 1976.

Swanson, D. W., & Dinello, F. A.   Severe obesity as a habituation syndrome: Evidence during a starvation study.   Archives of General Psychiatry, 1970, 22, 120-127.

Szasz, T. S.   The myth of mental illness: Foundations of a theory of personal conduct.   New York: Hoeber, 1961.

Szasz, T. S.   Ideology and insanity.   New York: Doubleday, 1970.

Tahaka, V. The alcoholic personality. Finnish Foundation for Alcohol Studies Report: Helsinki, 1966.

Talland, G. A., & Kasschau, R. Practice and alcohol effects on moto skill and attention: A supplementary report on an experiment in chronic intoxication and withdrawal. Quarterly Journal of Studies on Alcohol, 1965, 26, 393-401.

Talland, G. A., Mendelson, J. H., & Ryack, P. Tests of motor skills In J. H. Mendelson (Ed.), Experimentally induced chronic intoxication and withdrawal in alcoholics. Quarterly Journal of Studies on Alcohol, Supplement No. 2, 1964, 53-73.

Tamerin, J. S., & Mendelson, J. H. The psychodynamics of chronic inebriation: Observations of alcoholics during the process of drinking in an experimental group setting. American Journal of Psychiatry, 1969, 125, 886-889.

Ullmann, L. P., & Krasner, L. (Eds.). Case studies in behavior modification. New York: Holt, Rinehart and Winston, 1965.

Ullmann, L. P., & Krasner, L. A psychological approach to abnormal behavior. Englewood Cliffs, New Jersey: Prentice Hall, 1969.

Vanderpool, J. A. Alcoholism and the self-concept. Quarterly Journal of Studies on Alcohol, 1969, 30, 59-77.

Vogel-Sprott, M. Alcohol effects on human behavior under reward and punishment. Psychopharmacologia, 1967, 11, 337-344.

Vogel-Sprott, M. Defining "light" and "heavy" social drinking: Research implications and hypothesis. Quarterly Journal of Studies on Alcohol, 1974, 35, 1388-1392.

Vogler, R. E., Compton, J. V., & Weissbach, T. A. Integrated behavior change techniques for alcoholics. Journal of Consulting and Clinical Psychology, 1975, 43, 233-243.

Wall, J. H. Alcoholism: A medical responsibility. Medical Records, New York, 1953, 47, 497-500.

Wanberg, K. W., & Knapp, J. A multidimensional model for the research and treatment of alcoholism. The International Journal of the Addictions, 1970, 5, 69-98.

Ward, R. F., & Faillace, L. A. The alcoholic and his helpers: A systems view. Quarterly Journal of Studies on Alcohol, 1970, 27, 620-635.

Wayner, M. J., & Carey, R. J. Basic drives. *Annual Review of Psychology*, 1973, 24, 53-80.

Weissman, A. Elicitation by a discriminative stimulus of water-reinforced behavior and drinking in water-satisfied rats. *Psychonomic Science*, 1972, 28, 155-156.

Williams, R. J. The genetogrophic concept-nutritional deficiencies and alcoholism. *Annals of the New York Academy of Sciences*, 1954, 57, 794-811.

Williams, R. J., & Brown, R. A. Differences in baseline drinking behavior between New Zealand alcoholics and normal drinkers. *Behaviour Research and Therapy*, 1974, 12, 287-294.

Williams, T. K. *The ethanol induced "loss of control" concept in alcoholism*. Unpublished Doctoral dissertation. Western Michigan University, 1970.

Wilson, G. T., & Lawson, D. M. Expectancies, alcohol, and sexual arousal in male social drinkers. *Journal of Abnormal Psychology*, 1976, 85, 587-594.

Wilson, G. T., Leaf, R., & Nathan, P. E. The aversive control of excessive drinking by chronic alcoholics in the laboratory setting. *Journal of Applied Behavior Analysis*, 1975, 8, 13-26.

Witkin, H. Psychological differentiation and forms of pathology. *Journal of Abnormal Psychology*, 1965, 70, 317-336.

Witkin, H., Karp, S., & Goodenough, D. Dependence in alcoholics. *Quarterly Journal of Studies on Alcohol*, 1959, 20, 493-505.

Wooley, O. W., & Wooley, S. C. The experimental psychology of obesity. In T. Silverstone & J. Fincham (Eds.), *Obesity: Pathogenesis and management*. Lancaster: Medical and Technical Publishing Co., 1975.

Wooley, S. C. Physiologic versus cognitive factors in short-term food regulation in the obese and non-obese. *Psychosomatic Medicine*, 1972, 34, 62-68.

Zuckerman, M., Weary, R. S., & Brustman, B. A. Sensation-seeking scale correlates in experience (smoking, drugs, alcohol, "hallucinations" and sex) and preference for complexity (designs). *Proceedings of the 78th Annual Convention of the American Psychological Association*, 1970.

ALTERNATIVE SKILLS TRAINING IN ALCOHOLISM TREATMENT

Peter M. Miller

Department of Behavioral Medicine, Hilton Head Hospital

Hilton Head Island, South Carolina

ALTERNATIVE SKILLS TRAINING IN ALCOHOLISM TREATMENT: AN OVERVIEW

Forms of behavior therapy in the treatment of alcoholism were reported as early as 1928. Only recently, however, have behavior therapists become fully involved in the complexities of alcoholism assessment and treatment. Behaviorists have developed a keen interest in alcohol abuse for a variety of reasons. Historically, alcohol problems were largely abandoned by professionals espousing traditional therapeutic orientations. The alcoholism literature was, up until recently, filled with page after page of theories, case reports, and descriptions of treatment programs with few well-controlled experimental studies. The majority of "facts" regarding alcohol abuse and its treatment were unsubstantiated and based solely on unsystematic clinical observations and anecdotes. This "state-of-the-art" provided a challenge for behavior therapists whose main raison d'etre is the development and systematic evaluation of specific clinical procedures to accomplish specific behavioral goals.

Behavior therapists have been responsible for the development of new alcohol abuse assessment procedures (Miller, 1977), techniques to teach controlled drinking as opposed to complete abstinence (Sobell and Sobell, 1973a), and various new therapeutic techniques to modify drinking behavior (Miller, 1976). One of the most recent trends in behavior therapy with alcoholics is alternative skills training. Specifically, this involves teaching interpersonal, emotional, and cognitive skills that can serve as alternatives to abusive drinking.

A basic assumption underlying this treatment approach is that alcoholics have a limited repertoire of non-drinking skills required to cope with specific social, emotional, and cognitive precipitants of heavy drinking. For example, the alcoholic may drink excessively when confronted with marital or interpersonal problems. He is deficient in other, more appropriate, responses to these situations such as assertiveness or problem solving skills. Indeed, assertiveness deficits appear to be a major factor in drinking among many alcoholics. Miller and Eisler (1977) found alcoholics to be significantly less assertive than non-alcoholics when dealing with interpersonal encounters requiring expressing differences of opinion, sticking up for one's rights, or expressing anger. In addition, alcoholics who were least assertive were likely to drink more alcohol than those who were most assertive. Alcohol intake often serves a functional purpose in providing alcoholics with more flexibility in their interpersonal behavior. Thus, after a few drinks, an alcoholic wife may be able to express anger toward her husband. However, alcohol will affect her judgment and the intensity and timing of her reaction so that she may become hostile and assaultive rather than assertive. In this regard, alcohol abuse may be both (1) a response to emotionality engendered by situations requiring behaviors in which the alcoholic is deficient and (2) an attempt to cope with these situations by means of increased behavioral spontaneity brought on by alcohol. Indeed, in studying interpersonal interactions of the alcoholic via direct observational methods, Steinglass and his associates (Steinglass, Weiner, and Mendelson, 1971) concluded that alcohol enables an alcoholic to engage in various dyadic and inter-group roles that are unavailable to him when sober.

Two recent experimental investigations illustrate the relationship between social skills deficits and drinking. Miller, Hersen, Eisler, and Hilsman (1974) exposed alcoholics and non-alcoholics to a series of staged interpersonal encounters requiring assertive responses. Scenes were described to the subject and he was instructed to respond as he would if actually confronted with the situation in his everyday life. An example scene was:

Description: You had to work late at the office tonight. You are proud of the fact that you have not had a drink in a month. You have no desire to drink. You feel tired from your long day work. As you enter your house your wife greets you at the door and says in an angry tone, "Where have you been all this time! You've been drinking again. I can smell it on your breath. What do you have to say for yourself."

When the subject responded, the experimenter in the role-played scene would counter with an antagonistic response such as "Oh, stop making up excuses. I can always tell when you've been drinking."

As a control procedure, all subjects were also exposed to a non-stressful conversational period. Immediately after stressful and non-stressful procedures, all subjects were given the opportunity to drink alcohol via an operant drinking task. The results indicated that alcoholics drank significantly more alcohol after being exposed to these stressful social situations than the non-stressful ones. Moderate social drinkers, on the other hand, drank very little after this exposure.

A similar study by Marlatt, Kosturn and Lang (1975) is more directly related to alternative skills training. These investigators deliberately annoyed and angered heavy drinking college students and allowed some subjects to retaliate against the insult and prevented others from retaliating. Subjects who were angered and not allowed to retaliate consumed the most alcohol on a taste rating task. Subjects who could retaliate drank very little alcohol. Retaliation to naturally-occurring annoyances in the form of assertiveness may also lessen drinking in these situations.

In addition to interpersonal skills, it seems likely that alcoholics would also benefit from training in emotional and cognitive skills training. While emotions such as anxiety are not considered to be as important in the development and maintenance of abusive drinking as was once believed (Cappell and Herman, 1972), there are often occasions when excessive drinking follows periods of boredom, tension, depression or loneliness. Certainly the relationship between these emotions and episodes of drinking is a complex one related to the individual's expectations regarding the way in which alcohol will effect him and the type of situation arousing the emotion (Allman, Taylor, and Nathan, 1972; Briddell and Nathan, 1976). In any event, the development of an emotional skill, such as the ability to become completely relaxed, may assist alcoholics to refrain from drinking in particular situations. It appears that certain populations of problem drinkers such as females and younger males with a brief history of abusive drinking may especially benefit from such training (Miller, 1976).

Finally, self-control and cognitive skills may also be important alternatives to alcohol abuse. Although the use of self-management training and cognitive behavior modification are increasing, little has been reported with abusive drinkers (Thoresen and Mahoney, 1974). Behaviorists have yet to study the cognitive statements of alcoholics that lead to the ultimate decision to consume alcohol to excess. For example, the tendency (1) to think in absolute terms (e.g., "Well, I really blew it by having that one drink. Oh well, I'm back on it again."), (2) to feel cheated or "different" by not drinking (e.g., "It's not fair. Everybody else can drink. Why can't I?"), and (3) to concentrate only on the immediate satisfying effects of alcohol (e.g., "One drink will make me feel so much better. I've had a rough day at the office. A

drink will calm me down.") increase the alcoholic's chances of abus
ing alcohol. While much of this type of skills training is based
on conjecture, some reports are available which suggest that this
may be a fruitful area of clinical research investigation.

Let us now become more specific and discuss the interpersonal,
emotional, and cognitive skills that have proven useful with alcoholics.

INTERPERSONAL SKILLS

Assertiveness

Clinically, assertiveness training has gained wide popularity
in alcoholism treatment programs. Assertiveness refers to the direct expression of personal rights and feelings such as differences
of opinion, anger, love, negative replies to unreasonable requests,
and dissatisfaction with the infringement of one's rights (Miller,
1976). The general technique of assertiveness training can be illustrated in the following case study reported by Eisler, Hersen,
and Miller (1974). The patient was a 34-year-old divorced male
with a long history of alcohol abuse. He has been abstinent for a
number of months and was working as a night clerk in a small motel.
On the basis of his excellent job performance he was promoted to
the position of motel manager. His inability to handle the responsibilities of this position due to his lack of assertiveness resulted in a resumption of heavy drinking. Abusive drinking episodes were directly precipitated by such situations as (1) his
inability to confront housekeeping personnel regarding their inadequate cleaning of the motel rooms; (2) his inability to refuse to
buy unnecessary items from salesmen who pressured him; and (3) his
inability to deal with unreasonable complaints of motel guests. In
this case short-term abstinence (in terms of 3 to 4 months) was
reinforced by job satisfaction and promotions. However, long-term
maintenance of abstinence was threatened by the patient's limited
alternatives to interpersonal situations accompanying sobriety and
occupational success. In fact, at the time of treatment, the patient felt that, because of this pattern, long-term sobriety was
unattainable to him.

Six typical work-related situations were chosen for assertiveness training. An example of such a scene follows:

A guest comes to you complaining that his room was not cleaned
very well by the housekeeping personnel. You check other room
and find that they have not been cleaned properly. You mention this fact to the maid and she says "I cleaned the rooms a
well as I can. I'm too busy to be as neat as you would like."

Prior to training the patient was requested to role play these scenes with a treatment assistant. All role-played sequences were videotaped and subsequently rated on the following behaviors: (1) eye contact; (2) compliance with unreasonable requests; (3) affect; and (4) requests for the interpersonal partner to modify his/her behavior (e.g., "I would like you to vacuum these rooms every day"). During this assessment session the patient's typical response to the situations was to become very anxious, look at the floor or ceiling, sound very uncertain and apologetic (e.g., "I... ah...I'm sorry to have to ask you this...ah...but...ah...would you think about...ah...vacuuming these rugs a little more. Don't get me wrong. The rooms look great but...ah...some of the guests think the rugs are a little dirty.") The patient also complied with most of the unreasonable requests from salesmen or guests and rarely asked the other person in these situations to change his/her behavior.

Training consisted of role-playing scenes with the patient and providing him with specific instructions on assertive responses. Sequentially, he was instructed to increase eye contact, decrease compliance, increase more appropriate voice tone and facial expression, and increase behavioral requests. As the patient practiced these new response components, feedback was provided by therapists on the quality of his performance. After several training sessions significant increases were observed in duration of eye contact, quality of affect, and frequency of requesting a change in the behavior of the role model. Frequency of compliance decreased to the extent that the patient did not comply with any unreasonable demands. Through this training the patient successfully learned an alternative skill to be used in place of drinking and after treatment he was much better able to cope with troublesome situations. Unfortunately, most drinking by alcoholics is triggered by a more varied set of circumstances and training in several types of skills is required.

While assertiveness training as a general skill to be used in various interpersonal situations is often the focus of training, treatment geared toward specific encounters typically faced by alcoholics as a group has also been reported. In particular, due to the nature of the alcohol abuse problem, all alcoholics must learn to refuse offers of alcoholic beverages. In this regard, training can be applied in a group setting using other patients in role-playing scenes and more skilled patients as trainers of less skilled ones. Using a single case design experimental methodology, Foy, Miller, Eisler and O'Toole (1976) recently demonstrated the use of assertiveness training to teach alcoholics to refuse drinks effectively. Two chronic alcoholic patients who previously had difficulty in resisting social pressure to drink were the subjects of this study. For each patient three scenes were constructed which depicted encounters which would occur in their everyday lives. For example, one scene was:

> "You're at your brother's house. It's a special occasion and
> your whole family and several friends are there. Your brother
> says, 'How about a beer?'".

During pre-treatment baseline sessions each scene was described
separately by one of the experimenters. Two other experimenters
role-played each scene with each patient and served as "pushers",
trying to persuade him to have a drink. The pushers used various
persuasive remarks and rebuttals to counter the patient's initial
refusal to accept a drink. Such arguments as "One drink won't hurt
you.", "A real man should be able to handle his liquor!", "Just
have one; it'll make you feel better.", and "What kind of a friend
are you? We used to be great drinking buddies." These interactions lasted for two minutes for each scene. All responses of the
patient were videotaped and subsequently rated. After three baseline sessions, nine sessions of training conducted over a two-week
period were initiated. During the first three sessions, training
focused on teaching the patient to request that the pushers refrain
from asking him to take a drink in the present situation or in the
future. Videotaped modeling of this behavior, focused instructions, behavioral rehearsal, and verbal feedback on performance
were used to change the patient's response pattern. During the
next three sessions patients were taught to offer an alternative
(e.g., "Let's have a cup of coffee instead") and change the subject
by introducing an entirely different topic of conversation. In the
final three sessions patients were taught to look directly at the
pushers when responding to them and to respond firmly, with an ex-
pressive affect. The videotaped model was particularly useful in
teaching appropriate affect since the patient's attention could be
focused on facial expressions, voice tone and volume, and hand
gestures. After training was completed a three month posttreatment videotaped follow-up session was conducted. Figure 1 illustrates progress during and after treatment. It shows that changes
in target behaviors occurred only after training. These positive
changes, moreover, were maintained to the three month follow-up
session. Detailed self-report data revealed that both patients
used these new skills quite effectively. Both patients reported
increased feelings of confidence, self-esteem, and control over
their drinking.

Miller (1975) has also reported the use of this training when
controlled social drinking as opposed to abstinence is the treatment goal. Training techniques were similar to the ones described
above except that patients were also given practice at refusing the
second or third drink offered to them. After extensive training,
some patients were primed with two beers to simulate conditions
under which they would be refusing a third drink. Refusal behaviors were maintained under these "wet" conditions and patients used

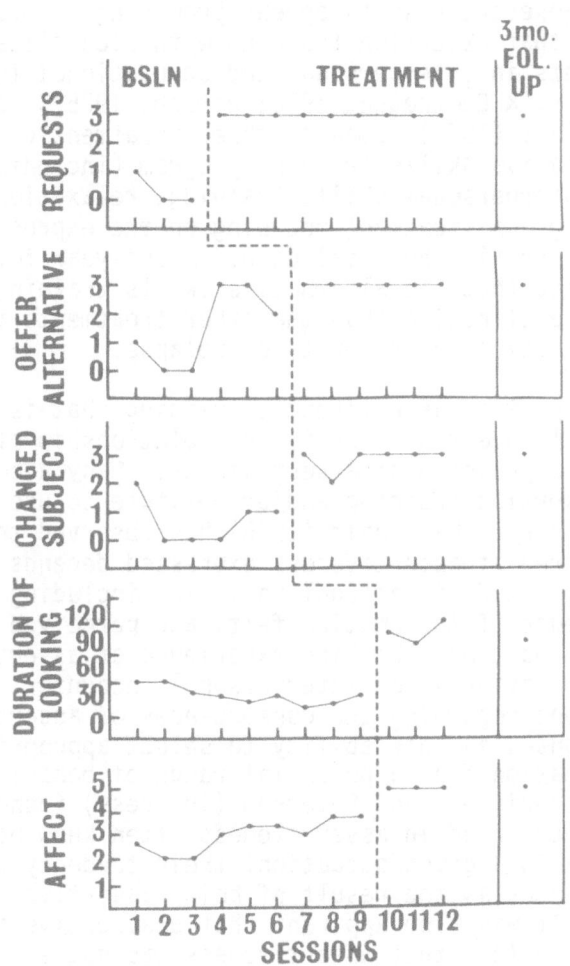

Figure 1: Changes in five target behaviors during treatment and at three-month follow-up (Foy, Miller, Eisler & O'Toole, 1976).

this skill effectively to control alcohol use. These skills were particularly useful with these patients since they were all young males who drank mostly in small social groups.

Group outcome studies evaluating the overall long-term effectiveness of assertion training with alcoholics have only recently been reported. However, results appear promising. Recent clinical studies indicate that assertion training with alcoholics leads to marked improvements in interpersonal and occupational functioning (Adinolfi, McCourt, & Geoghegan, 1975; Hirsch, 1975). Chaney, O'Leary, and Marlatt (1977) compared three treatment groups receiving (1) an alternative skills training program (including assertiveness training, interpersonal skills training, relaxation training), (2) a discussion group treatment focusing on the expression of feelings, and (3) routine hospital care. A one-year follow-up assessment revealed that the alternative skills training package wa significantly more effective than the other treatments in terms of decreasing the duration and severity of relapse.

One other aspect of assertiveness training that is often overlooked in clinical investigations is the relationship of cognitive variables to the expression of assertiveness. Many alcoholic patients resist assertion training and/or hesitate to use these skills when necessary. MacDonald (1975) has observed that the direction and degree of assertiveness expressed depends on a variety of perceptual and situational variables including the degree of intimacy, valence of the emotion felt, and perceived status of the interpersonal partner. Lacking experience at assertiveness, the alcoholic's behavior in an interpersonal encounter is related to his expectations regarding the consequences of assertive versus unassertive responses and his ability to select appropriate responses to a situation from a potential range of behaviors. Indeed, Eisler, Frederiksen, and Peterson (in press) found that while individuals inexperienced in assertiveness often know how to respond assertively in a given situation, their tendency to expect negative consequences as the result of being assertive inhibited their behavior. It was also apparent that unassertive individuals were more likely to feel that assertiveness was not a socially appropriate interpersonal response. While such cognitions are eventually overcome through experience at being "successful" in assertive encounters, the initial training frequently necessitates modifications in cognitions and expectancies. This is a particularly important issue with alcoholics which is not usually reported in clinical research studies on assertiveness.

Assertiveness, then, may be a very useful skill for alcoholics This area is ripe for clinical investigations into the questions of (1) which types of problem drinkers might benefit most from assertion training, and (2) which are the best methods to generalize training to the real world and to maintain assertive responding over time.

Marital Skills

Deficits in marital and parenting skills are frequently associated with excessive drinking. The following case illustrates the manner in which an alcoholic and his wife were taught (1) assertiveness skills, (2) problem solving skills, (3) more positive interactional skills, and (4) behavioral contracting skills.

The alcoholic husband was a 49 year-old married male who was hospitalized for alcoholism treatment. The client's history of alcohol abuse extended over the past ten years and had resulted in numerous arrests, automobile accidents, and marital problems. Prior to hospitalization he had been consuming from one pint to one fifth of vodka per day. Most of his drinking occurred at home, in his car, or at social functions.

While the client was able to maintain his job as a packer in a glass container company, the stability of his marriage was threatened. The couple had been married 29 years and the wife had attempted a variety of maneuvers to deal with her husband's excessive drinking (e.g., threatening, protecting, nagging, ignoring). Prior to the husband's hospitalization she had decided to divorce him if he did not agree to receive treatment. On this basis the client expressed a desire to change in order to save his marriage.

The husband's treatment involved a comprehensive behaviorally-oriented inpatient program (Miller, Stanford, and Hemphill, 1974), followed by periodic outpatient visits. Initially, marital counseling sessions were scheduled weekly in the hospital with the husband visiting the home each weekend.

Prior to initiation of counseling a behavioral assessment of the couple's marital interaction was obtained. Detailed evaluation is essential since both the goals and process of behavioral counseling are highly explicit. Three separate methods of assessing the disordered components of this marriage were used. First, direct observations of marital interactions were obtained through a series of interviews with the couple. These interviews, held both conjointly and separately, helped to delineate problem areas in the home environment as perceived by the husband and wife. Both partners were in general agreement that their current problems revolved around the husband's drinking, the husband's lack of responsiveness to the wife, the wife's nagging, and a lack of positive reciprocal verbal and non-verbal interactions in the home.

Second, the couple was videotaped for twenty minutes while conversing about various problems and non-problem areas. The counselors were not present to allow for a more "natural" interaction. Details regarding the setting, instructions, and equipment for this videotaping are presented elsewhere (Eisler, Hersen, and Agras, 1973). Ratings of this videotape indicated a minimum of positive interactional statements (only two during the session), an excess of non-goal oriented alcohol-related conversation (15 of the 20 minutes were spent on past drinking episodes, with the wife using threats in an attempt to prevent their future occurrence), and a total absence of positive suggestions to improve the relationship. The duration of talking during the session was approximately twice as long for the wife as for her husband.

Third, on two separate occasions the couple was provided with an audio tape recorder and instructed to record their mealtime interaction at home. These tapes were then rated for number of interruptions, number of positive comments, and number of requests for the partner to change his or her behavior. As far as content was concerned, the couple discussed a variety of problem and non-problem areas. The ratings indicated that the husband spoke little, not making any of the types of statements that were being rated. The wife, however, interrupted her husband approximately 30 times during each 40-minute tape, making between 3 and 5 positive comments, and requesting only 2 changes in her husband's behavior.

On the basis of these assessments the following general goals of counseling were established: 1) to increase the couple's ability to express themselves more directly and to solve mutual problems more efficiently; 2) to increase positive interactional patterns (the majority of their statements to one another were either neutral or negative; 3) to decrease conversations regarding negative incidents in the past; 4) to provide each partner with positive skills needed to increase more desirable behaviors in the other (the couple tended to use coercion, e.g., nagging and threats in this regard). In addition, the wife specifically requested that her husband: a) abstain from all alcoholic beverages; b) talk to her more frequently about his feelings, and c) take her out to a restaurant and/or movie more often. In turn, the husband wanted his wife to: a) decrease her nagging about his past and "possible" future drinking, b) sit and watch television with him on some evenings, and c) engage in pleasant conversations with him.

Skills necessary to attain these goals were then taught during conjoint sessions with a male and a female counselor present. Use of two such counselors often facilitates certain behavioral techniques such as modeling adaptive interactional patterns for the couple.

For example, the counselors modeled negotiations for a mutual agreement regarding Antabuse intake by the husband and cessation of nagging by the wife. In the presence of the couple the counselors roleplayed this situation, demonstrating compromise ("I'll agree to take Antabuse each day if you'll agree to quit bringing up my drinking in the past"), appropriate use of direct, assertive problem solving skills (e.g., offering concrete solutions), negotiating a written contract, and, subsequent to successful negotiations, the use of positive comments (e.g., "Frank, I can see now that you're really trying to change") to reinforce the other partner. More positive adaptive marital skills were also taught via videotaped feedback and roleplaying, together with feedback and social reinforcement from the counselors. The couple was initially taught to increase very simple non-verbal interactional behaviors such as eye contact, smiling, listening attentively, and touching. Simple positive statements such as "I feel good when you say that," or "You look very attractive today" were encouraged and reinforced by the counselors. Simple instructions, periodic prompts, and behavioral rehearsal fostered these new patterns quite rapidly.

Observations of the couple during counseling sessions and self-reports of home interactions revealed that these positive behavior patterns increased significantly (from only an occasional positive verbal or non-verbal response from the wife to a minimum of two to three per day from each partner). The couple also learned ways to utilize these new skills in a functional manner. For example, it was demonstrated that positive verbal and non-verbal reactions applied contingently to the partner's desirable behavior could increase the likelihood of these interactions in the future. The couple was instructed to practice these new patterns at home via half-hour sessions held on the days in which the husband was home.

Through assertive training (Hersen, Eisler, & Miller, 1973), the couple was taught more direct ways of communicating with one another. Videotape feedback and behavioral rehearsal proved most effective in teaching this ability to directly express personal rights and feelings.

In this regard, the twenty-minute videotape of the couple's interaction that was used initially for assessment purposes was shown to them on several occasions. Concomitantly, specifics of their deficit responses in the components of assertiveness were pointed out to them. They were then requested to practice discussing problem situations requiring assertiveness with particular attention to eye contact, appropriate affect, compromising, and requesting their partner to alter specific behavior patterns. These new behaviors were rehearsed repeatedly with feedback being provided by the counselors. These skills were used primarily to solve mutual problems. Various problem areas were discussed with the couple and they were provided with guided practice in solving them during the sessions. For example, one difficulty involved the couple's being frequently asked by their son to care for their granddaughter. After discussing this matter the couple agreed that:

a) they would limit such "babysitting" services for their own con-
venience and b) the husband would assist the wife in caring for the
child. In fact, during the session, the husband expressed the fact
that he enjoyed playing with the youngster but refrained from doing
this since he felt his wife preferred to spend time alone with the
child. The wife was previously unaware of her husband's feelings
about this matter and appreciated his offer of assistance.

   In an attempt to expand and generalize the changes occurring
within the session, mini-contracts were written for the couple at
the end of each weekly session. These contracts, signed by both
partners, specified one to two behavioral goals to be accomplished
by each partner during the husband's weekend home visit. For example, one contract illustrated in Figure 2 specified that the wife
was to refrain from mentioning alcohol or any related topic (such
behavior had frequently precipitated drinking episodes in the past)
and the husband was to take his wife out to dinner and to a movie
on Saturday evening. Social reinforcement was provided by the
counselors for the couple's compliance with these goals.

   At times more extensive reinforcement was necessary if compliance did not occur. For example, in the early sessions the wife
found it very difficult to refrain from alcohol-related conversations with her husband. During home visits, she would constantly
remind him of his promise to quit drinking and frequently expected
him to reaffirm this promise. Such behavior angered the husband
and made it more difficult for him to comply with his portion of
the weekly contracts. A survey of possible reinforcers for the
wife's behavior changes revealed that she often called the female
counselor or requested to talk with her privately. During these
conversations, which were admittedly very positive for her, she
discussed various personal feelings and concerns. After the third
counseling session, the wife was told that these individual conversations would only be allowed if she had complied with her con-
tractual goal for that week. This contingency served as a powerful reinforcer which modified the wife's behavior dramatically.
Prior to discharge from the hospital the husband signed a contract
with his wife agreeing that he would take Antabuse daily in her
presence if she would refrain from reminding him to take this medication.

   The couple continued marital counseling sessions on a biweekly
outpatient basis for three months and on a monthly basis for an
additional three months. Treatment was terminated at six months
after hospital discharge, with a follow-up interview occurring at
nine months. Based upon self-reports, reports from the wife, and
the client's work record, the patient has remained completely abstinent from alcohol for nine months and has taken Antabuse each
day. The couple report a more positive, enjoyable marital relationship. The wife has also reported (at both six and nine month

WEEKLY MARITAL CONTRACT

The undersigned, Frank and Wilma B., enter into the following agreement with each other. The terms of this agreement include the following:

1. During this weekend Wilma agrees not to mention any of Frank's past drinking episodes or possible drinking in the future.

2. Wilma will be allowed one infringement of this agreement per day provided that she immediately terminates her alcohol-related conversation contingent upon Frank's reminding her of this agreement.

3. On Friday, Saturday and Sunday afternoons or evenings Frank agrees to take Wilma out of the house for the purpose of a shopping trip, dinner, movie, or a drive, depending on her choice.

4. Wilma's agreement to refrain from alcohol-related conversation is binding only if Frank fulfills his agreement stated under term number 3.

5. Frank's agreement to take Wilma out each day of the weekend is binding only if Wilma fulfills her agreement stated under term number 1.

6. The terms of the contract are renewable at the beginning of each day so that failure of one partner to fulfill his or her part of the agreement on any one day breaks the contract for that day only.

_____

Frank B.

_____

Witness

_____

Wilma B.

Figure 2

follow-ups) that she and her husband go out together in the evening at least once a week and that, at least three times per week, they sit together in the evenings and discuss mutual concerns. The husband reports that his wife no longer mentions his past drinking. Although, during the first three months after discharge, the wife did mention it on six separate occasions, she has not done so in the past six months. The husband also reports that his wife is more pleasant, providing him with frequent positive responses (e.g., compliments) during the week. A number of mutual problems have arisen which the couple has handled directly and effectively. For example, six months after discharge, the wife angered her husband by initially agreeing to watch television with him for two hours in the evening but instead spending her time engaged in other activities. She excused her behavior by stating that she was attending to essential housekeeping chores. After two of these episodes the husband confronted his wife with her avoidance of their agreement (in the past he avoided dealing with such problems) and a contract was negotiated between the two. The wife agreed to watch television with her husband for one hour each evening in return for "feeling" talk (discussions regarding positive and negative feelings regarding work, home, friends) from him during at least three half-hour sessions per week. This agreement quickly solved the problem to the extent that these reciprocal behaviors became habitual and the contract was no longer needed after a few weeks.

Reports of these marital improvements were corroborated by a videotaped interaction in the hospital prior to discharge, direct observations of the couple's interactions at follow-up contacts, and one report (at nine months) from the couple's children. All of these data indicated marked increases in positive verbal and non-verbal interactions, more direct solution to problems, a total absence of threats and coercion, and an absence of alcohol-related conversation.

In a similar case study, Eisler, Miller, Hersen, and Alford (1974) taught an alcoholic husband skills to enable him to cope with marital arguments that triggered his drinking episodes. Prime problems involved the husband's inability to express opinions to his wife on matters of disagreement and to deal with family decisions firmly and decisively. Interpersonal skills training of both verbal and non-verbal behaviors consisted of videotape feedback, specific instructions, behavioral rehearsal, and verbal reinforcement. Training resulted in significant changes in the husband's assertiveness and drinking behavior. Breath alcohol levels taken weekly for six weeks prior to and six weeks after training decreased from a mean of .08% to .02%. In addition, ratings of videotaped interactions between the couple indicated marked improvements in marital problem-solving abilities.

While many comprehensive behavior therapy programs for alcoholism include marital skills training as part of a multifaceted treatment package (Azrin, 1976; Sobell & Sobell, 1973a), evaluations of this approach to training as the primary treatment modality are scarce. In one of the few such investigations of marital skills training, Hedberg and Campbell (1974) found behavioral marital counseling to be a very effective treatment technique. Through behavioral rehearsal, couples were taught to use feedback, positive reinforcement, assertiveness, and behavioral contracting to improve their marital interactions. At a one-year follow-up, 74% of 15 alcoholics had been successful in achieving their treatment goal (i.e., either complete abstinence or moderate, controlled drinking). In the same study, control groups receiving either systematic desensitization, covert sensitization, or electrical aversion therapy showed only 67%, 40%, and 0% success, respectively.

We conclude that teaching marital skills may be a very important aspect of behavioral alcoholism treatment programs. Clinical research in this area has only recently begun, with many investigators using elaborate, intensive intervention strategies (Paolino and McCrady, 1976; Steinglass, Davis, and Berenson, 1975).

Relaxation Training

Various forms of relaxation training have been used with problem drinkers to provide them an alternative to alcohol abuse in stressful emotional situations. While, at face value, the use of relaxation training with alcoholics appears to rely on the validity of the questionable tension-reduction hypothesis, this is not necessarily so. This is an important point since the rather simplistic tension reduction hypothesis as an explanatory concept for problem drinking has received little experimental support (Cappell & Herman, 1972). However, there is evidence (Allman, Taylor, & Nathan, 1972; Miller, Hersen, Eisler, & Hilsman, 1974) to support the notion that, under certain stressful circumstances, many alcoholics drink to relieve anxiety and tension on an aperiodic basis. Thus, while anxiety reduction may not satisfactorily explain all episodes of alcohol abuse, learning alternative ways to cope with anxiety may provide alcoholics with a viable alternative in some situations. For example, relaxation techniques may be helpful in dealing with the phenomenon of "craving" or anxiety associated with attitudinal expectancies regarding "loss of control." It may also be that younger, less chronic alcoholics would benefit more from relaxation training than more chronic alcoholics. Indeed, anxiety reduction appears to be a much less frequent consequence of excessive drinking in the latter group (Nathan and Lansky, 1978).

Besides its role in anxiety-reduction, relaxation may also serve as a delay tactic to bridge the gap between the initial thought, sight, or smell of alcohol and the decision to drink.

Self-management procedures have been used in this same fashion. A
Decision Delay Technique (DDT) can provide the alcoholic with the
time needed to evaluate his decision to drink more fully and perhaps
decide not to drink. Unfortunately, the exact manner in which relaxation training and other self-management procedures are effective
with alcoholics has yet to be determined.

In any event, forms of relaxation training have been reported
to reduce alcohol consumption. In a very early report, Blake (1967)
reported a 23% abstinence rate using aversion therapy along but a
46% abstinence rate when relaxation training was added. Marlatt,
Pagano, Rose, and Marques (1976) exposed heavy drinking college
students to one of four conditions: (1) meditation, (2) progressive muscle relaxation, (3) an attention-placebo condition in which
subjects were simply instructed to rest each day, and (4) a no
treatment control group. Drinking was monitored by daily drinking
diaries and by a laboratory analogue drinking task. The regular
practice of relaxation resulted in lowered alcohol consumption over
a six-week period of time. However, simply resting each day also
reduced alcohol intake. There was a slight tendency for the meditation group to continue using relaxation more consistently and over
a longer period of time. These results remain equivocal since attention-placebo factors were not successfully ruled out.

More recently, investigators at Baltimore City Hospital
(Strickler, Bigelow, and Wells, 1976) used relaxation training to
reduce tension associated with drinking-related stimuli. Subjects
were abstinent alcoholics who were instructed to use relaxation
while listening to an audiotape recording of a problem drinker in a
barroom, arguing with himself about having a drink. EMG recordings
of frontalis muscle tension levels indicated that subjects were
able to significantly lower tension in response to alcohol-related
cues. Control group subjects did not modify their tension responses. Using this procedure, abstinent alcoholics may be able to
cope with tension-producing stimuli which lead to a resumption of
heavy drinking.

Relaxation induced by EMG biofeedback is a relatively recent
but rapidly growing treatment technique with alcoholics. Basically,
the procedure teaches patients to control muscle tension in the
frontalis muscle of the forehead by providing continuous feedback
of muscle activity in the form of a visual or auditory signal.

Recent studies (Steffen, 1974; Steffen, Taylor, and Nathan,
1974) at Rutgers University examined the relationship between EMG
biofeedback training and alcohol consumption. Results indicated
that alcoholic subjects reached significantly lower blood alcohol
levels following EMG biofeedback training procedures than after a
placebo training procedure. It is interesting to note that, while
blood alcohol levels were decreased, absolute amount of alcohol

consumed did not change. Thus, biofeedback modified the pattern of drinking perhaps by slowing it down, but not altering total amount.

Unfortunately, clinical evaluations of biofeedback are only now being completed. Whether biofeedback adds substantially to relaxation training has yet to be determined.

Self-Management Skills

The ability to manage and regulate one's own behavior consistently is an important element in controlling any maladaptive habit pattern. Self-regulation or self-management involves the active manipulation of events that influence behavior. This involves a two-step process of (1) recognizing specific events that influence a particular behavior pattern and (2) manipulating those events to lessen their influence. These events can be either environmental (e.g., a party at which everyone is drinking heavily) or cognitive (e.g., "One night of drinking won't do me any harm. Besides, my liver can't be as bad off as Dr. Clarke says it is"). Therefore, self-management skills might involve the manipulation of situational factors or cognitive processes that either precede or follow alcohol abuse. Many alcoholics, especially chronic ones, are deficient in self-management skills and are almost completely dependent on externally-imposed reinforcement systems to control their drinking (Heilbrun and Norbert, 1972).

Unfortunately, this important clinical and research area has received very little attention in the alcoholism field. For example, little is known regarding the nature of these deficits among alcoholics. Clinically, alcoholics seem amazingly naive regarding the specific antecedents and consequences of their drinking and usually attribute their behavior to vague "cravings" which they feel unable to control. In fact, self-help treatment programs often reinforce this naivete by labeling specific antecedents of alcohol consumption (e.g., a marital argument) as "excuses" for drinking but not valid causes of the drinking. Thus, the alcoholic's self-evaluation process in regard to precipitants of abusive drinking is discouraged.

Perhaps, then, we are dealing with a deficit in the ability to discriminate relevant from irrelevant events that influence drinking behavior. On the other hand, alcoholics must also have the skills necessary to modify these events once they have identified them. Insight, even behavioral insight, is not enough.

While alcoholism clinicians have virtually ignored self-management training, behavioral clinicians in obesity treatment have found such training to be essential for successful long-term weight control (Stunkard, 1972). Unfortunately, alcoholism treatment,

even behavioral treatment, has emphasized external control of drinking behavior. The alcoholic has been hospitalized, medicated, or
conditioned via aversion therapy with little stress placed on
teaching him control over his own behavior. This is true in spite
of the fact that self-control is more likely to have long term effects than control by others. The advantage of the "magic pill"
approach, of course, is in its placebo effects and in the fact that
alcoholics are often looking for a quick and easy solution to their
problems. There is currently a trend, as the focus of this paper
indicates, for behavioral treatment to be more skills-training-
oriented. Indeed, alcoholism treatment is slowly becoming more
flexible and innovative as attested to by the current interest in
new therapeutic modalities and the increasing acceptance of controlled, social drinking as a viable treatment goal for some alcoholics.

    Self-management of alcohol consumption involves the ability
to (1) self-monitor a behavior pattern, (2) rearrange environmental
cues which trigger drinking, (3) rearrange social and environmental
consequences of excessive drinking, and (4) rearrange cues and consequences through cognitive changes (Miller and Mastria, 1977).

    Self-recording of drinking behavior is an important skill in
controlling abusive drinking, particularly when controlled drinking
as opposed to abstinence is the treatment goal. Teaching patients
to monitor their drinking behavior serves several purposes. First,
it provides a detailed, day-by-day analysis of functional relationships between excessive drinking and specific antecedent and consequent events. This information is essential in developing and
evaluating a treatment plan. Second, it provides the patient with
increased awareness of the frequency and quantity of his drinking.
Third, self-monitoring has been shown to effect behavior change
(Kazdin, 1974). Thus, merely by monitoring his behavior the problem drinker is likely to decrease his drinking. Sobell and Sobell
(1973b) have devised an Alcohol Intake Sheet on which the patient
records specific information on each drinking episode (e.g., type
of beverage, time, circumstances). Even abstinent alcoholics can
benefit from self-monitoring by using it to increase their use of
newly-learned alternative skills. Recording episodes of assertiveness, for example, serves to reinforce assertive behavior and increase the likelihood of its occurrence on future occasions.

    Miller (1976) reports two case studies in which alcoholics
were taught to rearrange environmental events to decrease the
likelihood of drinking. One patient was a traveling salesman who
drank excessively only while away from home on business trips.
Self-monitoring of drinking behavior revealed that he was most
likely to abuse alcohol when (1) he was alone in his motel room on
a sales trip (particularly the night before he was to meet a cus-
tomer) and (2) he was with a group of customers at dinner and they

were encouraging him to have a drink. With the assistance of the therapist the client devised the following self-management plan:

(1)   He agreed to arrange his future schedules so that he would arrive for an out-of-town business meeting very close to the time of the beginning of the meeting. In this way he avoided lengthy stays in his motel room with little to do.

(2)   Whenever possible he agreed to conclude his business in one day and return home that same evening.

(3)   Whenever possible he agreed to excuse himself from out-of-town cocktail gatherings.

(4)   He agreed to practice and use prepared verbal statements to refuse offers of alcoholic beverages from customers.

Since his excessive drinking occurred in such circumscribed situations, the client successfully modified his drinking pattern using these relatively straightforward self-management techniques. In essence, the patient learned to decrease his opportunities for drinking by avoiding situations in which he was most likely to drink.

The second case involved a 55-year-old chronic alcoholic binge drinker. Drinking binges were precipitated by the availability of a substantial amount of money. Once the patient had accumulated approximately $100 he would quit his job and begin a two to three week drinking binge. Between binges he was able to refrain from alcohol completely. The patient agreed to regulate his drinking by depositing all but a small amount of each weekly salary into a bank account that required 90 days notice for withdrawal of funds. He arranged with his employer to send this money to the bank prior to receiving his pay checks. He also agreed to carry as little money as possible with him from day to day. This strategy completely eliminated lengthy periods of binge drinking. At a one-year follow-up the patient was still working for the same employer and had engaged in only three very brief two-day drinking episodes.

Finally, a report by Mertens (1964) illustrates the use of both environmental and cognitive self-management. Patients were initially taught to rearrange visual alcohol stimuli in the environment to make these cues less prominent. For example, a patient might take a different route home from work to avoid passing a series of bars and liquor stores. Patients also learned to vividly imagine the ultimate negative consequences of drinking (e.g., developing mouth cancer, cirrhosis of the liver, loss of employment) and the ultimate positive consequences of sobriety (e.g., a more satisfying marriage, a feeling of personal accomplishment, more energy, improved sexual performance) whenever urges to drink alcohol occurred.

Unfortunately, most of these reports of the use of self-management techniques are case studies and program reports with little long-term follow-up data.

Further clinical research into the effectiveness of self-management training with alcoholics is definitely warranted.

Conclusions

In conclusion, while alternative skills training in the treatment of alcoholism has a relatively short history, it appears to have a very promising future. Alternative social, emotional, and self-control skills not only allow the alcoholic to avoid excessive drinking but also enable him to obtain more satisfaction from a sober life. Such satisfaction, in turn, helps to maintain sobriety. In this regard, Antabuse, aversion therapy, and operant conditioning via social and environmental engineering would be considered insufficient for the long-term success of alcoholic individuals. Recent follow-up evaluations of behavioral treatment programs emphasizing skills training techniques by Mark and Linda Sobell (Sobell and Sobell, 1973) and Nathan Azrin (Azrin, 1976; Hunt & Azrin, 1973) have demonstrated the effectiveness of this approach. More clinical follow-up studies are now needed focusing on specific alcoholic populations. Alternative skills treatment packages must be matched to the needs of individual alcoholic subgroups to maximize therapeutic effectiveness.

Finally, the alternative skills approach may have a significant impact on prevention of alcoholism. Children, especially those at high risk for the development fo alcoholism, seldom learn how to relax, be assertive, or manage their own behavior in a systematic manner. Preparing children with these skills prior to their exposure to alcohol may decrease the chances that they will become problem drinkers.

In any event, the challenge is there for clinicians and researchers to utilize and evaluate alternative skills training as a promising treatment modality in alcoholism programs.

REFERENCES

Adinolfi, A. A., McCourt, W. F., & Geoghegan, S. Group assertiveness training for alcoholics. Journal of Studies on Alcohol, 1975, 37, 311-320.

Allman, L. R., Taylor, H. A., & Nathan, P. E. Group drinking during stress: Effects on drinking behavior, affect, and psychopathology. American Journal of Psychiatry, 1972, 129, 669-
  678.

Azrin, N. H. Improvements in the community reinforcement approach
  to alcoholism. Behaviour Research and Therapy, 1976, 14, 339-
  348.

Blake, B. G. A follow-up of alcoholics treated by behavior therapy. Behaviour Research and Therapy, 1967, 5, 89-94.

Briddell, D. W., & Nathan, P. E. Behavior assessment and modification with alcoholics: Current status and future trends. In
  M. Hersen, R. M. Eisler, & P. M. Miller (Eds.), Progress in
  Behavior Modification. New York: Academic Press, Volume 2,
  1976.

Cappell, H., & Herman, C. P. Alcohol and tension reduction: A
  review. Quarterly Journal of Studies on Alcohol, 1972, 33,
  33-64.

Chaney, E. F., O'Leary, M. R., & Marlatt, G. A. Skill training
  with alcoholics. Unpublished manuscript, University of
  Washington School of Medicine, Seattle, Washington, 1977.

Eisler, R. M., Frederiksen, L. W., & Peterson, G. L. The relationship of cognitive variables to the expression of assertiveness. Behavior Therapy, in press.

Eisler, R. M., Hersen, M., & Agras, W. S. Effects of videotape and
  instructional feedback on nonverbal marital interaction: An
  analog study. Behavior Therapy, 1973, 4, 551-558.

Eisler, R. M., Hersen, M., & Miller, P. M. Shaping components of
  assertive behavior with instructions and feedback. American
  Journal of Psychiatry, 1974, 30, 643-649.

Eisler, R. M., Miller, P. M., Hersen, M., & Alford, H. Effects of
  assertive training on marital interaction. Archives of
  General Psychiatry, 1974, 30, 643-649.

Foy, D. W., Miller, P. M., Eisler, R. M., & O'Toole, D. H. Social
  skills training to teach alcoholics to refuse drinks effectively. Journal of Studies on Alcohol, 1976, 37, 1340-1345.

Hedberg, A. G., & Campbell, L. A comparison of four behavioral
  treatments of alcoholism. Journal of Behavior Therapy and
  Experimental Psychiatry, 1974, 5, 251-256.

Heilbrun, A. B., & Norbert, N. Self-regulatory behavior in skid row
    alcoholics. Quarterly Journal of Studies on Alcohol, 1972, 33,
    990-998.

Hersen, M., Eisler, R. M., & Miller, P. M. Development of assertive
    responses: Clinical, measurement, and research considerations.
    Behaviour Research and Therapy, 1973, 11, 505-521.

Hirsch, S. M. Experimental investigations of the effectiveness of
    assertion training with alcoholics. Doctoral Dissertation,
    Texas Technological University, Lubbock, Texas, 1975.

Hunt, G. M., & Azrin, N. H. A community reinforcement approach to
    alcoholism. Behaviour Research and Therapy, 1973, 11, 91-104.

Kazdin, A. E. Self-monitoring and behavior change. In M. J.
    Mahoney and C. E. Thoresen (Eds.), Self-control: Power to the
    person. Belmont, California: Brooks/Cole, 1974.

MacDonald, M. L. Teaching assertion: A paradigm for therapeutic
    intervention. Psychotherapy: Theory, Research, and Practice,
    1975, 12, 60-67.

Marlatt, G. A., Kosturn, C. F., & Lang, A. R. Provocation to anger
    and opportunity for retaliation as determinants of alcohol
    consumption in social drinkers. Journal of Abnormal Psycho-
    logy, 1975, 84, 652-659.

Marlatt, G. A., Pagano, R. R., Rose, R. M., & Marques, J. K. The
    effects of meditation and relaxation upon alcohol consumption
    in male social drinkers. Unpublished manuscript, University
    of Washington, 1976.

Mertens, G. C. An operant approach to self-control for alcoholics.
    Paper presented at symposium on "Alcoholism and Conditioning
    Therapy" at the American Psychological Association, September,
    1964.

Miller, P. M.   Behavioral Treatment of Alcoholism. New York:
    Pergamon Press, 1976.

Miller, P. M. Assessment of addictive behaviors. In A. Ciminero,
    K. Calhoun, & H. Adams (Eds.), Handbook of behavioral asses-
    sment. New York: John Wiley and Sons, 1977.

Miller, P. M. Training responsible drinking skills in veterans.
    Paper presented at the American Psychological Association,
    1975.

Miller, P. M., & Eisler, R. M. Assertive behavior of alcoholics: A descriptive analysis. Behavior Therapy, 1977, 8, 146-149.

Miller, P. M., Hersen, M., Eisler, R. M., & Hilsman, G. Effects of social stress on operant drinking of alcoholics and social drinkers. Behaviour Research and Therapy, 1974, 12, 67-72.

Miller, P. M., & Mastria, M. A. Alternatives to alcohol abuse. Champaign, Illinois: Research Press, 1977.

Miller, P. M., Stanford, A. G., & Hemphill, D. P. A social-learning approach to alcoholism treatment. Social Casework, 1974, 55, 279-284,

Nathan, P. E., & Lansky, D. Management of the chronic alcoholic: A behavioral viewpoint. In J. P. Brady & H. K. Brodie (Eds.), Controversy in Psychiatry. New York: Saunders, 1978.

Paolino, T. J., & McCrady, B. S. Joint admission as a treatment modality for problem drinkers: A case report. American Journal of Psychiatry, 1976, 133, 222-224.

Sobell, M. B., & Sobell, L. C. Individualized behavior therapy for alcoholics. Behavior Therapy, 1973, 4, 49-72(a).

Sobell, L. C., & Sobell, M. B. A self-feedback technique to monitor drinking behavior in alcoholics. Behaviour Research and Therapy, 1973, 11, 237-238(b)

Steffen, J. J. Electromyographically induced relaxation in the treatment of chronic alcohol abuse. Journal of Consulting and Clinical Psychology, 1974, 43, 275-279.

Steffen, J. J., Taylor, H. A., & Nathan, P. E. Tension reducing effects of alcohol: Further evidence and some methodological considerations. Journal of Abnormal Psychology, 1974, 83, 542-547.

Steinglass, P., Davis, D. I., & Berenson, D. In-hospital treatment of alcoholic couples. Paper presented at the American Psychiatric Association, May, 1975.

Steinglass, P., Weiner, S., & Mendelson, J. A systems approach to alcoholism: A model and its clinical application. Archives of General Psychiatry, 1971, 24, 401-408.

Strickler, D., Bigelow, G., & Wells, D. Electromyograph responses
    of abstinent alcoholics to drinking-related stimuli: Effects
    of relaxation instructions. Paper presented at the meeting of
    the Association for the Advancement of Behavior Therapy, New
    York, December, 1976.

Stunkard, A. J. New therapies for eating disorders: Behavior
    modification of obesity and anorexia nervosa. Archives of
    General Psychiatry, 1972, 26, 391-398.

Thoresen, C. E. & Mahoney, M. J. Behavioral self-control. New
    York: Holt, Rinehart, and Winston, Inc., 1974.

TREATMENT FOR MIDDLE INCOME PROBLEM DRINKERS

Ovide Pomerleau, Michael Pertschuk, David Adkins, and
Eugene d'Aquili

Center for Behavioral Medicine, Department of Psychiatry
University of Pennsylvania

Various psychological procedures for treating problem drinking
have been proposed over the years (Pomerleau, 1977). Few suggestions have been as controversal as behavioral treatment using controlled drinking--the idea the alcoholics might be able to learn to
moderate their drinking as a result of therapy. Interest in controlled drinking was started by reports of "spontaneous" moderation
in former alcoholics (Davies, 1962) and has received considerable
momentum from the lack of scientific support for the theoretical
basis for mandatory abstinence (Lloyd and Salzberg, 1975; Pomerleau,
Pertschuk, and Stinnett, 1976). While much careful investigation
still needs to be done, initial results on controlled drinking have
been promising.

Lovibond and Caddy (1970) were among the first to attempt moderate drinking as the goal of therapy. In their procedure, Austral-
ian outpatient alcoholics were taught to discriminate different
blood alcohol levels; in subsequent practice drinking sessions,
painful electric shocks were presented when blood alcohol exceeded
a pre-determined problem level. At follow-up (18 months later),
40% of treated patients were drinking moderately, in contrast to
control patients (given non-contingent shocks) whose drinking exceeded criterion levels even during the treatment phase.

In addition to procedures using punishing stimuli (aversive
stimuli such as shock), some researchers have explored positive reinforcement contingencies. Investigators at Baltimore City Hospital
employed a variety of positive reinforcement procedures to inculcate
abstinence or moderation with in-patients. Among the contingencies
studied was time out from positive reinforcement as a schedule consequence for drinking, access to an enriched environment as a reward

for abstinence or moderate drinking (Cohen, Liebson, Faillace, et al., 1971; Pickens, Bigelow and Griffiths, 1973).

The most elaborate test of controlled drinking was performed at the Patton (Cal.) State Hospital (Sobell and Sobell, 1976). Alcoholic in-patients who qualified for controlled drinking were randomly assigned to behavioral treatment with moderation as a goal or traditional treatment with abstinence as a goal; patients who did not qualify for controlled drinking were assigned to behavioral treatment with abstinence as a goal or traditional treatment with abstinence as a goal. At follow-up, two years after treatment, the controlled drinking group functioned best on several indicators, including number of days abstinent. When "abstinent days" and "controlled drinking days" were combined, the percentage of patients "functioning well" 80% or more of the time was as follows: 79% in the behavioral moderation group versus 22% in its traditional control group; 54% in the behavioral abstinence group versus 21% in its traditional control group.

At present there has been little systematic investigation of the use of positive reinforcement techniques in out-patient settings In addition, few studies have treated middle income problem drinkers --people who are still employed, have intact families, and have not yet encountered serious health problems because of drinking. Each of these variables represents a promising point of departure for treatment research: Positive reinforcement techniques are generally more acceptable to patients than negative ones. Out-patient treatment minimizes disruption of employment and other life patterns and, because residence and meals are not required, therapy can be extended less expensively over sufficient time to encounter those difficulties in living which contribute to problem drinking. Treating people while they still have something to lose by continued excessive drinking represents an important innovation in the management of alcohol problems, for it may make possible the prevention of late-stage alcoholism and the personal devastation that often accompanies it. Finally, the major deficit in current alcoholism treatment research is the scarcity of studies comparing maximized, clinically-relevant treatment procedures. In response to these concerns, the present research compared outcome for a multi-component positive reinforcement procedure emphasizing moderation and traditional group encounter therapy emphasizing abstinence, treating self-selected middle-income problem drinkers on an out-patient basis. Since the experimental findings are described in some detail by Pomerleau, Pertschuk, Adkins, et al. (in press), the present report will concentrate on the clinical implications of the research

METHOD

Subjects

Problem drinkers were recruited through physican referrals and announcements of treatment availability in the local media. The project was conducted in Philadelphia, a major metropolitan area of several million people. Criteria for inclusion in the study included expressed willingness to attend treatment and follow-up sessions, capacity to follow instructions, and absence of marked psychopathology as judged by an interview.

There were 32 subjects treated in the present study. Seven problem drinkers were screened but then referred to another treatment modality (four for treatment of depression, one for psychosis, and two for open-ended group support for abstinence). An additional seven received a screening interview and were accepted for the study but did not attend the first treatment session or pay a fee.

On a random basis, 18 subjects were assigned to behavioral treatment and 14 were assigned to traditional treatment. The procedure assigned problem drinkers to one or the other treatment group until six or seven participants were recruited, rather than trying to accumulate sufficient subjects to form two groups at one time. This approach was developed to minimize the delay between screening and treatment, as previous experience had shown that attrition increased with the waiting period.

Treatment was carried out at the Center for Behavioral Medicine of the Department of Psychiatry, University of Pennsylvania, a unit specializing in developing methods for preventing disease through behavior change (Pomerleau, Bass, and Crown, 1975). Therapy was conducted in a large room seating 15 comfortably; a private office was available nearby for individual conferences.

Procedure

Behavioral and traditional therapy were conducted with groups of from 3 to 7 problem drinkers in 90 minute sessions meeting once a week for three months and five additional sessions at increasing intervals over nine months following treatment.

The behavioral treatment procedure (Pomerleau and Pertschuk, Note 1) consisted of four consecutive phases: (1) Baseline--including the screening interview and first therapy session; (2) Reduction of drinking or cravings--the second through fifth session; (3) Behavior therapy for problems contributing to excessive drinking-- the sixth through the ninth session; and (4) Maintenance of thera-

peutic gain--the tenth through twelfth session and five follow-up
sessions. The techniques used were those which the research literature had shown to be effective on some aspect of the problem of alcoholism. The sequence of instructions was arranged to facilitate
transition from one phase to another.

In the baseline phase, the problem drinker was interviewed; he
also completed demographic forms and questionnaires on drinking
history. The treatment procedure was explained at this time. A
prepaid treatment fee was required (on a sliding scale from $85 to
$500, based on ability to pay). A "commitment fee" (Chapman, Smith,
and Layden, 1971) of up to $300 was also requested. The fee could
be earned back in its entirety by following treatment instructions,
i.e., by (a) keeping records (refunds based on completeness, not
content); (b) coming to treatment with no detectable breath alcohol
(measured by the Alcohol Level Evaluation Road Tester; Borg-Warner,
DesPlaines, Illinois); (c) carrying out selected non-drinking activi
ties as corroborated by a self-designated monitor; and (d) attending
follow-up. Participants were told which activities were required
but not when refunds would be given. During the treatment, an
intermittent schedule of reinforcement was chosen to increase resistance to extinction (Ferster and Skinner, 1957) in follow-up,
when monetary reinforcement was given more predictably but less
frequently. If the participant dropped out, all fees were forfeited

In the first week of treatment, participants were asked to make
no special effort to modify their drinking but to keep a detailed
record of consumption, drink by drink. The situations which led to
excessive drinking were identified by examining the circumstances
under which drinking took place. The principal exception to the
above procedure was in the case of participants who required immediate detoxification when seen at screening; in the first week of
treatment, they were typically abstinent and recorded circumstances
contributing to craving for alcohol rather than actual drinking.

Active therapy began in the reduction phase. Participants
designated specific daily quotas for the coming week as well as
final goals for treatment. The emphasis was on gradual, steady improvement rather than abrupt change (response shaping and stimulus
fading--Keller and Schoenfeld, 1950). Moderate drinking was permitted as a goal (1) if the participant requested it as the principal reason for treatment at that time, (2) if the participant had
exhibited some degree of control in recent drinking, and (3) if
there were no medical contraindications to continued alcohol consumption. For participants attempting to learn controlled drinking,
the final goal was three days of abstinence per week, consumption
of no more than three ounces (88.7 cc.) of absolute alcohol on days
when drinking was authorized, and consumption of no more than ten
ounces (296 cc.) of alcohol per week (the "3-3-10" rule). Participants for whom abstinence was the goal used similar methods but set

subgoals more stringently so that abstinence could be reached
within two weeks; subsequently, they recorded cravings on their
daily records. Quotas were publicly-stated and social reinforcement for success was provided by therapists as well as participants
(Hunt and Azrin, 1973).

During the first weeks of reduction, participants used various
stimulus control and contingency management techniques (Mahoney and
Thoresen, 1974) to specify appropriate and inappropriate drinking
circumstances, to delay and interfere with habitual drinking patterns, and to enhance non-drinking in designated situations. Among
the techniques used were sipping rather than gulping drinks, not
drinking at certain times or in certain places, keeping glass filled
with a non-alcoholic drink on social occasions, storing quantities
of non-alcoholic beverages while not stockpiling liquor, and pre-
planning--writing down a specific drinking plan prior to known problem situations. These techniques were derived from research in
which the drinking patterns of alcoholics and social drinkers were
studied objectively (Nathan and O'Brien, 1971). Desire to drink was
handled, in part, through covert conditioning--pairing craving with
the imagined negative consequences of sustained excessive drinking
(Cautela, 1970).

In the behavior therapy phase, increased emphasis was placed on
identifying and, subsequently, modifying emotional situations such
as anger, anxiety, depression, and marital discord which might precipitate or contribute to excessive drinking. Standard behavioral
procedures including assertion training, desensitization, and deep
muscle relaxation training were made available on an individual
basis as needed (Lazarus, 1971; Wolpe, 1969). Resocialization
training through modeling (Bandura, 1969) and family counseling
(Cheek, Franks, Laucius et al., 1971; Miller, 1972) were also provided.

In the final, consolidation phase of treatment, participants
were encouraged to develop alternative activities to replace problem
drinking, such as hobbies, physical exercise, courses, etc. which do
not involve alcohol. New social companions were suggested if necessary. The importance of discovering that pleasure and satisfaction
could be derived from various activities without drinking was
stressed. Toward the end of treatment and the beginning of the
follow-up phase, record keeping was phased out gradually on an op-
tional basis. Participants were prepared for decreased formal support from therapy by encouraging them to call one another or the
therapists during times of difficulty in the follow-phase. Occasional loss of control was defined as the failure of a technique
rather than as personal failure. At such times, participants were
given support and were told to reinstate record keeping. They were
also shown how to deal with contributory circumstances such as depression or personal setbacks by appropriate corrective action.

Anticipation and a realistic perspective on the vicissitudes of living were emphasized. At all times, personal responsibility in the
acceptance of the consequences of maladaptive behavior was stressed.

Traditionally-oriented treatment was also conducted in small
groups over the same length of time as behavioral treatment by therapists who were committed to it. The treatment procedure was explained in the screening interview. Fees for treatment were on a
sliding scale; they ranged from $5 to $30 per session for 17 sessions, based on ability to pay, with payment due before each session. No prepaid commitment fees were required nor were refunds
awarded. No monetary penalty for dropping out was specified.

The first or introductory phase of traditional treatment, consisting of three sessions, was devoted to developing a sense of
group cohesion and mutual trust among participants and therapists.
The importance of total abstinence was emphasized. Controlled
drinking was allowed in a few participants only under the special
condition that it was the sole basis for entering treatment at that
time and was not medically contraindicated. Participants were encouraged to discuss their problems (with alcohol and in other aspects of living) within a supportive group milieu.

During this time, therapists observed various denial patterns
emerging from participants' conceptualizations of personal problems.
Since denial of drinking problems is considered a major barrier to
recovery, considerable attention was paid to various denial patterns which ranged from the obvious to the subtle--for instance, an
alcohol abuser with years of excessive drinking and associated liver
damage might assert that he is not an alcoholic, or a problem drinke
who has temporarily stopped drinking and is willing to call himself
an alcoholic, but indicates in various ways that he does not believe
there is anything seriously wrong with him.

The second, confrontation phase of this treatment lasted six or
seven sessions and utilized all material developed in earlier sessions. Therapists taught participants to recognize denial patterns
in others and to confront them effectively. Participants each took
a turn on the "hot seat" for 20 to 30 minutes in successive sessions
(Fehr, 1976), during which they were the focus of observation, comment, and analysis by all other group members and therapists. Confrontation by the group often generated anger, sadness, guilt, and
other strong emotions which, within the context of a carefully-con-
trolled group situation, was felt to be conducive to generating insight into the nature of personal denial mechanisms, the latter an
essential component of the traditional view of the healing process.
In general, selection for the "hot seat" arose out of the group
process, though at times therapists made assignments based on their
knowledge of group interaction and alliances. At the same time,
participants were encouraged to discuss personal or family problems

which they felt contributed to their drinking.

The third, resolution phase lasted two to three sessions. At this time, attempts were made to channel the intense emotions generated in the preceding phase of treatment into productive, future-directed activity. As in phase one, positive feelings of group solidarity and mutual support were encouraged. Adjunctive psychotherapy for depression, anxiety, family problems, etc. was provided as needed.

During the five follow-up sessions extending over nine months, supportive therapy for non-drinking continued to be made available.

Therapists

A doctoral-level therapist was responsible for all phases of both treatment programs; a subdoctoral co-therapist assisted during treatment and was the principal therapist during follow-up. One of two doctoral-level therapists, a licensed psychologist or a board-eligible psychiatrist, led behavioral treatment in successive groups; both had several years of general clinical experience. The behavioral co-therapist held a masters degree in psychiatric nursing. Traditional treatment was led by a board-certified psychiatrist who had specialized in the treatment of alcoholism for several years and was in private practice; the traditional co-therapist held a masters degree in psychiatric social work. Traditional therapists were not familiar with the behavioral treatment procedures.

RESULTS

A comparison of pre-treatment characteristics of the treated population as a whole and behavioral and traditional subjects separately is given in Table 1. It reveals no statistically significant differences between behavioral and traditional participants in any pre-treatment category. The table also shows that behavioral participants registered a non-significant 18% reduction in drinking between the week before screening and the first week of treatment (baseline) while traditional participants showed a significant 46% reduction during this interval (Wilcoxon Test, p < .01; two tailed).

Drinking level as a function of time for those behavioral and traditional participants who remained in treatment is shown in Figure 1. Statistical comparisons (which exclude dropouts) indicate

Table 1

Pre-treatment Characteristics (Medians)

	All Subjects	Behavioral Participants	Traditional Participants
Age (years)	44	45	41.5
Education (years)	16	16	15.5
Duration of problems with alcohol (years)	8.5	9.5	7.8
Number of prior treatment attempts	2	2.5	1
Delay between screening and treatment (weeks)	3	3	3
Percent Male	69	56	86
Amount of alcohol (100% ethanol) consumed week prior to screening (oz.)	28.5	25.5	31.4
(cc.)	843	754	928
Amount of alcohol consumed in baseline (oz.)	19.8	21.0	16.8
(cc.)	586	621	497
Number of participants	32	18	14

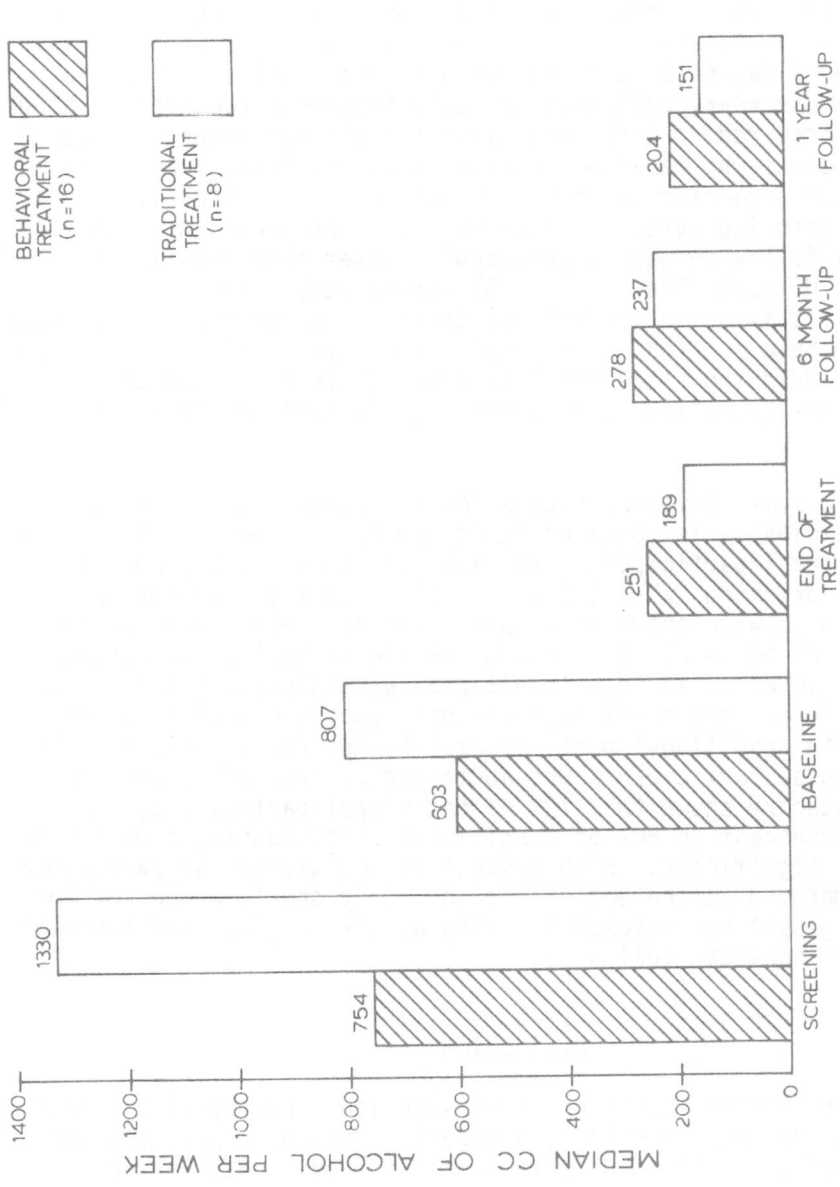

Figure 1: Median cc. of ethanol per week at various points in time for the 16 behavioral and 8 traditional participants who remained in treatment. Dropping out was defined as missing at least the last session of treatment and not attending subsequent follow-up.

that traditional subjects remaining in treatment consumed significantly more than remaining behavioral subjects at the time of initial screening (Mann-Whitney U-test, $\underline{p}$ < .05; two tailed). There
were, however, no significant differences in consumption between
members of the two treatment groups at baseline, during subsequent
treatment, or at follow-up. Using subjects as their own controls,
drinking level decreased significantly from initial screening to
the first anniversary follow-up for both treatment procedures though
reductions occurred at different times for the two groups. Thus,
for participants who remained in behavioral treatment, consumption
rates between screening and baseline were not significantly different but drinking rates the last week of treatment and at the firs
anniversary follow-up were significantly lower than baseline (in
both cases, Wilcoxon Test, $\underline{p}$ < .005; one-tailed). By contrast,
participants staying in traditional treatment consumed significantly
less during baseline than at screening (Wilcoxon Text, $\underline{p}$ < .05; two-
tailed) and drinking rates the last week of treatment and at the
first anniversary follow-up were not significantly different than
at baseline.

The one-year follow-up status for all behavioral and traditional participants is compared in Figure 2. It shows that somewhat
fewer behavioral participants abstained but considerably more reduced their drinking (χ^2 = 2.93, $\underline{p}$ < .1); slightly more were unimproved compared with traditional participants. Behavioral participants who reduced level of drinking consumed a smaller percentage
of previous level (35%) than traditional participants (50%). Significantly fewer behavioral participants dropped out of treatment
compared with traditional participants (Fisher Exact Test, $\underline{p}$ < .05)
Combining those participants who abstained or reduced to form an
improved category shows that 72% of behavioral participants improved, in contrast to 50% of traditional participants, a difference
that is not significant. With respect to recidivism, of participant
who were improved at the end of therapy, only one (a member of the
behavioral group) had resumed drinking at his original pre-treatment
level at the one-year follow-up.

DISCUSSION

The significant reduction in consumption shown by traditional
participants during screening and baseline periods apparently was
the result of an emphasis on abstinence in traditional therapy
(during the screening process); behavioral participants did not
display a significant reduction in drinking during this interval.
Traditional participants who stayed in treatment reported significantly more drinking at screening than did behavioral participants;
there were no significant differences in consumption levels between
treatment groups at any other subsequent follow-up point.

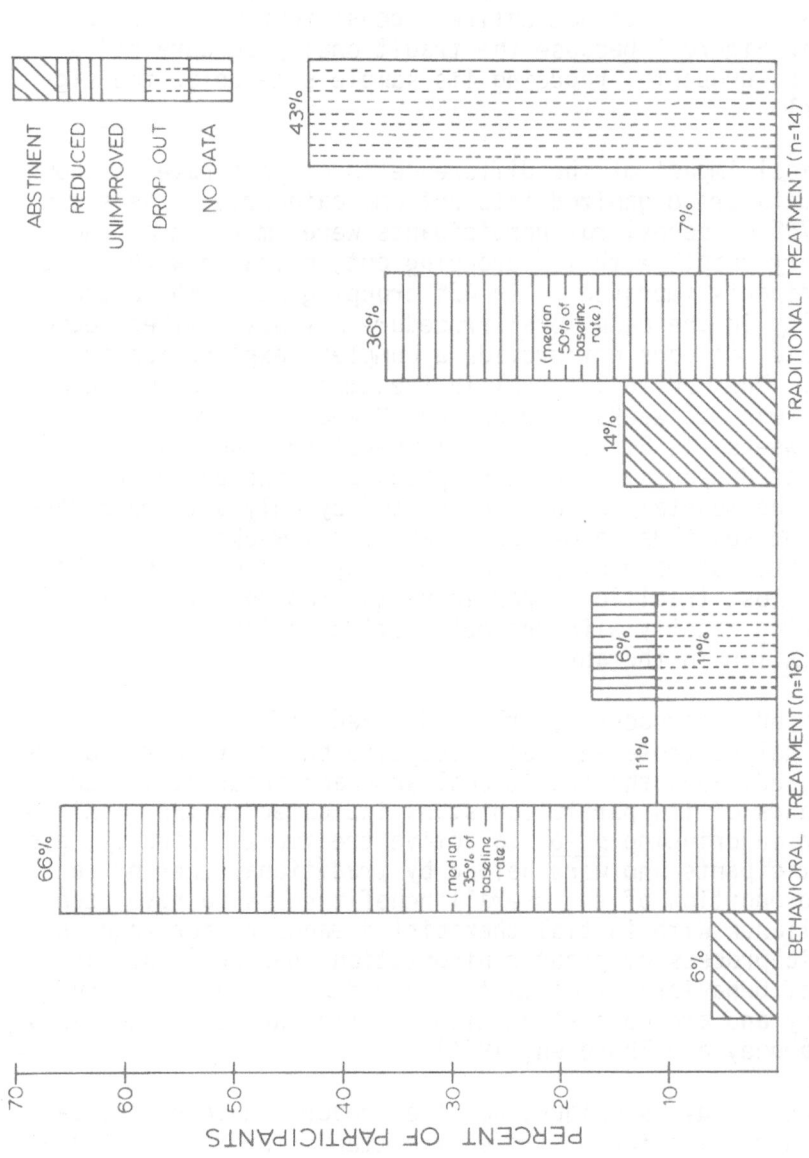

Figure 2: Percent in various outcome categories for 18 behavioral and 14 traditional participants at the one-year follow-up. A participant was classified as unimproved if drinking at the same level or greater than baseline levels. One behavioral participant, for whom data could not be obtained at the anniversary follow-up, was classified in the dropout (no data) category.

Additional significant changes occurred at different times within each treatment group. Traditional subjects reported a significant reduction in drinking between screening and baseline periods whild behavioral participants showed a significant reduction between the end of treatment and the first anniversary follow-up. The statistical analysis is not entirely consistent with the graphic presentation in Figure 1 because the traditional procedure had a larger number of dropouts, reducing its sample size below the significance level.

The critical impact of the difference in dropout rate is shown clearly when data are organized into outcome categories. As shown in Figure 2, 72% of behavioral participants were improved at the year anniversary point, with 11% dropping out, compared with 50% of treated participants improved, with 43% dropping out. While the monetary penalty in the behavioral procedure may have helped reduce the dropout rate, it does not provide a complete explanation of what happened. The median interval in treatment for the two behavioral dropouts was 5.5 weeks (1 week and 10 weeks), compared with a median of 9.0 weeks (1, 8, 9, 9, and 11 weeks) for the six traditional treatment dropouts. Fees were prepaid in the behavioral procedure and the monetary penalty decreased by only a third during the treatment phase; thus, a median of about six weeks in treatment seems an adequate estimate of central tendency. In contrast, the point at which most traditional participants dropped out is nine weeks, coinciding exactly with the culmination of intense interpersonal confrontation in therapy.

The therapeutic process clearly had mixed effects in traditional treatment; it helped those who were receptive but drove out those wh were not. In addition, the traditional approach apparently made mor immediate demands of its participants, as indicated by the significant reduction in drinking prior to active treatment. Finally, ever for those participants who were helped by traditional therapy, a considerable proportion of the overall benefit seems to have been based on compliance with initial therapist demands rather than on the therapeutic process of group confrontation, per se. The favorable effects of behavioral treatment occurred during and, presumably, through therapy and are consistent with the findings on contingency management (Mahoney and Thoresen, 1974).

The present study is unique; no other reports have been published in which results for maximized treatments for middle income problem drinkers are examined. Because of this, comparisons with other studies in the literature must be made with care. The research by the Sobells on controlled drinking (1976) is perhaps most similar to ours. Major points in common are that behavioral and conventional hospital treatments were compared, subjects volunteered for the study, and subjects had reasonably favorable prognoses (Sobell, Note 2). A difference is that treatment was conducted in an in-patient setting and that the conventional state hospital therapy used was probably less than maximized. The general thrust of the Sobell's findings is quite consistent with the present results --in particular, the observation that behavioral treatment with moderation as a goal produced greater improvement than traditional, abstinence-oriented treatment. Recent reviews of the behavioral treatment literature (Miller and Barlow, 1973; Nathan and Briddell, 1977) support these conclusions.

A few words about methodology are also in order. Among the key problems in alcoholism treatment research at present is determination of valid measures of outcome. While the issue is far from settled, some general guidelines seem to be emerging (Sobell, 1978). The present study shows clearly how measures like daily alcohol consumption before and after treatment can lead to a distorted view of efficacy. Organizing the same data into continuous categories produced a much clearer picture of outcome, for it differentiated non-drinkers from moderate or excessive drinkers. In addition, since information about drinking was often unobtainable from dropouts, these participants would have been excluded from an analysis based solely on drinking levels. By incorporating participants who dropped out, a more comprehensive report was made possible.

The importance of accounting for all participants should not be underestimated. The present study indicates the status of all problem drinkers who interacted with the program. Too many studies only provide data for those subjects who completed treatment, ignoring those problem drinkers who dropped out or refused treatment. Without such information, valid conclusions concerning the general effectiveness of a particular treatment are limited and may, in many cases, be impossible. A major source of bias which cannot be glossed over is that dropping out of treatment is usually associated with lack of progress (Sobell, 1978). Thus, in the present investigation, the lower dropout rate for behavioral treatment is as much an indicator of clinical effectiveness as its larger proportion of participants in the improved category.

The present study explored several major treatment variables in combination. While the preliminary objective of developing an adequate methodology for demonstrating differences between two complex clinical procedures was met, many questions still remain unanswered. The main experiment compared clinical outcome for a multi-component behavioral procedure which emphasized moderation and a traditional group confrontation procedure which emphasized abstinence. Among the treatment variables which were common to the procedures were that the middle income participants formed strong relationships with the professional therapists and each other and were exposed to similar enthusiasm by therapists for their particular approach. Both treatments were conducted with the same frequency on an outpatient basis over the same time period.

The following observations can be made with respect to some of these similarities. First, while lower income problem drinkers have been treated extensively with traditional group encounter techniques, comparable experience is lacking on self-help methods. Preliminary work with other participants at the Center for Behavioral Medicine suggests that, while middle income participants respond well to self-management procedures, lower income clients require more explicit conditioning procedures and goals, such as those used by Lovibond and Caddy (1970) and Sobell and Sobell (1976). Second, while the Rand Report (Armor, Polich, and Stambul, 1976) reported uniform remission rates among treatment modes using professional and non-professional therapists (and even between purportedly different therapeutic methods), more carefully controlled investigations will be needed to settle this issue. As a final note, providing therapy on an out-patient basis was quite satisfactory for both behavioral and traditional approaches, having the advantage of being sufficiently intensive during the treatment phase and adequately extensive during the follow-up phase.

Among the treatment variables which differentiated the procedures were the following: First, traditional treatment was psychodynamic and insight-oriented and did not employ behavioral techniques such as stimulus control or contingency management to modify drinking habits or to deal with problems which contributed to excessive drinking. Second, the chief change vehicle for traditional treatment was group confrontation (social punishment), in contrast to various behavioral contingencies including group support and monetary rewards (positive reinforcers) in behavioral treatment. Third, the emphasis in traditional treatment was on abstinence, whereas in behavioral treatment it was on moderation.

With regard to the first set of variables, the multi-component behavioral procedure was more effective than traditional therapy in several ways and the improvement observed seemed to be attributable to the therapeutic techniques employed. In a sense, the experiment was conducted with a bias against demonstrating a difference, for traditional therapy was considerably more extended than is usually the case and was conducted by experienced, highly-trained professional therapists. Thus, the attempt to maximize traditional therapy probably made it less representative of conventional treatment.

The relative contribution of positive reinforcement and punishment contingencies to the outcome for the two procedures cannot be fully determined. But, as has been discussed above, group confrontation of denial apparently increased the dropout rate in traditional treatment. It is interesting to note also that monetary rewards had a measurable effect on those behavioral participants who remained in treatment. Excluding dropouts from the analysis, behavioral participants attended a median of 92% of scheduled treatment sessions, compared with 87% for traditional participants; in follow-up, attendance rose to 100% for behavioral participants but decreased to 80% for traditional participants. The probable explanation for this is that, though conditions were relatively similar between treatment, as a whole, and follow-up for traditional participants, behavioral participants were given monetary refunds contingent on attending follow-up sessions. The difference between behavioral and traditional treatment is even more pronounced if participants who dropped out are taken into account.

Finally, the role of the goal of therapy cannot be isolated from other treatment variables in the present study because of the small number of participants involved and the differences in the complex procedures being compared. Preliminary evidence in the literature, however, suggests that behavioral treatment is more effective in producing either abstinence or moderation than conventional therapy (Sobell and Sobell, 1976) and that traditional treatment is not well-suited to fostering moderation (Popham and Schmidt, 1976). While the present results are not inconsistent with these findings, they do not constitute a definitive test. A particular concern at present is the relative recidivism rate between the two goals. A productive approach might be to compare outcome in follow-up for a sufficient number of problem drinkers who either moderated or abstained after behavioral treatment. A higher non-improvement rate for one or the other might help settle the question of whether controlled drinkers are more likely to relapse than abstainers. With the present state of knowledge, it is no more appropriate to blame recidivism on an attempt at controlled drinking than it is to designate a period of abstinence as the cause of subsequent relapse.

In conclusion, the present research examined several underexplored variables in alcoholism treatment and found that a multi-component behavioral procedure was superior to intensive traditional group confrontation for middle income problem drinkers. We hope that other clinicians and researchers will find the results of sufficient interest to explore these techniques and methods more fully by replication and further extension.

ACKNOWLEDGMENT

The present investigation was supported in part by the Governor's Council on Drug and Alcohol Abuse, Commonwealth of Pennsylvania. The opinions and conclusions stated in the present paper are those of the authors and are not to be construed as official or as necessarily reflecting policy of the Governor's Council.

The authors wish to thank Gail Hough, Kathryn Ratner, Ceil Hirsh, and Leah Walcoff for their assistance on the project.

REFERENCE NOTES

1. Pomerleau, O. F. & Pertschuk, M. Behavioral treatment of alcohol abuse using multiple techniques. Paper presented at the meeting of the Association for the Advancement of Behavior Therapy, San Francisco, 1975.

2. Sobell, L. C. Personal communication, March, 1977.

REFERENCES

Armor, D. J., Polich, J. M. & Stambul, H. B. Alcoholism and Treatment. Santa Monica, Cal.: Rand Corporation, 1976.

Bandura, A. Principles of Behavior Modification. New York: Holt, Rinehart, and Winston, 1969.

Cautela, J. R. Covert reinforcement. Behavior Therapy, 1970, 1, 33-50.

Chapman, R. F., Smith, J. W. & Layden, T. A. Elimination of cigarette smoking by punishment and self-management training. Behavior Research and Therapy, 1971, 9, 255-264.

Cheek, F. E., Franks, C. M., Laucius, J. & Burtle, V. Behavior modification training for wives of alcoholics. Quarterly Journal of Studies on Alcohol, 1971, 32, 456-461.

Cohen, M., Liebson, I. A., Faillace, L. A., & Allen, R. P. Moderate drinking by chronic alcoholics. Journal of Nervous and Mental Disease, 1971, 153, 434-444.

Davies, D. Normal drinking in recovered alcohol addicts. Quarterly Journal of Studies on Alcohol, 1962, 23, 94-104.

Fehr, D. N. Psychotherapy: Integration of individual and group methods. In R. E. Tarter & A. A. Sugerman (Eds.), Alcoholism: Interdisciplinary approaches to an enduring problem. Reading, Mass.: Addison-Wesley Publishing Co., 1976.

Ferster, C. & Skinner, B. F. Schedules of Reinforcement. New York: Appleton-Century-Crofts, 1957.

Hunt, G. M. & Azrin, N. H. A community-reinforcement approach to alcoholism. Behaviour Research and Therapy, 1973, 11, 91-104.

Keller, F. & Schoenfeld, W. N. Principles of Psychology. New York: Appleton-Century-Crofts, 1950.

Lazarus, A. A. Behavior Therapy and Beyond. New York: McGraw-Hill, 1971.

Lloyd, R. W. & Salzberg, H. C. Controlled social drinking: An alternative to abstinence as a treatment goal for some alcoholics. Psychological Bulletin, 1975, 82, 815-842.

Lovibond, S. H. & Caddy, G. Discriminated aversive control in the moderation of alcoholics' drinking behavior. Behavior Therapy, 1970, 1, 437-444.

Mahoney, M. & Thoresen, C. Self-Control: Power to the Person. Monterey, Cal.: Brooks/Cole, 1974.

Miller, P. M. The use of behavioral contracting in the treatment of alcoholism: A case report. Behavior Therapy, 1972, 3, 593-596.

Miller, P. M. and Barlow, D. H. Behavioral approaches to the treatment of alcoholism. Journal of Nervous and Mental Disease, 1973, 157, 10-20.

Nathan, P. E. and Briddell, D. W. Behavioral assessment and treatment of alcoholism. In B. Kissin and H. Begleiter (Eds.), The biology of alcoholism, Vol. 5. New York: Plenum Publishing Co., 1977.

Nathan, P. E. & O'Brien, J. S. An experimental analysis of the behavior of alcoholics and non-alcoholics during prolonged experimental drinking: A necessary precursor of behavior therapy? Behavior Therapy, 1971, 2, 455-476.

Pickens, R., Bigelow, G. & Griffiths, R. An experimental approach to treating chronic alcoholism: A case study and one-year follow-up. Behaviour Research and Therapy, 1973, 11, 321-325.

Pomerleau, O. F. Research priorities in alcohol studies: The role of psychology. Journal of Studies on Alcohol, special supplement, 1977.

Pomerleau, O. F., Bass, F. & Crown, V. The role of behavior modification in preventive medicine. The New England Journal of Medicine, 1975, 292, 1277-1282.

Pomerleau, O. F., Pertschuk, M., Adkins, D. & Brady, J. P. A comparison of behavioral and traditional treatment for middle income problem drinkers. Behavior Therapy, in press.

Pomerleau, O. F., Pertschuk, M. & Stinnett, J. A critical examination of some current assumptions in the treatment of alcoholism Journal of Studies on Alcohol, 1976, 37, 849-867.

Popham, R. & Schmidt, W. Some factors affecting likelihood of moderate drinking by treated alcoholics. Journal of Studies on Alcohol, 1976, 37, 868-882.

Sobell, M. B. & Sobell, L. C. Second year treatment outcome of alcoholics treated by individualized behavior therapy: Results. Behaviour Research and Therapy, 1976, 14, 195-215.

Sobell, L. C. Empirical assessment of alcoholism treatment outcome evaluation: Past, present and future. In G. A. Marlatt & P. E. Nathan (Eds.), Behavioral Assessment and Treatment of Alcoholism. New Brunswick, N. J.: Rutgers Center of Alcohol Studies, 1978.

Wolpe, J. The Practice of Behavior Therapy. New York: Pergamon Press, 1969.

STUDIES IN BLOOD ALCOHOL LEVEL DISCRIMINATION

Peter E. Nathan

Rutgers University

New Brunswick, New Jersey

In recent years the contention that unwavering abstinence constitutes the only legitimate treatment goal for alcoholism has come under increasing scrutiny. Acceptance of a single abstinence-oriented treatment goal for all alcoholics has been challenged by findings indicating that some alcoholics can acquire and maintain patterns of limited social drinking without any accompanying "loss of control" over intake (e.g., Davies, 1962; Pattison, 1968). The apparent success of alcoholism treatment programs having controlled drinking as an explicit treatment goal (e.g., Lovibond and Caddy, 1970; Pomerleau et al., in press; Sobell and Sobell, 1973) has also contributed to reconsideration of treatment goals for some alcoholics.

A number of studies exploring the utility of controlled drinking treatment approaches have incorporated blood alcohol level (BAL) discrimination training into their programs. In general, the authors of these studies (e.g., Caddy and Lovibond, 1976; Lovibond and Caddy, 1970; Miller, in press; Vogler, Compton and Weissbach, 1975; Vogler, Weissbach, Compton, and Martin, 1977) have maintained that training in BAL discrimination allows the alcoholic to monitor his level of intoxication and that this ability can be incorporated within a treatment program designed to aid the alcoholic to maintain more moderate BALs.

The first study of BAL discrimination training was reported by Lovibond and Caddy (1970). The goal of that study was to establish accurate BAL discrimination abilities in alcoholic subjects and thence to induce a discriminated conditioned aversion to high BALs (defined as those over 65 mg/%). Of 44 alcoholic subjects accepted into the treatment program, 31 were assigned to an experimental group and 13 to a control group.

Experimental subjects were trained in BAL discrimination in the first phase of treatment. At the beginning of the single 90- to 120-minute training session, subjects were given a scale describing the typical behavioral effects of different BALs. They then ingested an alcohol and fruit juice mixture, then were asked to examine their subjective experiences as a basis for estimating BAL. Every 15-20 minutes of the session, subjects received Breathalyzer tests, estimated their BAL, then were given immediate feedback on actual BAL.

Conditioning procedures were implemented during the second, treatment, phase of the study. Subjects consumed their preferred alcoholic beverage at an experimenter-determined rate designed to raise BALs to approximately 65 mg/% by the end of 90 minutes. As before, subjects made BAL estimates, then received accurate feedback every 15-20 minutes. However, when BALs rose above 65 mg/%, as they were programmed to do, painful (4-7 mA) electric shocks were delivered to subjects' faces on a partial reinforcement schedule. Subjects were required to continue drinking throughout the entire session. Treatment lasted 6-12 sessions; subjects received between 30 and 70 shocks in all. Control subjects received the same BAL discrimination training but, throughout the aversive conditioning phase of the study (which lasted only three sessions), shocks were administered on a random basis.

Although Lovibond and Caddy did not report data on acquisition rates of BAL discriminations, they did report that "after a single training session, errors in excess of + .01% (10 mg/%) rarely occur (p. 440)." Nonetheless, because pretraining data on BAL estimation accuracy were not reported and the discrimination accuracy of a control group was not assessed, it cannot be concluded unequivocally that subjects' post-training accuracy was due to training per se. Further, because BAL estimates by subjects were never made in the absence of BAL feedback, it is not possible to be certain that post-training estimation accuracy was maintained when feedback was removed.

In a follow-up comparison of these experimental and control subjects a greater pre-post reduction in alcohol intake was reported for the former group than for the latter. Lovibond and Caddy attributed the greater improvement in experimental subjects' drinking patterns to a change in their motivation to drink to high BALs, produced as a result of the pairings of shock with ostensibly readily-discriminated BALs.

Along with a too-ready acceptance of the efficacy of electrical aversion conditioning with alcoholics (cf., Nathan and Briddell, 1977; Wilson, Leaf and Nathan, 1975), the validity of conclusions drawn by Lovibond and Caddy on the basis of their outcome data can be questioned. First, outcome was assessed on the basis of self-report data whose validity and reliability were not measured. Second, by virtue of the differential attrition rates of the two groups (control subjects: 61%; experimental subjects: 10%), it appears that the control procedure was probably not a very convincing one. Hence, post-treatment group differences might well have been due to such non-specific factors as differential expectancies for improvement and other demand characteristics than to the specific procedures employed. Finally, because the authors never established that BAL discrimination training via internal cues had both taken place in the first place and been maintained over time, it is difficult to accept this aspect of their intervention as responsible for the short-term improvement in drinking pattern they reported.

In an effort both to replicate and to extend Lovibond and Caddy's exploration of the blood alcohol level discrimination training paradigm, Silverstein, Nathan and Taylor (1974) designed a study initially including four male "gamma" alcoholics who participated as inpatients in both phases of a two-part, 36-day study. The goal of the first phase of the study (which lasted 10 days) was to examine some of the factors involved in training alcoholics to estimate BAL accurately. Drinking was programmed in five 2-day cycles such that BAL rose on the first day of a cycle to 150 mg/%, then fell overnight and over the next day to zero. During the first (baseline) two-day cycle, subjects estimated BAL approximately ten times per day without receiving feedback on accuracy. During the following three 2-day cycles, subjects were continuously alerted to the emotional and physical correlates of changing levels of blood alcohol while receiving feedback 1) after each BAL estimate; 2) after 50% of their estimates; and 3) after 50% of their estimates with positive reinforcement delivered contingent on accurate BAL estimation. During the final 2-day cycle of this phase of the study, which represented a return to baseline conditions, subjects were again required to make their BAL estimates in the absence of training, feedback, or contingent reinforcement.

During the second phase of the study (which lasted 26 days), three of the four subjects who had participated in the study's first phase were trained to drink to, then maintain, a prescribed BAL (80 mg/%). Three converging behavioral shaping procedures were utilized for this purpose: 1) Responsibility for control over drinking was gradually shifted from the experimenter to the subject; 2) The range of positively-reinforced BALs was successively narrowed closer and closer to the target BAL of 80 mg/%; 3) All reinforcement and feedback were gradually faded-out over the nearly four weeks of this phase of the study.

Data from the first phase of the study showed that the most powerful factor influencing BAL estimation accuracy was, simply, the presence or absence of accurate feedback on blood alcohol level. Whether this feedback was continuous or intermittent, or accompan- ied or unaccompanied by reinforcement for accuracy was unimportant; estimation error scores during the three training cycles were uni- formly lower than those during the initial pre-training or conclud- ing post-training baseline periods of this phase of the study. During the second, control-training phase of the study, subjects were able effectively to control their drinking - to maintain BAL within the prescribed range - but only so long as feedback on BAL was provided. Degree of control decreased significantly when feedback was removed (during the post-experimental baseline asses- sment period) (see Figure 1).

These data by Silverstein and his colleagues called into ques- tion Lovibond and Caddy's explanation of their successful treatment outcomes - that subjects maintained the ability to discriminate BAL from internal cues through the follow-up period. Though the Silver- stein study affirmed that alcoholics can learn to discriminate BAL with considerable accuracy and, following acquisition of that skill, confine their drinking behavior to narrowly-defined limits, both abilities were significantly attenuated once external feedback of accurate BAL was removed.

Specifically relevant to the results of the studies by Lovibond and Caddy and Silverstein, Nathan, and Taylor are data from a clinical report by Paredes, Jones and Gregory (1974), who trained a single alcoholic subject to discriminate BAL. Although he did, in fact, learn to monitor rising and falling BALs, whether he did so on the basis of training in the use of subjective experience for this purpose or from the accurate feedback on BAL he was continu- ously provided throughout training could not be determined from the research design employed. What was clear, however, was that the subject did not attain the high degree of estimation accuracy, even after approximately 50 hours of training, either group of al- coholics studies previously (by Lovibond and Caddy, and Silverstein and his colleagues) attained. Studies in which BAL discrimination training comprised a component of a comprehensive behavior therapy package have also been reported. Most of these studies employed BAL discrimination training to establish discriminated aversions to BALs above a certain level (for example, Caddy and Lovibond, 1976; Miller, in press; Vogler, Compton and Weissbach, 1975; Vogler, Weissbach, Compton and Martin, 1977; Wilson and Rosen, 1975). Over- all, these studies have not contributed to a fuller understanding of the issues discussed in this paper thus far. For one thing, they were not designed to permit inquiry into the actual acquisition patterns of BAL discrimination. For another, assessment of these acquired discriminations following termination of training was not among the studies' goals. Finally, the authors of most of these studies presumed both that alcoholics can acquire accurate BAL dis- crimination abilities and that these discriminations can be estab-

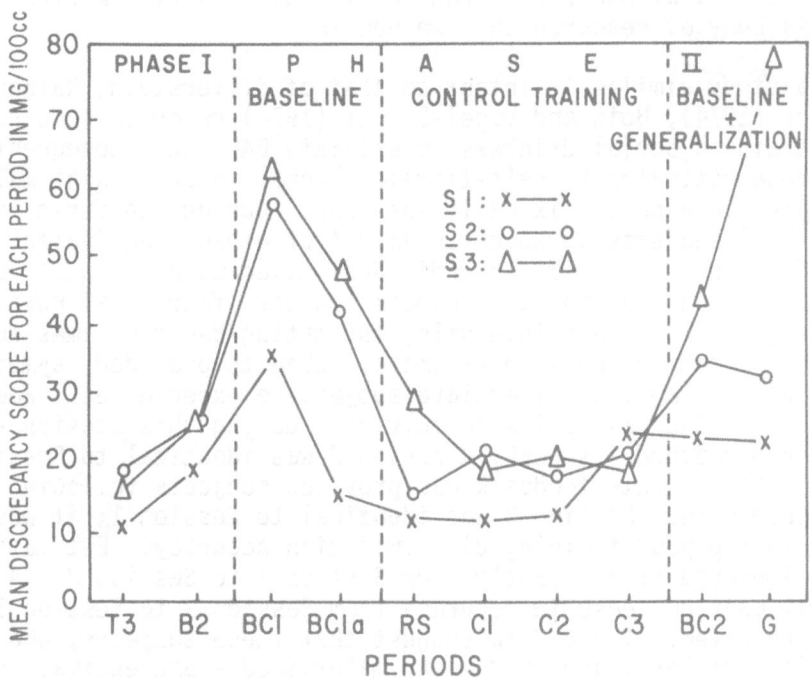

Figure 1: Blood alcohol level estimation accuracy
during phase II of the first BAL estimation study.
Estimation accuracy again improved markedly when
feedback on BAL was reintroduced at the beginning
of control training (From Silverstein, Nathan,
& Taylor, 1974).

lished on the basis of sensory awareness training alone. However, neither of these latter assumptions had been proven when these studies were undertaken. Data bearing on the BAL discrimination ability of alcoholics at the time these treatment studies were planned were both sparse and equivocal.

A series of investigations exploring these issues with both alcoholic and non-alcoholic subjects were then initiated; some were conducted at the Alcohol Behavior Research Laboratory, Rutgers University, others, at other research facilities in North America. It is to this body of research that we now turn.

In a study similar in intent to that of Silverstein, Nathan and Taylor (1974), Bois and Vogel-Sprott (1974) reported some success in training social drinkers to estimate BAL and, subsequently, to use these estimates to self-titrate alcohol intake. Nine males participated in each of six daily sessions. During the first three sessions, all subjects consumed an amount of ethanol equivalent to 135 ml of alcohol for a 150 lb individual mixed with an equal volume of "7-up." During Session 1, subjects consumed four equal portions of the drink at 20-minute intervals, estimating BAL ten times during that time. Accompanying each estimate, subjects provided "symptom reports" describing their immediate subjective experiences. Feedback of actual BAL was delivered only once during this session - when BAL had reached its peak. Session 2 was identical to Session 1 except that accurate feedback was provided subjects following each BAL estimate. Session 3 was identical to Session 1; it was designed to tap post-training discrimination accuracy. Estimation accuracy improved significantly from Session 1 to Session 2. A non-significant decrease in accuracy from Session 2 to Session 3 was also reported. These data suggest that these subjects, who were social drinkers, acquired - and maintained - BAL estimation accuracy on the basis of internal cues. However, BAL feedback was provided during Sessions 1 and 3, and amount and rate of drink consumption were identical all three sessions. As a result, subjects may have linked these external cues to the feedback provided in Session 2 and, in that way, learned to formulate subsequent estimates on this basis, rather than on the basis of internal cues.

Sessions 4, 5, and 6 were designed to assess subjects' ability to maintain discrimination accuracy when some of the external cues previously provided them were modified by altering the manner in which the drinks were constituted. Thus, drinks given in Session 4 averaged 100 ml of ethanol (instead of 135) and 100 ml of "7-Up" (instead of 135). Subjects were not given ethanol during Sessions 5 and 6 as before but, instead, their preferred alcoholic beverage. In addition, during all three of these sessions, rate and amount of alcohol consumed were determined by the subjects themselves. During all three sessions, subjects were required to select a target BAL (between 40 and 60 mg/%) and, on the basis of internal sensations,

stop drinking when this level had been reached. Ten minutes following the decision to stop drinking, subjects were asked to give a report on subjective sensations, then to make a BAL estimate. Ten
minutes later, a second estimate and symptom report were obtained
and BAL feedback was again provided. Five subsequent symptom reports and estimates were then taken at ten-minute intervals.

Results of these experimental manipulations were that estimation errors increased moderately, though non-significantly, from
Session 3 to Session 4. Estimation accuracy improved markedly,
however, from Session 4 to Session 6. Although estimation accuracy
apparently improved through these three sessions, the improvement
may actually have been more apparent than real. First, although
subjects were informed that their Session 4 drinks would differ
from those previously consumed, these differences were not in fact
substantial. Actually, if subjects had been discriminating the
strength of their drinks, they might have programmed their Session
4 drinking to parallel that which characterized previous sessions.
As a result, BAL estimates could have been made on the basis of
feedback first delivered during Session 2. Likewise, estimates
during Sessions 5 and 6 may have been guided by subjects' familiar-
ity with their "customary" drinks and by the limited feedback
which was available during these sessions rather than by the internal cue training provided earlier.

One of the most adequately-controlled studies in this area was
reported by Huber, Karlin and Nathan (1976). Its principal finding
was that non-alcoholics can acquire and maintain accurate BAL discriminations whether provided internal or external cue training.
Thirty-six non-alcoholic college students participated in three
day-long experimental sessions. In each, subjects consumed a total
of seven ounces of vodka mixed with tomato juice randomly distributed across six drinks. In an initial session, measures of pre-
training estimation accuracy were obtained. In the second, subjects were assigned - on the basis of pre-training accuracy scores
- to one of three training groups: Internal training only (I);
External training only (E); or Internal and External training (I+E).
Internal training was designed to teach subjects to focus on changes
in mood and bodily sensations as a basis for identifying changes in
BAL. External training relied on a programmed learning booklet
designed to teach subjects relationships between the amount and
frequency of alcohol intake and changes in BAL.

In order to disguise the alcoholic content of drinks during
Sessions 2 and 3, subjects were required to gargle before every
drink with an anesthetic mouthwash. In both of these sessions,
subjects also estimated the alcoholic content of their drinks in
order to permit assessment of discriminability of drink strength.
During training, BAL estimates, made seven times, were immediately
followed by feedback. Subjects who received external training were

told immediately prior to each estimate what the actual alcoholic
content of the immediately preceding drink was; they were to use
this information and the formulae taught them in the programmed
learning booklet to estimate BAL. Subjects who received internal
training were not told the alcoholic content of their drinks during
training; they were to formulate BAL estimates on the basis of internal sensations and feelings. Subjects in the I+E group made two
sets of estimates: one based on external cues, the other, on internal cues.

In the third, test, session of this study, subjects made four
BAL estimates; no feedback was given following any of these estimates. Prior to these estimates, half the subjects in each group
were told the actual alcoholic content of their drinks, the other
half were not. An analysis of variance revealed that all subjects
significantly improved BAL estimation accuracy during training,
then maintained this improved accuracy in the third session. This
improvement was independent of the kind of training provided and
whether or not subjects had been told the alcoholic content of
their drinks in the third session. These results suggested that
these non-alcoholic subjects could use internal and external cues
equally well to estimate BAL. However, a word of caution regarding
interpretation of these data is also necessary. Despite the elaborate procedures employed to disguise the strength of drinks, subjects were able to discriminate the various drink dosages to a
limited extent; as a consequence, discriminability of drink doses
may have played some role in subjects' improved post-training BAL
estimation accuracy.

Given, though, that these data and those reported earlier suggest that non-alcoholics can, in fact, learn to discriminate BAL on
the basis of changes in mood and bodily sensations, it was still an
open question whether alcoholics could learn to make BAL discriminations on the basis of the same cues. Data from the study by
Silverstein and his colleagues suggested that alcoholics and non-
alcoholics might well differ on this basis. No study of alcoholic
subjects, however, had yet explored the differential efficacy of
BAL discrimination training methods focussing on external and internal cues. Instead, research in this area, almost entirely
clinical, had attempted to train alcoholics to discriminate BAL
only via internal cues - despite the dearth of evidence that alcoholics can in fact do so. This research lacuna required a direct
comparison of the effectiveness of external and internal training
methods with alcoholics, a comparison which would afford evidence
as to the optimal mode of training BAL discrimination skills for
treatment purposes. To fill this research void, alcoholic subjects
were selected for a partial replication of the Huber et al. (1976)
study which, like it, was conducted at the Alcohol Behavior Research Laboratory, Rutgers University. The replication also served
as a test of the hypothesis first advanced by Huber and his coworker

that the effects of tolerance to ethanol may be reflected in differences between alcoholics' and non-alcoholics' abilities to utilize internal and external cues to discriminate BAL.

Like the 1976 study of nonalcoholics by Huber and his colleagues, subjects in this study (by Lansky, Nathan, and Lawson, in
press) participated in three day-long experimental sessions. The
internal and external training methods used in this study were
identical to those used by Huber et al. The study differed from
that of Huber et al. in the following ways: First, only two of the
six experimental conditions of the earlier study were replicated:
Internal and External training, alcoholic content of third session
drinks Known. Second, one extra BAL estimate was obtained in all
sessions in order to increase the number of analyzable data points.
Third, there were slight differences in the programming of drinks
(the interval between drinks was five minutes longer) and BAL estimates (the interval between drinks and estimates was five minutes
shorter). Fourth, the scale on which subjects based their BAL estimates was modified in order to make it more suitable for alcoholic
subjects. Finally, while training followed baseline assessment and
testing followed training in the former study by, respectively, one
to two weeks and three days, all sessions in this study were separated from each other by one day. The study's design is shown in
Figure 2.
Two separate groups of four chronic alcoholic subjects lived
in the Alcohol Behavior Research Laboratory for two-week periods.
During those periods, one group of four was given training in BAL
discrimination via internal cues while the other received external
cue training. All subjects participated in three experimental sessions, each separated by a day on which no experimental activities
took place. An initial baseline session was designed to obtain
pre-training measures of all subjects' BAL discrimination accuracy.
Subjects were given six drinks containing a total of seven ounces
of 80-proof vodka over a three-hour period; they were required to
estimate BAL four times. Blind as to actual alcoholic content of
drinks, subjects were nonetheless required to estimate the amount
of alcohol contained in each drink. During Session 2, the training
session, subjects who received external training were given programmed learning booklets detailing BAL-dose relationships; those
receiving internal training were to focus on the physiological and
affective concomitants of different BALs. During this session subjects were again administered seven ounces of 80-proof vodka and
required to estimate drink strength. They estimated BAL eight
times in the course of the session and were given feedback on actual BAL following each estimate. During this and the final experimental session, subjects gargled with an anesthetic mouthwash
in order to mask the strength of drinks. Session 3, designed to
assess post-training discrimination accuracy, was identical to
Session 1 except that subjects were told the alcoholic content of
each drink and were required to gargle with the mouthwash.

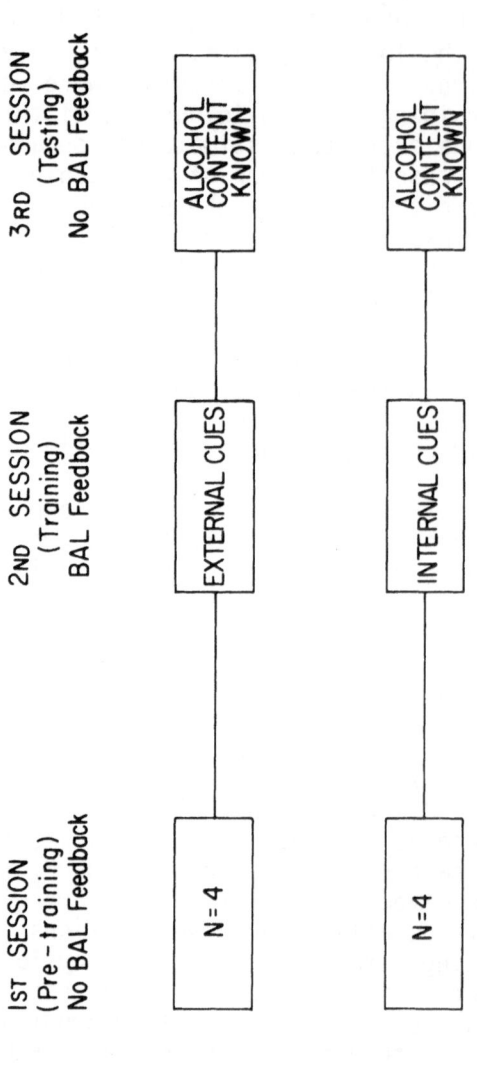

Figure 2: In this study, two comparable groups of chronic alcoholics each received BAL estimation training, one with external cues to BAL, the other with internal cues (From Lansky, Nathan, & Lawson, in press).

Prior to training, as Figure 3 shows, both groups of alcoholics were unable to estimate BAL with accuracy. During training (Session 2), when veridical BAL feedback was available, estimation accuracy increased significantly and equally for both groups: both groups of alcoholics acquired the ability to estimate actual BAL and to follow the changes which took place in BAL over the course of the session. Results of the third (test) session revealed, however, that, once feedback of actual BAL was removed, only the externally-trained alcoholics maintained the ability accurately to estimate BAL and to follow changes in actual BAL.

It was concluded that, unlike the non-alcoholics Huber et al. (1976) studied, the alcoholics in this study were not able to learn to discriminate BAL effectively on the basis of internal feelings and sensations, although they could do so by referring to external cues. These findings, then, supported an hypothesis first made by Silverstein, Nathan, and Taylor (1974), subsequently refined by Huber, Karlin, and Nathan (1976) - that alcoholics have a fundamental deficit in the ability to discriminate blood alcohol level on the basis of internal cues. Huber and his colleagues had suggested, in addition, that the relative inability of alcoholics to monitor internal cues may be a function, at least in part, of shifting levels of tolerance experienced by them during their lengthy drinking histories. As a result of these varying tolerance levels, discrete sets of internal cues had likely become associated with many BALs, not just one (as with most social drinkers). Other hypotheses to account for alcoholics' apparent inability to discriminate BAL on the basis of internal cues include inherited dysfunction of internal receptors and the effects of sustained high levels of alcohol in the blood on the sensitivity of receptors.

The first two of these hypotheses were tested in a recent study (Lipscomb and Nathan, 1978) eithin this program of research at the Alcohol Behavior Research Laboratory. Twenty-four male Rutgers University undergraduates were selected to fall into four experimental groups on the basis of usual drinking pattern (heavy versus light) and familial alcoholism (present in close biological relatives versus absent). Subjects were also tested on a standing steadiness measure before and after consuming alcohol, then divided into high and low tolerance groups based on changes in standing stability under alcohol. Because standing steadiness is an extremely sensitive measure of intoxication, it has also been suggested as a potentially valuable measure of tolerance (Moskowitz, Daily and Henderson, 1974).

Subjects participated in a three-session blood alcohol level discrimination training program which utilized only internal cue training. During Session 1, the baseline session, subjects consumed alcohol in six programmed doses and made eight estimates of

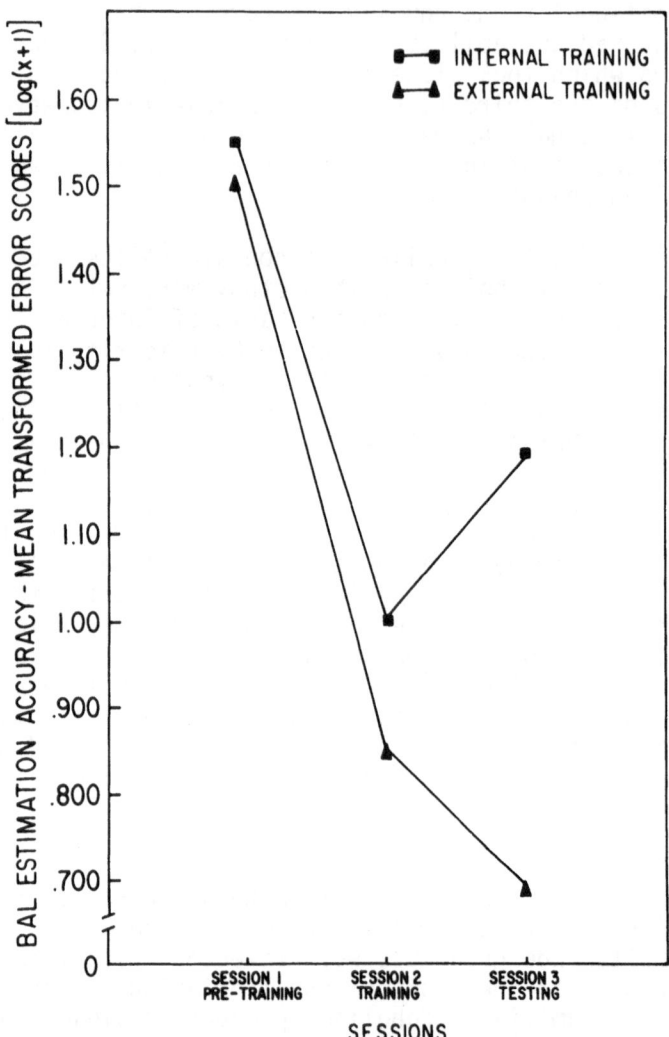

Figure 3: Mean transformed BAL estimation error
scores for the two groups of alcoholics. The
figure shows that, when feedback on BAL was
withdrawn in Session 3, the externally-trained subjects maintained their level of accuracy
while the subjects trained internally deteriorated markedly (From Lansky, Nathan, &
Lawson, in press).

intoxication without training or feedback on actual BAL. Accurate BAL feedback following each estimate and internal cue training were provided in Session 2. During the session's training sequence, subjects identified bodily sensations and feelings with the aid of relaxation instructions and adjective checklists, then matched these with the BAL feedback provided. Subjects made BAL estimates without training or feedback during Session 3, the test session. A week separated each of the three sessions.

Results showed that groups differing in drinking pattern or familial alcoholism did not differ in ultimate BAL estimation accuracy following internal training. By contrast, when subjects were grouped according to tolerance, "low tolerant" subjects (those whose body sway sober and drunk differed markedly) were found to have been significantly more accurate in their Session 3 estimates than "high tolerant" subjects (those whose body sway sober and drunk differed very little). An analysis of covariance indicated that this effect could not be accounted for by pre-training differences, suggesting that low tolerant subjects were better able to use internal training.

These results suggest that previously observed differences between alcoholics and non-alcoholics in ability to monitor internal cues of intoxication may be related to the development of tolerance by alcoholics. This finding further reinforces the veiw that internal cue training is not an effective means of teaching BAL discrimination to alcoholics; it should be discontinued as a component of controlled drinking treatment programs. Instead, future research should investigate the therapeutic use of an alternative mode of BAL discrimination, one that utilizes externally-based cues.

ACKNOWLEDGMENT

Preparation of this manuscript and support for some of the research discussed in it were enabled by NIAAA Grant No. AA00259.

REFERENCES

Bois, C. & Vogel-Sprott, M. Discrimination of low blood alcohol level and self-titration skills in social drinkers. Quarterly Journal of Studies on Alcohol, 1974, 35, 86-97.

Caddy, G. R. & Lovibond, S. H. Self-regulation and discriminated aversive conditioning in the modification of alcoholics' drinking behavior. Behavior Therapy, 1976, 7, 223-230.

Davies, D. L. Normal drinking by recovered alcohol addicts. Quarterly Journal of Studies on Alcohol, 1962, 23, 94-104.

Huber, H., Karlin, R., & Nathan, P. E. Blood alcohol level discrimination by non-alcoholics: The role of internal and external cues. Journal of Studies on Alcohol, 1976, 37, 27-39.

Lansky, D., Nathan, P. E., & Lawson, D. Blood alcohol level discrimination by alcoholics: The role of internal and external cues. Journal of Consulting and Clinical Psychology, in press

Lipscomb, T. H. & Nathan, P. E. Effect of family history, of alcoholism, drinking pattern, and tolerance on blood alcohol level discrimination. Manuscript in preparation, 1978.

Lovibond, S. H. & Caddy, G. R. Discriminated aversive control in the moderation of alcoholics' drinking behavior. Behavior Therapy, 1970, 1, 437-444.

Miller, W. R. Behavioral treatment of problem drinking: A comparative outcome study of three controlled drinking therapies. Journal of Consulting and Clinical Psychology, 1978, 46, 74-86

Moscowitz, H., Daily, J. & Henderson, R. Acute Tolerance to Behavioral Impairment by Alcohol in Moderate and Heavy Drinkers. Report to the Highway Research Institute, National Highway Traffic Safety Administration, Department of Transportation, Washington, D. C., April, 1974.

Nathan, P. E. & Briddell, D. W. Behavioral assessment and treatmen of alcoholism. In B. Kissin & H. Begleiter (Eds.), The Biolog of Alcoholism, Volume 5. New York: Plenum Press, 1977.

Paredes, A., Jones, B. M., & Gregory, D. An exercise to assist alcoholics to maintain prescribed levels of intoxication. Alcohol Technical Reports, 1974, 2, 24-36.

Pattison, E. M. A critique of abstinence criteria in the treatment of alcoholism. International Journal of Social Psychiatry, 1968, 14, 268-276.

Pomerleau, O. F., Pertschuk, M., Adkins, D., & Brady, J. P. A comparison of behavioral and traditional treatment for middle income problem drinkers. Journal of Behavioral Medicine, in press.

Silverstein, S. J., Nathan, P. E., & Taylor, H. A. Blood alcohol level estimation and controlled drinking by chronic alcoholics Behavior Therapy, 1974, 5, 1-15.

Sobell, M. B. & Sobell, L. C. Alcoholics treated by individualized behavior therapy: One year treatment outcome. Behaviour Research and Therapy, 1973, 11, 599-618.

Vogler, R. E., Compton, J. V., & Weissbach, T. A. Integrated behavior change techniques for alcoholics. Journal of Consulting and Clinical Psychology, 1975, 43, 233-243.

Vogler, R. E., Weissbach, T. A., Compton, J. B., & Martin, G. T. Integrated behavior change techniques for problem drinkers in the community. Journal of Consulting and Clinical Psychology, 1977, 45, 267-269.

Wilson, G. T., Leaf, R. C., & Nathan, P. E. The aversive control of excessive alcohol consumption by chronic alcoholics in the laboratory setting. Journal of Applied Behavior Analysis, 1975, 8, 13-26.

Wilson, G. T. & Rosen, R. Training controlled drinking in an alcoholic through a multi-faceted behavioral treatment program: A case study. In J. D. Krumboltz & C. E. Thoresen (Eds.), Counseling Methods. New York: Holt, Rinehart, & Winston, 1975.

ALTERNATIVES TO ABSTINENCE: EVIDENCE, ISSUES AND SOME PROPOSALS

Mark B. Sobell

Vanderbilt University

Scientific studies of alcohol problems were largely nonexistent prior to the decade of the 1960's. Nevertheless, despite the absence of a scientific basis, a set of popular, highly influential conceptualizations of alcohol problems developed. These popular views derived from the only available source, the life experiences of persons with alcohol problems and the clinical experiences of those who had to deal with them. These views, recently described by Pattison, Sobell and Sobell (1977) as the traditional model of alcoholism, in turn became the determining principles upon which treatment procedures were formulated. In fact, it has been argued by some that these beliefs have become so reified by workers in the alcohol field that defense and proselytization of the traditional model is characterized by an evangelical religious fervor (Pattison et al., 1977; Robinson, 1972; Verden and Shatterly, 1971). Because of such a commitment by many, alcohol problems have come to be recognized as extremely serious, pervasive health problems worthy of intensive scientific study. Less positively, the widespread acceptance of traditional concepts has also provided the foundation for divisive controversy.

Considering the origins of traditional concepts of alcohol problems, it is only reasonable to expect that scientific findings will sometimes necessitate a reformulation of those ideas. In recent years, the need for such changes has become increasingly apparent (see Pattison et al., 1977). In large part, these change processes seem to characterize problems in which the alcohol field is currently enmeshed.

From an historical perspective, it is probably valid to describe the field of alcohol studies as presently enduring some

serious dilemmas. One such quandry involves debate about the ability of some individuals who have had drinking problems to successfully recover from those problems without being totally abstinent.
This particular aspect of the changing conceptualizations of alcoho
problems is the topic of this chapter.

Several reviews (e.g., Gerard and Saenger, 1966; Lloyd and
Salzberg, 1975; Pattison, 1976; Pattison, Headley, Gleser and
Gottschalk, 1968; Pattison et al., 1977; Sobell and Sobell, 1975)
have documented that there exists a vast amount of evidence demonstrating that some individuals, identified as "alcoholic," have
been able to recover successfully from drinking problems without
fully abstaining from alcoholic beverages. That prolific literature will not be reviewed here and issues regarding definition and
measurement of nonabstinent outcomes will receive only brief attention. However, various characteristics of the total body of evidence will be discussed. Also, two studies purported to constitute
negative evidence for alternatives to abstinence will be reviewed
in some detail since they have not heretofore been scrutinized with
the same intensity as studies which constitute positive evidence.
Next, both the nature of nonabstinent outcomes and the available
evidence regarding variables associated with such outcomes will be
examined in some depth. Then, several issues regarding the veracity of the evidence and how traditionalists have attempted to explain the findings will be discussed. Finally, several proposals
will be presented regarding the nature of successful treatment outcomes.

EVIDENCE

The Evidence for Successful Nonabstinent Outcomes

Pattison et al. (1977) have recently reviewed 74 studies
which reported that some identified alcoholics has recovered from
drinking problems without an adherence to total abstinence. Since
that review was compiled, several additional studies have reported
similar outcomes (Armor, Polich and Stambul, 1976; Orford, Oppenheimer and Edwards, 1976; Vogler, Weissbach, Compton and Martin,
1977; Caddy, Note 1; Harris and Walters, Note 2; Pomerleau,
Pertschuk, Adkins and Brady, Note 3; Pomerleau, Pertschuk and
Stinnett, Note 4). Taken collectively, these studies are of considerable interest, since the majority have been follow-up reports
of treatment which strongly emphasized a uniform treatment goal of
total abstinence. Moreover, subjects in the majority of studies
were explicitly identified as chronic, or gamma, alcoholics
(Jellinek, 1960). This is particularly important, since gamma alcoholics, by definition, have experienced "loss of control" and
physical dependence on alcohol.

Also of interest, nearly half of the total studies reporting
successful nonabstinent outcomes were conducted outside the United
States. As a whole, the studies report outcomes for subjects in
both inpatient and outpatient treatment programs, as well as subjects who never received any formal treatment. In fact, only about
one fourth of the studies specifically used a nonabstinent treatment
goal for some subjects. Such studies have been very recent, and
were first reported from Japan (Mukasa and Arikawa, 1968; Mukasa,
Ichihara and Eto, 1964), South Africa (Lazarus, 1965), and Australia
(Caddy, 1972; Lovibond and Caddy, 1970), before recently proliferating in the United States. The treatment modality primarily associated with nonabstinent outcomes has been behavior therapy.

Before examining the characteristics of such outcomes in
greater detail, it should be noted that an equally impressive body
of literature documents that, within a laboratory setting and given
appropriate incentives, even severely chronic alcoholics can drink
in a highly controlled fashion. This evidence has been reviewed
elsewhere (Lloyd and Salzberg, 1975; Marlatt, 1978; Mello, 1972;
Miller, 1976; Nathan and Briddell, 1977; Pattison et al., 1977).

On the Nature of Successful Nonabstinent Outcomes

The literature reporting successful nonabstinent outcomes
(sometimes referred to as "controlled drinking," "nonproblem
drinking," "social drinking," or by various other terms) demonstrates that such results are best considered as a set of outcomes,
rather than a uniform entity. For instance, these outcomes have
been described in terms of arbitrary limits upon consumption (Armor
et al., 1976; Popham and Schmidt, 1976; Sobell and Sobell, 1972,
1973a, b, 1976), arbitrary criteria for blood alcohol concentration
as estimated by subjects (Lovibond and Caddy, 1970; Vogler, Compton
and Weissbach, 1975; Vogler et al., 1977), clinical impressions
(Gerard and Saenger, 1966), evaluation by significant others
(Orford, 1973; Orford et al., 1976), and case descriptions (Davies,
1962; Davies, Scott and Malherbe, 1969) among other methods. Such
outcomes have included drinking behaviors reported as ranging from
extremely minimal and infrequent, to daily drinking of limited
quantities, to occasional intake of excessive but not debilitating
amounts of alcohol. Indeed, the variety of outcomes reported seem
to incorporate a degree of heterogeneity strikingly similar to that
which characterizes so-called "normal" drinking. The few studies
which report individual subject data in greater detail (e.g., Orford, 1973; Sobell and Sobell, 1973a, b, 1976, in press), furthermore, make it clear that such a range of outcomes applies not only
between studies, but also to subjects within studies.

If one attempted to define "successful" nonabstinent outcomes
simply in terms of quantity of ethanol consumed, therefore, the

task would be fruitless. However, these varied outcomes do share
an important component: <u>adverse</u> <u>consequences</u> <u>which</u> <u>result</u> <u>from</u>
<u>drinking</u> <u>define</u> <u>drinking</u> <u>problems</u> and, consequently, whether an
outcome is labeled "successful." Adverse consequences can affect
many different areas of life health--physical, psychological, socioeconomic, etc.--but they must be identifiable. A more complete
definition of consequences might also involve consideration of the
<u>potential</u> consequences which an individual risks by drinking. A
caution is in order here, however, in that such an orientation
could easily become unrealistic and lacking in utility.

In a related regard, Pattison, in several publications
(Pattison, 1976; Pattison, Coe and Rhodes, 1969; Pattison <u>et al</u>.,
1968), has suggested that the "meanings" - functions - of drinking
are important determinants of how drinking should be evaluated.
He suggests (Pattison, 1976) that drinking by former alcoholics can
be trichotomized as either (1) "attenuated drinking"--the amount
and frequency of drinking are decreased, but adverse consequences
follow the drinking, (2) "controlled drinking"--a person's consumption of alcohol is self-limited, although the person is evaluated
as having a high potential for drinking in ways and circumstances
which are likely to result in adverse consequences, and (3) "normal
drinking"--drinking is not engaged in for purposes which could
reasonably be considered likely to result in adverse consequences.
Thus, Pattison suggests that drinking has a different "symbolic
meaning" for the normal drinker than for the controlled drinker.
As will be seen, such a differentiation has both its assets and
liabilities.

On the beneficial side, it seems prudent to pay greater attention to measuring the "meaning" of drinking when evaluating nonabstinent outcomes. Seemingly, these aspects of drinking could be
operationalized in terms of measuring, for instance, urges to
drink, thoughts about drinking, the circumstances (antecedent
stimuli) determining such events, the circumstances which accompany
actual drinking, and the consequences which follow it. These
facets of drinking and associated behaviors have seldom been assessed in studies reporting nonabstinent outcomes, including
studies employing behavioral methods.

Similarly, however, one might speculate that abstinent outcomes are subject to the same variety of interpretations. That is,
a person may be abstinent, yet experience repeated urges to drink,
and have a high likelihood of drinking in a pathological manner
should conducive circumstances arise. In a classic paper, Gerard,
Saenger and Wile (1962) examined the life styles of 50 abstinent
alcoholics. On the basis of extensive interviews with the subjects
as well as pretreatment and posttreatment data, they concluded that
only five (10%) of the subjects could be classified as "Independent
Successes." Although the remaining 45 subjects were also abstinent

Gerard et al. classified 27 (54%) as "Overtly Disturbed"--subjects who sustained abstinence although they were assessed as being psychologically unstable, 12 (24%) as having "Inconspicuous Inadequate Personalities"--these were "ex-patients whose total functioning is characterized by meagerness of their involvement with life and living" (p. 89), and 6 (12%) as "AA Success"--these subjects were described as being..."as dependent on AA as they were before on alcohol...(and having)...little or no social life apart from AA" (p. 90).

If data similar to those gathered by Gerard et al. were collected from the general population, some persons would likely be identified who have suffered no identifiable drinking problems, are very conscious of their drinking, or drink in circumstances where adverse consequences are likely to eventually occur, others who suffer infrequent and minimal adverse consequences of ostensibly normal drinking would also be found, and still others pay virtually no apparent attention to their drinking and do not suffer adverse consequences. Thus, it is possible that, by Pattison's definition, "normal" drinking is not a normative behavior in most societies. Although no studies have investigated "normal" drinking in this much detail, it seems quite likely that the "symbolic meaning of drinking" for the average drinker is more like Pattison's "controlled drinking" category than his "normal drinking" category, except that the average drinker is perhaps less aware of the functions which drinking can serve and lacks a history of adverse consequences from drinking.

This critical distinction - between drinking per se and the functions of drinking - raises issues pertinent to abstinent as well as nonabstinent outcomes. Surely some nonabstinent outcomes incorporate factors of high risk and have a greater potential for relapse to problem drinking than do other nonabstinent outcomes. Similarly, however, one can expect that abstinent outcomes are subject to the same liabilities. Put succinctly, current abstinence does not guarantee future abstinence nor does current nonproblem drinking guarantee continuation of that status. Unfortunately, the available data preclude an evaluation of both abstinent and nonabstinent outcomes in terms of the potential of each for relapse to drinking problems.

The Evidence Against Successful Nonabstinent Outcomes

Upon surveying the literature, it rapidly becomes evident that there is actually very little scientific evidence which demonstrates the impossibility of successful nonabstinent outcomes. The majority of evidence exists in the form of clinical anecdote rather than empirical research and generally is similar in content to Ruth Fox's critique of Davies' 1962 study (see Davies, 1963): "My own practice covers many hundreds of alcoholics, and though I

have never been in a position to do a follow-up, I do not know of a
single patient of mine who has been able to resume normal drinking"
(p. 117). This kind of argument will be considered later in this
paper when issues, rather than evidence, are reviewed. Recently,
however, spokesmen for the National Council on Alcoholism cited
studies by Pittman and Tate (1972) and Ewing and Rouse (1976) as
demonstrating that nonproblem drinking outcomes are not possible
and as incorporating relatively stronger methodological structures
than recent reports of successful nonabstinent outcomes (specifically, the Rand Report) (The Alcoholism Report, Note 5). Cohen
(1977) has also cited the report by Ewing and Rouse as a major research finding which casts doubt upon reports of nonproblem drinkin
outcomes. Inasmuch as the critics of studies which have reported
nonproblem drinking outcome have been opposed to publication of
such findings (National Council on Alcoholism, Note 6), it seems
appropriate and useful to subject studies cited in support of those
criticisms to the same scrutiny as the studies which have been at-
tacked.

      Pittman and Tate (1969, 1972) reported the results of a follo
up investigation of 255 individuals who had been treated for alcoholism at a detoxification center from 1962 to 1964. Some subjects
participated in an extensive outpatient program following detoxification while the remaining subjects received only inpatient care.
Pittman and Tate reported that they "found no patients who had returned to what may be called 'normal social drinking' during the
follow-up period" (p. 188). Upon closer scrutiny, however, these
results are subject to alternative interpretations.

      The validity of the Pittman and Tate study is called into ser
ious question as a result of several methodological deficiencies.
For instance, although follow-up interviews were stated as having
begun "one year after the person's initial discharge from the treat
ment facility" (p. 185), the reader is later informed that "the
length of the follow-up period covered in the study ranged from
nine to 32 months, the median being 12.9 months" (p. 186). Thus,
subjects sometimes had to recall events which had transpired over a
lengthy interval, thus increasing the possibility of inaccuracy in
subjects' self-reports.

      Another serious methodological and definitional problem in
the Pittman and Tate study is the considerable ambiguity with which
they discuss their lack of evidence for normal drinking outcomes.
Although several of their outcome variables were well defined and
quantified (e.g., median weekly outcome, frequency of bi-weekly
attendance at AA meetings, employment status, place of residence,
etc.), the authors inexplicably provide no quantitative data at all
regarding the drinking behavior of their subjects, although they
report that the majority of subjects were drinking less following
treatment than prior to treatment. Emrick (1974) has noted that it

is not uncommon for alcohol treatment outcome evaluation studies to
fail to report quantitative data regarding drinking outcome, beyond
generally describing subjects as abstinent, improved, or unchanged.
In their study, Pittman and Tate described a "moderated" drinking
outcome as simply involving at least seven months of abstinence
during the one-year followup period. With regard to their conclusion that no subjects were engaging in "normal social drinking,"
nowhere did the authors define what they meant by "normal social
drinking." Obviously, considerable disagreement might be found
even among experts concerning what constitutes a "normal social
drinking" pattern. Thus, the lack of definition and quantification
of this most important variable--drinking behavior--forces the
reader to either accept or dispute the author's judgment without
benefit of evidence.

The second study cited by the National Council on Alcoholism
as refuting the possibility of nonproblem drinking by former alcoholics, reported by Ewing and Rouse (1976), deserves detailed consideration because it purports to demonstrate a failure at inculcating "controlled" drinking patterns in alcoholics. Regarding
internal validity, the study is plagued with several design problems. First and foremost, the study was not a controlled investigation (no control subjects or control treatments were used) and involved a highly selected group of subjects. More specifically,
Ewing (Note 7) initially reported having embarked on a clinical
trial because of his frustration with traditional abstinence-oriented treatment methods. For instance, he stated: "The fact is that
it is a rare alcoholic who will remain abstinent indefinitely and
if we can offer him a ray of hope, perhaps we should" (p. 12).

Ewing and Rouse invited subjects to participate in a trial
study which had an announced objective of inculcating controlled
drinking patterns. Subjects were only accepted for the study if
they had a history of failure with Alcoholics Anonymous and expressed an unwillingness to accept treatment approaches requiring
total abstinence. Furthermore, subjects were excluded from the
study if they had a history of previous extended periods of abstinence or previous successful affiliation with Alcoholics
Anonymous. Although the authors considered their subject population
to be relatively nonselective, that description would seem to be at
considerable variance with Ewing's 1972 statement that: "So far I
have only accepted those who have failed with Alcoholics Anonymous
and with total abstinence goals, or have rejected such approaches
for the time being" (Note 7, p. 9). In essence, therefore, this
subject population could be described as highly recalcitrant to
additional treatment and of generally poor prognosis. Among the
more interesting case histories discussed by Ewing and Rouse (1976),
two of the original six subjects who were considered as having completed treatment had been diagnosed as schizophrenic and received
concurrent psychiatric treatment while in the program, and an

additional subject was described as an indigent female who was co-
erced by her case social worker to attend treatment sessions invol-
untarily.

The treatment methods used by Ewing and Rouse are also vulner
able to serious criticism. Subjects, accompanied by their spouses
if possible, attended weekly outpatient group meetings where they
had informal discussions with the investigators while they engaged
in a drinking situation. Major therapeutic techniques used by
Ewing and Rouse (1973, 1976) included aversive conditioning for
excessive drinking (Lovibond and Caddy, 1970), learning drinking
skills, maintaining a drinking record, including spouses at treatment sessions, and modeling (therapists sometimes drank with subjects during group sessions, thereby providing a model for appropriate drinking). The nature of the group discussions which occurred is unclear from published reports. For instance, in
Ewing's first description of the method (Note 7), he stated that
"no effort is made to invoke dynamic psychotherapeutic methods,
although some patients continue in therapy with other psychiatrists
while in the program" (p. 10). This general description was repeated by Ewing and Rouse (1973), albeit slightly modified: "No
attempt was made to develop formal group psychotherapy in these
sessions but undoubtedly communications of a therapeutic nature
occurred" (p. 70). However, in their recent summary report (Ewing
and Rouse, 1976), the authors state that "...we included group
therapy, couples therapy, modeling and other techniques to be described, because we wanted to make our therapy as powerful as possible so that it might have the greatest chance of success" (p. 124).

With regard to the blood alcohol concentration discriminatior
training used by Ewing and Rouse, Caddy (1975) has elsewhere criticized the clinical use of this method in the absence of a thorough
behavioral counseling program as an inappropriate use of the
Lovibond and Caddy (1970) methodology and a procedural variation aļ
to have little influence on drinking behavior.

Several additional factors also cast doubt upon the validity
of this clinical trial. For instance, by way of explaining a rathe
high dropout rate, Ewing (Note 7) stated that "sometimes the comments of professionals or members of Alcoholics Anonymous have led
patients to abandon the technique without giving it a fair trial"
(p. 9). Given the basic methodological problems summarized above,
this study would surely be inadequate as an evaluative test of any
treatment approach, much less one intended to examine the feasi-
bility of alternatives to abstinence. Moreover, complementing the
design faults of the study, the procedures used to evaluate treatment outcome stand unmatched in their vulnerability to critical
appraisal.

In their recent summary paper, Ewing and Rouse (1976) report

treatment outcome for 35 subjects, with follow-up periods ranging
from 27 to 55 months per subject. Only 14 of these subjects had
completed as many as six treatment sessions; all were considered by
the authors to be total treatment failures. Ewing and Rouse (1976)
described their reasoning in this regard as follows:

> Although it is true that some people who are 'social' or
> 'controlled' drinkers may sometimes drink too much, it is
> our experience that they do not drink in the uncontrolled
> and self-destructive way of chronic alcoholics. Thus, we
> believe that the only satisfactory outcome for a program
> that is trying to inculcate controlled drinking is for
> the patient to maintain such controlled drinking pat-
> terns indefinitely. Only by doing so can the patient
> demonstrate that his drinking is like nonalcoholics (p.
> 132).

In other words, instead of assessing subjects' cumulative functioning over the entire follow-up interval, Ewing and Rouse chose to
categorize subjects' outcomes according to the poorest single day
experienced by each subject at any time post-treatment. According
to this criterion, an individual who had been totally abstinent
over 54 months of the follow-up interval following a minor drinking
episode the first month after treatment would have been evaluated
as having the poorest possible outcome score. At this point, it
should be noted that several studies (e.g., Davies, Sheperd and
Myers, 1956; Gerard and Saenger, 1966; Gillies, Laverty, Smart and
Aharan, 1974; Orford, 1973; Parades, Hood, Seymour and Gollob,
1973; Pattison et al., 1968; Pittman and Tate, 1972) have indicated
that even alcoholics who are not necessarily recalcitrant to abstinence-oriented treatment are very likely to drink to excess within
the first year following traditional treatment, with typical abstinence rates ranging from 5% to 15% at the end of the first year.
Accordingly, by Ewing and Rouse's criteria, all programs, including
abstinence-oriented ones, would have to be judged as failures. In
conclusion, in assessing the impact of their study, we can only
strongly support Ewing and Rouse (1976) in their assertion that:
"Based on our experience with these patients and a long-term follow-
up, we have concluded that, in our hands at least, further attempts
to inculcate controlled drinking by such methods are unjustified"
(p. 134, italics added).

Summary of Existing Evidence

In 1966, Gerard and Saenger concluded that: "It might be proper logically and clinically to question or challenge the permanence
or stability of this behavior (nonproblem drinking outcomes), but to
deny its existence for a stated period of time is merely prejudice,
not science" (p. 110). In view of the much stronger evidence

currently available, it is clearly a fruitless exercise to debate
whether or not some alcoholics can ever successfully drink. Such
outcomes do occur and their frequency is astonishingly high when
one considers the lack of treatment services oriented toward or ac-
cepting of alternative goals. The more relevant questions, answers
to which will provide both humanitarian and scientific gains, have
to do with for which persons such a treatment outcome might be a
legitimate objective, how it might be best achieved, for whom it
might be a preferred treatment objective, what the characteristics
are of such recovery, and what the ramifications are of such a major conceptual change for service providers and potential recipient:
of these services. Fortunately, some evidence suggestive of predictors of successful nonabstinent outcomes has now accrued. This
evidence is reviewed in the next section of this paper.

Predictors of Successful Nonabstinent Outcomes

Existing evidence indicates that successful nonabstinent outcomes can be achieved. Therefore, the ability to predict for whom
such outcomes are possible and for whom they might be contraindicated would be very useful. Unfortunately, data on this issue have
been scarce.

In 1962, Davies evoked unprecedented commentary when he reported on 7 of 93 former alcoholics, all traditionally treated, who
had successfully engaged in normal drinking for periods ranging
from 7 to 11 years following their hospital discharge. Two aspects
of Davies' report relate to possible predictors of successful nonab
stinent treatment outcomes. First, all seven of Davies' subjects
reported they had begun to drink following an extended period of
abstinence, as long as one year in length. Second, substantial
changes had occurred in the life situations of six of the seven.
Four had changed occupations as a result of treatment, switching to
a vocation which involved less risk of drinking. Another two had
apparently resolved social and sexual relationship problems as a
precursor to their successful outcome. However, since similar life
change information was not provided for subjects who were either
abstinent or functioning poorly at follow-up, this early report
can only be considered a suggestive predictor of successful nonabstinent outcomes.

Gerard and Saenger (1966) gathered follow-up data on 767 alcoholics who had been treated by eight different treatment programs.
The eight programs provided geographically and socially disparate
subject populations. Five percent of all subjects had been prac-
ticing "controlled drinking" during the six months prior to the
follow-up interview; 13% had been abstinent. Groups of subjects
with differing outcomes were compared statistically on several pretreatment characteristics using Chi-square and t test analyses.

Gerard and Saenger reported no significant differences between abstinent and controlled drinking subjects in terms of age, sex, race, education, social stability, or prior duration of drinking problems. When successful subjects from the eight treatment programs were differentiated in terms of abstinence versus controlled drinking outcomes, however, the incidence of controlled drinking outcomes varied greatly, between 8% to 67%. It is likely that this finding reflected differences in subject populations, in treatment programs or in both. Gerard and Saenger also found that subjects who became controlled drinkers tended to have maintained less contact with their treatment program than subjects who had remained abstinent. Lastly, significantly more subjects who had become abstinent (74%) had received tranquilizers during treatment than subjects who became controlled drinkers (46%).

These two groups of subjects did not differ in outcome status as reflected by several adjunctive measures, including physical health, social health, interpersonal relationships, work adjustment, and work status. In all cases, subjects were evaluated for their functioning over a one-year follow-up period. This finding is consistent with one-year follow-up results reported by Pattison et al. (1968), who compared 11 abstinent and 11 normal drinker subjects who had been treated in outpatient alcohol treatment programs. However, it is at variance with findings reported by Pattison et al. (1969), who completed a follow-up evaluation of subjects from four different treatment programs. Those authors reported that subjects classified as "successful limited drinkers" reported improvement in other areas of life health which was "intermediate between the abstinent and the unsuccessful drinkers" (p. 483).

Case histories of 100 abstinent and 35 controlled drinking alcoholics provided by Gerard and Saenger (1966) yielded several interesting observations. Many of the abstinent subjects had experienced previous problems which convinced them that the cost of continued drinking had become too high for them to risk further drinking. Several controlled drinking subjects had undergone changes in their life situations similar to those described by Davies (1962), but few abstinent controlled drinking subjects overall had changed in many ways other than their drinking. Although controlled drinking subjects tended to have shifted their drinking pattern to include weaker beverages, they were characterized by Gerard and Saenger as being relatively unaware of determinants of their prior drinking problems and as attributing their successful outcomes largely to their own willpower. All but one of the 35 controlled drinking cases had had severe drinking problems at intake, including chronic intoxication, ill health, arrests, and work or family problems.

Davies et al. (1969) presented four case reports of controlled drinking subjects who had been hospitalized at one time with a

diagnosis of alcoholic psychosis. Follow-up periods ranged from 5
to 27 years. For all cases, it was suggested that substantial
changes in subjects' life situations had accompanied their acquisition of nonproblem drinking patterns; in three of the four cases
these changes were in terms of diminished responsibility for vocational or family affairs. As with Davies' (1962) first report,
however, comparable data for subjects of similar background who did
not become social drinkers were not presented. It is likely that
most individuals undergo substantial changes in life circumstances
over extended periods of time and this seems especially likely for
individuals who successfully recover from alcohol problems (i.e.,
by becoming abstinent or successfully nonabstinent). Unfortunately
it is difficult to assess the importance of changes in life circumstances as related to non-problem drinking outcomes from these reports.

Several other reports of successful nonabstinent outcomes published within the present decade have attempted to identify predictors of such outcomes. Orford (1973) reported outcome data for 100
males diagnosed as alcoholics at the time of their intake into an
experimental marital therapy program. Follow-up data were summarized for weekly periods; a controlled drinking week was defined as
any week a) with "at least one drinking occasion but never amountin
to five pints of beer or its equivalent per day" (p. 567) and b)
when the drinking during that week was described by the subject's
spouse as not unacceptable. Of 77 subjects for whom complete follow-up data were available, only 10 were abstinent during the entire one-year period following treatment and only three had been
drinking in a totally controlled fashion. The majority of subjects
had outcomes which were intermediate and included weeks of controll
drinking as well as weeks of uncontrolled drinking.

Orford also compared pretreatment behavior of 22 subjects who
were totally uncontrolled drinkers during the follow-up interval
with the behavior of 14 subjects who had engaged in "mainly controlled drinking." In contrast to the uncontrolled drinkers, controlled drinker subjects were less likely to have self-identified
themselves as alcoholic, less likely to have expressed a preference
for abstinence as a treatment goal, had reported fewer instances of
morning drinking, tremors, hallucinations, and time lost from work,
and were more likely to have thought that their drinking problem
was of very recent origin. An intriguing finding was that while
controlled drinker subjects described their own drinking problem as
being of more recent origin than did uncontrolled drinker subjects,
the wives of these two groups of subjects did not differ significantly in their estimates of the chronicity of their spouses' drink
ing problems. Though the two groups of subjects did not differ in
age, controlled drinker subjects had suffered slightly fewer family
difficulties from their drinking (p < .10).

A further analysis of the two-year follow-up data for this subject population (Orford et al., 1976) compared 11 subjects who were primarily abstinent with 10 subjects who were primarily controlling their drinking. In all cases, subjects' reports were fully substantiated by their spouses. Controlled drinker subjects had reported relatively fewer symptoms at intake (morning drinking, morning nausea, passing out when drinking, morning tremor, "loss of control," secret drinking, hallucinations), were less likely to have been physically dependent on alcohol, and had been treated less intensively. This latter finding is consonant with findings reported by Gerard and Saenger (1966). Both studies, however, reported outcome data from treatment programs which were explicitly abstinence-oriented. Thus, it cannot be determined whether the successful non-abstinent subjects needed less treatment, or whether they had simply availed themselves of less treatment because they were alienated by the abstinence orientation. Orford et al. also found that subjects who were mainly controlled drinkers had expressed greater pretreatment confidence in their ability to abstain from alcohol. Finally Orford et al. found that subjects' outcomes for the first 12 months of follow-up were good predictors of their status after 24 months of follow-up. This finding is consistent with other evidence showing that alcohol treatment outcome results are quite stable after a period of 12 to 18 months, although individual case variation does occur (Davies, Shepard and Myers, 1956; Gerard and Saenger, 1959; Gerard and Saenger, 1966; Gibbins and Armstrong, 1957; Sobell and Sobell, 1976; Caddy, Note 1).

Hyman (1976) investigated the functioning of former alcoholics 15 years after treatment. He followed five total abstainers, two primarily abstainers, and five moderate drinkers. Hyman reported that, compared to abstainers, moderate drinkers were employed more regularly, more often married, and had fewer arrests prior to treatment. He also reported that, in many cases, the moderate drinkers gained that status only after having suffered some additional drinking problems shortly following treatment. All of Hyman's subjects had been involved in abstinence-oriented treatment.

In another long term follow-up study, Levinson (Note 8, Note 9) reported five-year follow-up data for alcoholics who had been treated at the Donwood Institute in Canada. Of 115 subjects for whom data were available, 11 were categorized as controlled drinkers, defined as "reduced drinking and improvement in psychological functioning and social adjustment." No differences were found between controlled drinkers and the remainder of the sample in terms of age, sex, occupational level, education, employment record, drinking history prior to treatment, and estimated intellectual functioning. The only variable which significantly discriminated controlled drinkers from other subjects was that a greater percentage of controlled drinker subjects were married at the time they began treatment. Fewer than half the controlled drinker subjects had experienced major social changes in their lives prior to acquisition of their nonproblem drinking pattern, a finding at some variance with the hypothesis put forth by Davies (1962).

Reporting on a rather unique study, Popham and Schmidt (1976) summarized the results of an otherwise conventional treatment program which stressed moderate drinking as an acceptable outcome. The overall rate of success was not substantially different from that of traditional abstinence programs, except that a higher rate of nonproblem moderate drinking was reported by subjects. This finding, with data presented by Gerard and Saenger (1966), Pattison et al. (1969), and a recent report by Chaney, O'Leary and Marlatt (1978), suggests that program orientation - the extent to which the abstinence outcome is emphasized - may influence the outcomes of successful subjects. A lower incidence of limited drinking occurs when programs strongly emphasize the need for total abstinence, although a recent study by Pomerleau et al. (Note 3) suggests that such findings may be due in part to certain individuals (those more likely to engage in limited drinking) dropping out of traditional treatment.

"Moderate drinking" was defined by Popham and Schmidt as consumption averaging no greater than 2.5 oz of ethanol daily. Of 96 subjects evaluated one year following treatment, nine were classified as abstainers and 18 as moderate drinkers according to this criterion. Nine of the moderate drinkers typically drank less than twice a week, six reported drinking from two to three times per week, and three drank almost daily. While consumption by all moderate drinker subjects was usually less than two oz of ethanol on any given occasion, at least 11, and possibly 14, of the 18 moderate drinkers had become acutely intoxicated at least once during the follow-up interval. Popham and Schmidt also found that pretreatment consumption of alcohol was the best predictor of posttreatment alcohol consumption (r = +.61) and suggested, as a consequence, that "prime candidates for the moderate-drinking approach are patients whose pre-treatment consumption rate was below the median of the range" (p. 880). No significant difference was found between pretreatment beverage preference and treatment outcome; although the moderate drinkers were somewhat younger and better educated than the abstainers, these differences were also not significant. There was a trend (p < .10) for a greater proportion of the moderate drinkers to be female, however, and this difference may have mediated the apparent relationship between pretreatment and posttreatment alcohol consumption.

Vogler and his associates (Vogler, Compton and Weissbach, 1975; Vogler, Weissbach and Compton, 1977; Vogler, Weissbach, Compton and Martin, 1977) have reported follow-up results for two different populations of subjects, inpatient alcoholics and "problem

drinkers" - persons never hospitalized or diagnosed as alcoholic
who had nonetheless experienced legal, vocational or marital problems from drinking. A controlled drinking outcome was defined as:
(1) average monthly consumption of less than 50 oz of absolute
ethanol, and (2) no more than one drinking episode per month in
which a subject's self-estimated blood alcohol level exceeded .08%.
Using multiple stepwise regression and discriminant analysis techniques, Vogler and his colleagues reported that the best predictor
of controlled drinking outcome was pretreatment alcohol intake.
They also concluded that "the best candidate for moderation was the
less chronic, younger drinker with a relatively lower alcohol intake, a more stable vocational record, and no history of hospitalization for alcohol abuse or physical deterioration from drinking"
(Vogler, Weissbach and Compton, 1977, p. 31). However, the scope
of this conclusion remains uncertain because pretreatment alcohol
intake was by far the most substantial predictor of outcome. Added
variables of age, hospitalizations and job history only minimally
increased the amount of variance which could be accounted for by
pretreatment factors. However, the multiple regression analyses
performed by Vogler et al. incorporated a large number of independent variables (the exact number of variables was not reported,
but appeared to be at least 24). The multiple correlation obtained
in a multiple regression analysis tends to be systematically biased
upward as a function of the ratio of the number of independent variables in the regression equation to the sample size (Nunnally,
1967). Since it is unclear whether the reported multiple correlations were adjusted to control for this bias and considering the
minimal proportion of variance accounted for by variables other
than pretreatment alcohol intake, it seems unlikely that other pretreatment characteristics had substantial predictive value in these
studies.

Sobell, Maisto and Sobell (Note 10) have recently reported the
results of a step-wise multiple regression analysis of the relationship of several variables to treatment outcome during the final
quarter of a two-year follow-up of subjects treated in the Individualized Behavior Therapy (IBT) for alcoholics program (Sobell and
Sobell, 1972, 1973a, b, 1976, in press). Independent variables included three posttreatment variables (follow-up functioning for the
first, second and third six-month intervals of follow-up), treatment goal assignment (controlled drinking or nondrinking) and seven
pretreatment variables (age, education, occupational status, years
drinking problem, prior alcohol hospitalizations, previous withdrawal symptoms, and prior alcohol arrests). When multiple regression equations were computed using all 13 independent (predictor)
variables, early posttreatment functioning (predominantly, the
third six-months) was found to overshadow all other variables, typically accounting for from one half to three quarters of the total
outcome variance. This finding further attests to the relative
stability of outcome results after a twelve-month follow-up period

has passed.

Acknowledging the strong predictive value of earlier posttreatment functioning, posttreatment variables were then deleted from the regression equations in order to examine the relative contributions of within-treatment and pretreatment variables to the outcome data. This analysis revealed that assignment to a controlled drinking goal was the next most important determinant of outcome, accounting for approximately 20% of successful outcomes and 33% of controlled drinking outcomes. When within-treatment variables were also deleted from the regression equations, the only pretreatment factor which significantly predicted a controlled drinking outcome was prior alcohol-related hospitalizations: Subjects who had fewer hospitalizations were more likely to engage in controlled drinking. Controlled drinking, measured on a daily basis, was defined as consumption of six or fewer oz of 86-proof liquor or the equivalent in alcohol content.

Some additional findings from the IBT study are also relevant to the present discussion. First, subjects who had successful controlled drinking outcomes simultaneously engaged in substantially more abstinent days than other subjects over the follow-up period. Second, statistical analyses and individual subject drinking profiles demonstrated that when controlled drinking subjects drank, their controlled drinking typically did not lead to excessive drinking. Third, controlled drinking subjects also had better outcomes than other subjects on several adjunctive measures of outcome, including general adjustment, occupational and vocational status, residential status and stability, and physical health. Finally, an examination of individual subject drinking profiles suggested that controlled drinking tended to occur in their own home, with others present.

Recently, Pomerleau et al. (Note 3) reported initial findings from a controlled study in which outpatient problem drinkers were randomly assigned to either a multicomponent behavioral treatment program emphasizing moderation or a traditional treatment program emphasizing abstinence. Twelve-month follow-up data indicated that subjects were more likely to drop out of the traditional program. Also, while subjects in both groups reduced their consumption of alcohol, traditional treated subjects significantly reduced their consumption prior to therapy (apparently as the result of a screening interview), whereas behaviorally treated subjects significantly reduced their consumption during the course of treatment. At the end of the one-year follow-up period, 72% of the behaviorally treated subjects were either abstinent (6%) or had reduced their drinking below pretreatment levels (66%), with 11% dropping out; of traditionally treated subjects, 50% were either abstinent (14%) or had reduced consumption (36%), but 43% had dropped out. Between group differences were not statistically significant. Interestingly those subjects who dropped out of traditional treatment reported less pretreatment drinking the week before screening than those who continued in treatment ($p < .06$). In this regard, the authors reported that most traditionally treated subjects dropped out at a time which "coincides with the culmination of intense interpersonal confrontations in therapy."

Summary of Predictors of Successful Nonabstinent Outcomes

The studies just reviewed comprise the bulk of existing data concerning variables related to successful nonabstinent outcomes. Nearly all studies report correlational data and all but four (Popham and Schmidt, 1976; Sobell and Sobell, in press; Vogler et al., 1975; Vogler et al., 1977; Pomerleau et al., Note 3) consist of follow-up reports of abstinence-oriented treatment. Clearly, it would be valuable to have available more data derived from controlled experiments which provided for random assignment to treatment orientations and compared nondrinking, controlled drinking, and excessive drinking outcomes in terms of the same relevant factors (rather than simply reporting case characteristics for subjects who engaged in nonproblem drinking). However, despite the relative lack of controlled research on this topic, the studies which have been reviewed, as diverse and often methodologically weak as they are, offer some tentative suggestions regarding (1) the types of individuals most likely to adopt a successful pattern of nonproblem drinking, (2) the characteristics of such outcomes, and (3) directions for future research.

In terms of pretreatment variables predictive of successful nonabstinent outcomes, the findings of Orford (1973), Vogler (1977), and Sobell and Sobell (in press) suggest that individuals who have less serious drinking problems at entry into treatment are relatively more likely to acquire a posttreatment pattern of nonproblem drinking. This relationship is suggested by less pretreatment alcohol intake, less prior symptomatology (especially in terms of physical dependence) and a shorter self-reported history of drinking problems. Orford's results also suggest that several cognitive factors may be related to nonproblem drinking outcomes, but there is presently insufficient data upon which to base such a determination.

Evidence concerning the types of within-treatment factors which may be related to nonproblem drinking outcomes is nearly nonexistent. In one of the two studies which differentially assigned subjects to abstinent and nonabstinent treatment goals (Sobell and Sobell, in press), assignment to a controlled drinking goal was found to be significantly related to controlled drinking outcomes. In terms of nonproblem drinking outcomes following treatment in abstinence oriented programs, the findings of Gerard and Saenger

(1966), Popham and Schmidt (1976), and Chaney et al. (1978) suggest that individuals who attain such outcomes also tend to spend relatively less time in treatment. In a related context, Pomerleau et al. (Note 3) reported a higher drop out rate for subjects randomly assigned to abstinence-oriented conventional treatment as compared to subjects assigned to a multi-component behavioral treatment which emphasized moderation. At this time, however, it is not possible to determine whether these findings indicate that individuals who achieve successful nonabstinent outcomes have a lesser need for treatment or that their lessened involvement in treatment stems from other reasons (e.g., being alienated from treatment).

Finally, in terms of early posttreatment factors associated with nonproblem drinking outcomes, several additional observations are possible. First, Davies' (1962) early report suggested that an extended period of posttreatment abstinence might be a necessary prerequisite for a nonproblem drinking outcome. However, data from Hyman (1976), Popham and Schmidt (1976) and Sobell and Sobell (1972, 1973a, b, 1976, in press) indicate that a somewhat different pattern is typical: It often seems to be the case that subjects experience some drinking problems shortly following treatment and then overcome these problems. The nature of some of these problems and possible reasons for their occurrence are discussed later in this paper.

Second, several studies have suggested that the relationship between nonproblem drinking outcomes and improvement in other areas of life health is similar to the relationship between abstinence and other life health variables (Gerard and Saenger, 1966; Pattison et al., 1968; Sobell and Sobell, 1972, 1973a, b, 1976, in press): A successful drinking outcome is frequently but not necessarily associated with improvement in other areas of life health. It seems likely that the relationship between drinking and other life health variables will eventually be best summarized by an approach similar to that adopted by Pattison et al. (1969), whereby different patterns of life health improvement are found for different subject populations.

Lastly, again stemming from Davies' (1962) work, it has been suggested that a significant positive change in life circumstances sometimes precedes the onset of a successful nonabstinent pattern. However, data supporting this hypothesis have been in the form of case histories while controlled studies have not provided support for such a relationship. However, controlled studies have rarely focused on subjects and circumstances which could reflect such relationships.

The research evidence clearly indicates that successful nonabstinent outcomes do occur and that some progress has been made in identifying characteristics of persons for whom such outcomes are possible. Still, the topic of alternatives to abstinence continues to be fraught with controversy. In large part, these conflicts concern clinical issues and attitudes; further research data will undoubtedly aid in their eventual resolution. The remainder of this paper is devoted to a general discussion of several of these issues, as well as the formulation of some proposals which might help resolve existing differences of opinion among those in the alcohol field.

ISSUES, ATTITUDES AND CONTENTIONS

Arguments Against Nonabstinent Outcomes

Some of the criticisms of studies reporting nonproblem drinking outcomes derive basically from personal clinical experience. For instance, several respondents to Davies' 1962 article stated, simply, that such outcomes were not possible (Block, Fox, Esser, and Smith, in Davies, 1963). Such unsubstantiated observations are impossible to refute. Given the current evidence, however, it does seem likely to maintain contact with clinicians or programs strongly oriented toward abstinence (Chaney et al., 1978; Gerard and Saenger, 1966; Orford et al., 1976; Pomerleau et al., Note 3). Consequently, there would seem to be little likelihood of such clinicians having contact with or recognizing such persons.

Some traditionalists have tried to explain reports of nonproblem drinking by former alcoholics by alleging that anyone able to return to some type of limited nonproblem drinking, ipso facto, was never a "real" or "true" alcoholic (Block, Lemere, see Davies, 1963). Even after Davies' report and the subsequent publication of many dozens of reports documenting such outcomes for individuals, many of whom were previously physically dependent on alcohol, the argument of misdiagnosis continues to be evoked. For instance, Weisman (1975) has stated that "what may be involved when some individuals do return to controlled drinking is that they are obviously different from those who cannot, and that our skills in diagnosing alcoholics are sometimes wanting...The history of medical diagnosis is replete with many examples of false positives which only later turn out to be medical errors" (p. 4). Such reasoning is specious and tautological, since there is as yet no a priori way of identifying those individuals who can resume drinking. This circuitous logic, consisting essentially of post hoc relabeling of cases which do not conform to traditional expectations, unfortunately lacks any practical utility other than to preserve the belief system of its advocates.

An indirect way to discount evidence of successful nonabstinent outcomes is to suggest that these data merely reflect precursors to an eventual full alcoholic relapse (Esser, in Davies, 1963). In

other words, while it is sometimes acknowledged that some alcoholics
can drink without problems for limited periods of time, this recover
of control is viewed as merely temporary, before uncontrolled drinking ensues. Accordingly, it is prophesized that this type of drinking inevitably develops into a pattern of alcohol dependence. Besides the fact that empirical data contradict this assertion
(Pattison et al., 1977), this line of reasoning is equally applicable to abstinent outcomes. That is, it is well documented that many
alcoholics have successfully attained extended periods of abstinence, only to reinitiate a pattern of self-destructive drinking.
Obviously, arguments can be made to the effect that either outcome
--abstinence or nonproblem drinking--can be maintained over long
periods of time and that neither outcome precludes the possibility
of relapse.

The relative efficacy of any treatment objective can only be
determined by clinical research. Moreover, treatment effectiveness
will likely be at least partly a function of individual case circumstances and treatment characteristics. This same reasoning also
applies to the topic of relapse. The presently available data regarding abstinence outcomes indicate that individuals seldom become
totally abstinent following the completion of any treatment program. Similarly, some individuals treated according to nonproblem
drinking treatment objectives may also experience periods of excessive drinking. When abstinence is the goal and the client starts
to drink again, this does not mean that the treatment goal of abstinence must be discarded or that treatment has "failed." In such
cases, treatment is typically reinitiated, with the objective of
reacquisition of an abstinent state. This orientation to success
and failure is equally appropriate when individuals have been
treated with the objective of nonproblem drinking. Repeated episodes of problem drinking may indeed reflect a need to consider a
change in treatment goal, but they might occur no matter what the
treatment goal. Neither goal need be eternal!

Another strategy traditionalists have used to criticize reports of nonproblem drinking by former alcoholics has been to acknowledge such cases, but to treat them as exceptions to the rule:
As a "freak anomaly of human biochemistry or psychopathology"
(Thimann, see Davies, 1963, p. 325) or as cases of spontaneous
remission (Block, Smith, see Davies, 1963; Weisman, 1975). Block
(see Davies, 1963), for example, described such individuals as
"unique, they are different, they are rare" (p. 116).

Still another attempt to explain these data is the hypothesis
that "loss of control" does not occur until an alcoholic has consumed enough alcohol to exceed some minimal blood alcohol threshold (Kjølstad, Glatt, see Davies, 1963). This hypothesis has
been developed most fully by Glatt (1967). As discussed by
Pattison et al. (1977), however, the available evidence is that

physical dependence on alcohol follows massive intake of alcohol, usually over a period of a few days. Thus, while the threshold notion probably has some theoretical validity, it is unable to serve its originally intended purpose of countering reports of non-problem drinking.

The criticism of reports of successful nonabstinent outcomes which appears most valid derives from the fact that these events are newsworthy, not part of the public's conception of recovery from alcoholism. As a result, sensationalistic publicity on such findings might encourage experimentation by abstinent alcoholics, might even be used by some alcoholics as a justification for further drinking (Bell, Armstrong, Smith, see Davies, 1963). These criticisms are, unfortunately, apt; methods of minimizing the detrimental public impact of scientific findings will be discussed later in this paper. In the present context, however, two observations are relevant. First, the impact of such publicity derives from two sources. Basically, the content of any public media communication about scientific findings is necessarily a popularization of those findings. It is not unusual for second-hand accounts to contain erroneous or misleading statements. As Davies (1963) pointed out, however, "It is unfortunate that communications to scientific journals may be reproduced in newspapers, but this is surely not an argument against reporting clinical findings" (p. 331).

The second factor affecting the impact of publicity about nonabstinent outcomes, however, derives not from inaccurate scientific reporting but from the exaggerated reactions of persons who are ignorant of - or choose to ignore - the evidence under consideration. Such an instance occurred recently when the National Council on Alcoholism (Note 6) issued a press release stating that, "In the present state of our knowledge, we firmly believe and emphasize that there can be no relaxation from the stated position that no alcoholic may return with safety to any use of alcohol" (p. 1-2). Instead of a more temperate statement--an assertion, for example, that such outcomes may be rare, should not be expected to result from home remedies, and may be unwise in many cases--a categorical denial of any such outcomes was issued. Verden and Shatterly (1971) have noted that such a reaction may derive, in part, from the threat which nonproblem drinking outcomes may constitute to the philosophical recovery base of some abstinent alcoholics.

An issue related to the reactions of traditionalists to reports of successful nonabstinent outcomes concerns what Chafetz has characterized as a "paternalistic" attitude toward the alcoholic among workers in the alcohol field (The Alcoholism Report, Note 11). Chafetz, former Director of NIAAA, has observed that, ironically, it is often the very same individual who most loudly proclaims it his mission to remove the moral stigma of alcoholism who does not

trust the alcoholic to have access to knowledge about his/her disorder.

Arguments for Nonabstinent Outcomes

It was mentioned above that decisions regarding choice of alternatives to abstinence as treatment objectives basically involve
a balance of potential risks and benefits. Just as a multitude of
criticisms have been offered against alternatives to abstinence, so
too have arguments supporting the value of nonabstinence approaches
been put forth. Interestingly, with few exceptions (e.g., Brunner-
Orne, see Davies, 1963), these arguments have either been ignored
or overlooked by most critics of nonabstinence approaches. For
instance, most of the studies cited above that used the total abstinence criterion reported that treatment outcome successes (cases
of total abstinence) over a one year follow-up period are rare indeed. This unhappy outcome has a number of causes. For example,
alcohol problems are extraordinarily difficult to treat and highly
resistant to long-term change by any approach. At the same time,
total abstinence, while undeniably a positive outcome in most cases,
may be an unreasonably stringent, unrealistic treatment outcome
criterion for every alcoholic. This being so, it could be argued
that sole dependence on this criterion obscures recognition of
actual improvement in drinking with treatment. Another unfortunate
consequence of evaluating treatment efficacy in terms only of total
abstinence is that it may have helped create the popular view that
treatment for alcoholism seldom works.

Further, insistence upon abstinence from the beginning of
treatment may create a fatalistic orientation among service providers and consumers alike. Marlatt (1978), for instance, proposes
elsewhere in this book a cognitive-behavioral phenomenon - the
"Abstinence Violation Effect" which describes a common cognitive
reaction by an individual who consumes an initial drink if (s)he
has made a previous commitment to abstinence. Since Marlatt deals
with this topic at length in his paper, only its clinical implications will be discussed here. Put succinctly by Marlatt, they are
that "a single drink is enough to shatter the image of oneself as
an abstainer." Others (Engle and Williams, 1972; Merry, 1966;
Sobell, Sobell and Christelman, 1972) have similarly speculated that
the maxim, "One drink, then drunk," may function as a self-fulfilling
prophecy for the abstinent alcoholic. That is, how many alcoholics,
sincerely believing that they will begin a drinking binge if they
consume a first drink, continue to drink after taking a first drink
because they feel they must do so? Similarly, how many alcoholics
conclude they are not actually alcoholics when they are able to
consume one or two drinks without feeling a physical need to continue drinking?

In essence, what appears needed as prophylaxis are practical procedures to prepare individuals to deal effectively with such situations should they occur. For example, it would seem helpful to explain to individuals who have participated in abstinence-oriented treatment programs that, while an initial drink will not force them to continue to drink, there are several reasons why they might continue to drink as a result of such experimentation. It would seem advantageous for clients to be aware that any decision to take a first drink, second drink, and every drink thereafter is their own--just as they could make the decision to stop after one or two drinks or immediately to seek treatment.

Another very important reason why the criterion of total abstinence appears in need of modification is the contention that traditional treatment programs may have precluded many persons from seeking or obtaining early treatment for developing drinking problems - and inhibited the development of appropriate clinical services for such individuals. For instance, we have little empirical knowledge about how many persons consistently deny that they have a drinking problem until they have truly developed a serious physical dependence on alcohol and to what extent this denial is based on resistance to making a personal commitment to life-long abstinence. It is not unusual for those in the alcohol field to attribute the failure of clients to remain in treatment to the client's lack of "motivation" or "unwillingness to change." Might not this supposed deficiency in motivation simply reflect the rigidity of traditional ideas and service providers' inability to adequately treat a broad spectrum of persons with drinking problems?

It has already been mentioned that several studies have found evidence that persons who attain successful nonabstinent outcomes are likely to have only limited contact with traditional treatment programs. It has also been found that there is little participation by persons under 30 years of age in Alcoholics Anonymous (Leach and Norris, 1977), although national survey data suggest that males in their late 20's report the highest incidence of drinking problems (Cahalan and Room, 1974). Consider for a moment the circumstances of problem drinkers, those individuals traditionally thought of as in transition to heavy, dependent drinking. At present, there are no separately identifiable services for individuals who have mild or moderate drinking problems, the result being that they either go without services or are referred to programs oriented toward treating the chronic, physically dependent alcoholic. One might expect that a treatment program which was not totally abstinence oriented would be appealing to this population. Moreover, the available research evidence suggests that such individuals constitute the population most likely to achieve a successful nonabstinent outcome.

At the present time, it is unfortunately the case that traditional beliefs about alcoholism offer little reason for problem

drinkers to curtail their drinking, and, in fact, may provide them
with a basis for repeated attempts to prove that they are not "alcoholics." This is not to deny that abstinence may indeed be an
appropriate treatment objective for some of these individuals but
rather to state that abstinence is simply one possible alternative.
At this time the term "alcoholic," with all its negative connota-
tions, tends to be applied rather indiscriminantly to all varieties
of drinking problems. It seems likely that fewer persons would
deny present or potential drinking problems if the field were to
stop depicting the typical outcome of such problems as humiliating,
debasing, and requiring a major and permanent change in life style
simply to arrest the problem. A preferred orientation would be to
illustrate that recovery is possible and that one can minimize future handicaps by early, limited actions. At the present time,
this suggestion is admittedly based largely on extrapolation from
research reviewed earlier in this chapter; it has not yet been
evaluated empirically. However, it is reasonable to question the
utility of subjecting all individuals, even those who have relatively less serious drinking problems, to the traditional mandate
of total abstinence.

Two other reasons for recognizing the legitimacy of successful
nonabstinent outcomes are relevant to the present discussion. Perhaps the most compelling of these is the sheer mass of evidence
which demonstrates that such outcomes are possible and numerous.
It is surprising that such a convincing amount of evidence has for
so long been ignored or overlooked. Yet, although traditional
abstinence-constrained concepts have neither produced particularly
good rates of treatment success nor been supported in controlled
research studies, many insist that those concepts are not in need
of modification, since they have not yet finally been disproven.
On the one hand, no amount of positive evidence will suffice; on the
other, no amount of negative evidence is adequate. In the end, it
is the individual with drinking problems who would appear to suffer
most from this paradox.

A final argument for legitimizing nonproblem drinking outcomes
arises from the case histories of individuals who have attained such
outcomes (Sobell and Sobell, in press). For many individuals labeled "alcoholics," a serious impediment to the attainment of successful nonabstinent outcomes exists. Although the social stigma of
being an alcoholic is severe, an even more oppressive stigma can
result when an alcoholic makes known his/her belief that s(he) can
drink without experiencing further drinking problems. I know several individuals who have adopted successful nonproblem drinking
and, consequently, have encountered unexpected, unwarranted hostility and ostracism from family members and others in their community. For this reason, treatment programs which consider nonproblem
drinking to be a legitimate outcome must tell clients about the
negative attitudes held by the general populace about alcoholics

returning to some form of nonproblem drinking. For instance, such
clients should be made aware of and counseled about the possible
resistance they may encounter from friends, relatives, and unexpected sources, including physicians and judges, to their drinking in
a nonproblem manner. In light of the hostile reactions which sometimes occur, it seems likely that the individual who engages successfully in nonproblem drinking might well be reluctant to confide in
others about his past drinking problem. While there has been a
great deal of speculation about the so-called "hidden" alcoholic,
it is ironic that traditional concepts of alcoholism may actually
encourage a great many persons to be "hidden ex-alcoholics." Only
when it becomes acceptable for some persons, no matter how few, to
recover from serious drinking problems and successfully engage in
nonproblem drinking will we be able validly to determine the proportion of persons capable of achieving such a goal, how many individuals have already achieved such a status, and what their characteristics are.

Based on the increasing frequency of reports of nonproblem
drinking outcomes, awareness of those reports among service providers
and the general public, and the need to provide services for the
heretofore neglected population of problem drinkers, it appears
clear that traditional conceptualizations regarding the nature of
recovery from alcohol problems are in need of substantial revision.
In particular, service providers must become aware of the existing
data. It is also clear that now, more than ever, there is opportunity to develop an array of treatment services which can effectively, efficiently, and ethically meet the needs of persons with
alcohol problems.

A SUMMARY AND SOME PROPOSALS

A bimodal view of treatment outcomes for persons treated for
drinking problems - categorizing individuals only as abstinent or
drunk - is anachronistic; it may also actually be deleterious to
the conduct of treatment. For instance, this view of outcome virtually compels alcohol treatment providers to view any drinking
which occurs during or after treatment as constituting treatment
"failure." Interestingly, a rigidly binary view of recovery is
seldom found in other health-related fields. To this end, recovery
from depression or pneumonia or cardiac insufficiency is not viewed
as an all-or-none phenomenon but rather in terms of degrees of recovery. Recent empirical evidence, some reviewed in this chapter,
suggests the value of a similar view of recovery from alcohol problems: outcome should be evaluated as reflecting degrees of improvement or recovery.

When one examines the factors which define drinking problems,
it is apparent that these problems are defined by consequences of

drinking, not in terms solely of quantity of alcohol consumed.
This shift in focus, however, from the act to the consequences of
drinking suggests the use of a shared goal for all alcohol treatment
efforts. In particular, it suggests that a single treatment goal is
appropriate for treating all individuals with alcohol problems: a
reduction in drinking to a nonproblem level. For many, perhaps
most, such a reduction might only be achieved through total abstinence while, for others, it might be accomplished within the context
of nonproblem drinking.

Several caveats accompany this suggested common goal. First,
while it is appropriate to recognize alternatives to abstinence as
legitimate treatment objectives for some individuals, these alternatives to abstinence do not constitute a panacea. Neither should
legitimizing alternatives to abstinence be taken to imply that all
individuals with drinking problems, even most persons, can achieve
such an outcome, or that all, or even most, persons currently working in the alcohol field are equipped to pursue these goals with
their clients. However, a realistic perspective requires acceptance
of the fact that successful nonabstinent outcomes can occur. Given
that some individuals who have experienced alcohol problems can
apparently acquire a pattern of drinking without incurring further
problems, our efforts should now be directed at generating information which will help predict which kinds of treatment procedures
might be most appropriate for use with which types of clients to
achieve which kinds of outcomes.

Second, a failure to recognize alternative treatment objectives
may have caused some individuals with alcohol problems to be denied
efficacious treatment. The limited array of traditional treatment
approaches may also have caused some individuals to be reluctant to
become labeled as "alcoholic" and, more importantly, may have served
as a powerful deterrent to treatment for persons with less serious
drinking problems. Thus, service providers need to acknowledge
that successful nonabstinent treatment outcomes do occur.

However, while we can no longer afford to ignore alternatives
to abstinence, it is clear that nonproblem drinking is probably no
more or less likely than abstinence to be attained simply by self-
commitment. Obviously, striving toward nonproblem drinking could
be highly detrimental should an individual do so to justify confin-
ued problem drinking. Therefore, like any other therapeutic procedure or goal, nonproblem drinking should only be used by trained
and knowledgeable individuals, aware of the methods, benefits,
dangers and limitations inherent in such an approach.

A substantial amount of correlative clinical evidence has now
accrued which suggests that nonabstinent outcomes are more often
achieved by persons who have not become physically dependent on alcohol, have relatively lower pretreatment ethanol intakes than the

average chronic alcoholic, are less likely to be self-identified as having alcohol problems serious enough to warrant total abstinence, and have access to environmental resources likely to support a non-abstinent outcome. Clearly, although a great deal of further controlled research is needed to determine the types of individuals most likely to achieve successful nonabstinent outcomes, the current evidence appears to suggest that individuals who have less chronic alcohol problems are more likely to attain successful nonabstinent outcomes.

Persons with minor drinking problems constitute an extremely neglected, probably sizeable target population. Thus, it is apparent that there is a need to develop new, different, and less intensive and demanding treatment programs in order to provide services for problem drinkers.

It also appears that the time is ripe for adding a greater degree of sophistication to our assessments of treatment outcomes. Successful nonabstinent outcomes subsume a broad spectrum of outcomes, sometimes including drinking patterns which seem to have a high risk of incurring adverse consequences, just as some individuals who are abstinent are nevertheless constantly preoccupied with thoughts of drinking, find it necessary to structure their daily activities very carefully so as to avoid ready access to alcohol, and could generally be considered as highly vulnerable to relapse. Thus, it is important not only to develop more sophisticated indices of drinking outcomes but to apply a similar scrutiny to nondrinking (abstinent) outcomes.

Finally, the evidence reviewed in this chapter suggests that there is a crucial need for further controlled research investigating nonproblem drinking outcomes, how they are best achieved, and how they differ from other outcomes.

These proposals are advanced as tentative suggestions, in need of detailed examination and modification, as appropriate. However, I believe they identify some important issues concerning treatment and prevention of alcohol problems, issues about which we can no longer postpone debate.

REFERENCE NOTES

1. Caddy, G. R. Individualized behavior therapy for alcoholics; third year follow-up: Advantages and disadvantages of conducting double blind treatment outcome evaluation studies. Paper presented at the 23rd Annual Meeting of the Southeastern Psychological Association, Hollywood, FL, May, 1977.

2. Harris, R. N. & Walters, J. Outcome, reliability and validity
    issues of alcoholism follow-up. Paper presented at the 27th
    Annual Meeting of the Alcohol and Drug Problems Association of
    North America, New Orleans, LA, September, 1976.

3. Pomerleau, O. F., Pertschuk, M., Adkins, D. & Brady, J. P. Com
    parison of behavioral and traditional treatment for problem
    drinking. Paper presented at the Annual Meeting of the Associa
    tion for Advancement of Behavior Therapy, New York, NY, Decembe
    1976.

4. Pomerleau, O. F., Pertschuk, M. & Stinnett, J. Behavioral
    treatment of middle income problem drinkers: Outcome results.
    Paper presented at the 23rd Annual Meeting of the Southeastern
    Psychological Association, Hollywood, FL, May, 1977.

5. The Alcoholism Report, June 25, 1976, p. 3.

6. National Council on Alcoholism, press release, July 19, 1974.

7. Ewing, J. A. Behavioral approaches for problems with alcohol.
    Paper presented at Winter Meeting of the American Academy of
    Psychoanalysis, New York, NY, 1972.

8. Levinson, T. The Donwood Institute--a five year follow-up stud
    Presented at the 31st International Congress on Alcoholism and
    Drug Dependence, 1975.

9. Levinson, T. Controlled drinking in the alcoholic: A search
    for common features. Paper presented at the Third Internationa
    Conference on Alcoholism and Drug Dependence, Liverpool,
    England, April, 1976.

10. Sobell, M. B., Maisto, S. A. & Sobell, L. C. Multiple regressi
    analysis of variables predictive of alcohol treatment outcome a
    controlled drinking. Manuscript in preparation.

11. The Alcoholism Report, July 23, 1976, p. 2.

REFERENCES

Armor, D. J., Polich, J. M. & Stambul, H. B. Alcoholism and treat-
    ment. Report No. R-1739-NIAAA. Santa Monica, CA: Rand Cor-
    poration, 1976.

Caddy, G. R. Behaviour modification in the management of alcoholis
    Unpublished doctoral dissertation, University of New South
    Wales, Australia, 1972.

Caddy, G. R. How not to replicate: A commentary on Maxwell et al.'s (1974) replication of Lovibond and Caddy's (1970) "Discriminated aversive control" study. Behavior Therapy, 1975, 6, 710-711.

Cahalan, D. & Room, R. Problem drinking among American men. Monograph No. 7 of the Rutgers Center of Alcohol Studies, New Brunswick, NJ, 1974.

Chaney, E. F., O'Leary, M. R. & Marlatt, G. A. Skill training with alcoholics. Unpublished manuscript, University of Washington, 1978.

Cohen, S. A primer on alcoholism. (Review of Alcoholism: Its causes and cure. A new handbook, by H. Milt). Contemporary Psychology, 1977, 22, 460-461.

Davies, D. L. Normal drinking in recovered alcohol addicts. Quarterly Journal of Studies on Alcohol, 1962, 23, 94-104.

Davies, D. L. Normal drinking in recovered alcohol addicts. (Comment by various correspondents). Quarterly Journal of Studies on Alcohol, 1963, 24, 109-121, 321-332.

Davies, D. L., Shepard, M. & Myers, E. The two-year's prognosis of 50 alcohol addicts after treatment in hospital. Quarterly Journal of Studies on Alcohol, 1956, 17, 485-502.

Davies, D. L., Scott, D. F. & Malherbe, M. E. L. Resumed normal drinking in recovered psychotic alcoholics. The International Journal of the Addictions, 1969, 4, 187-194.

Emrick, C. D. A review of psychologically oriented treatment of alcoholism. Quarterly Journal of Studies on Alcohol, 1974, 35, 523-549.

Engle, K. B. & Williams, T. K. Effect of an ounce of vodka on alcoholics' desire for alcohol. Quarterly Journal of Studies on Alcohol, 1972, 33, 1099-1105.

Ewing, J. A. & Rouse, B. A. Outpatient group treatment to inculcate controlled drinking behavior in alcoholics. Journal of Alcoholism, 1973, 8, 64-75.

Ewing, J. A. & Rouse, B. A. Failure of an experimental treatment program to inculcate controlled drinking in alcoholics. British Journal of Addictions, 1976, 71, 123-134.

Gerard, D. L. & Saenger, G. Interval between intake and follow-up as a factor in the evaluation of patients with a drinking problem. Quarterly Journal of Studies on Alcohol, 1959, 20, 620-630.

Gerard, D. L. & Saenger, G. Out-patient treatment of alcoholism.
    Toronto: University of Toronto Press, 1966.

Gerard, D. L., Saenger, G. & Wile, R. The abstinent alcoholic.
    Archives of General Psychiatry, 1962, 6, 83-95.

Gibbins, R. J. & Armstrong, J. D. Effects of clinical treatment on
    behavior of alcoholic patients: An exploratory methodological
    investigation. Quarterly Journal of Studies on Alcohol, 1957,
    18, 429-450.

Gillies, M., Laverty, S. G., Smart, R. G. & Aharan, C. H. Outcomes
    in treated alcoholics. Journal of Alcoholism, 1974, 9, 125-
    134.

Glatt, M. M. The question of moderate drinking despite 'loss of
    control.' The British Journal of Addiction, 1967, 62, 267-274

Hyman, H. H. Alcoholics 15 years later. Annals of the New York
    Academy of Science, 1976, 273, 613-623.

Jellinek, E. M. The disease concept of alcoholism. New Brunswick,
    NJ: Hillhouse Press, 1960.

Lazarus, A. A. Towards the understanding and effective treatment
    of alcoholism. South African Medical Journal, 1965, 39, 736-
    741.

Leach, B. & Norris, F. L. Factors in the development of Alcoholics
    Anonymous (A.A.). In B. Kissin & H. Begleiter (Eds.), The
    biology of alcoholism, Vol. 5. New York: Plenum Press, 1977.

Lloyd, R. W., Jr. & Salzberg, H. C. Controlled social drinking:
    An alternative to abstinence as a treatment goal for some al-
    cohol abusers. Psychological Bulletin, 1975, 82, 815-842.

Lovibond, S. H. & Caddy, G. Discriminated aversive control in the
    moderation of alcoholics' drinking behavior. Behavior Therapy,
    1970, 1, 437-444.

Marlatt, G. A. Alcohol, stress and cognitive control. In P. E.
    Nathan, G.A. Marlatt, & T. Løberg (Eds.), Alcoholism: New dir-
    ections in behavioral research and treatment. New York: Plenum
    Press, 1978.

Mello, N. K. Behavioral studies of alcoholism. In B. Kissin &
    H. Begleiter (Eds.), The biology of alcoholism, Vol. 2. New
    York: Plenum Press, 1972.

Merry, J. The "loss of control" myth. Lancet, 1966, 1, 1257-1258.

Miller, P. M. Behavioral treatment of alcoholism. New York: Pergamon Press, Inc., 1976.

Mukasa, H. & Arikawa, K. A new double medication method for the treatment of alcoholism using the drug cyanamide. The Kurume Medical Journal, 1968, 15, 137-143.

Mukasa, H., Ichihara, T. & Eto, A. A new treatment of alcoholism with cyanamide (H_2NCN). The Kurume Medical Journal, 1964, 11, 96-101.

Nathan, P. E. & Briddell, D. W. Behavioral assessment and treatment of alcoholism. In B. Kissin & H. Begleiter (Eds.), The biology of alcoholism, Vol. 5. New York: Plenum Press, 1977.

Nunnally, J. C. Psychometric theory. New York: McGraw-Hill, 1967.

Orford, J. A comparison of alcoholics whose drinking is totally uncontrolled and those whose drinking is mainly controlled. Behaviour Research and Therapy, 1973, 11, 565-576.

Orford, J., Oppenheimer, E. & Edwards, G. Abstinence or control: The outcome for excessive drinkers two years after consultation. Behaviour Research and Therapy, 1976, 14, 409-418.

Paredes, A., Hood, W. R., Seymour, H. & Gollob, M. Loss of control in alcoholism: An investigation of the hypothesis, with experimental findings. Quarterly Journal of Studies on Alcohol, 1973, 34, 1146-1161.

Pattison, E. M. Nonabstinent drinking goals in the treatment of alcoholism. Archives of General Psychiatry, 1976, 33, 923-930.

Pattison, E. M., Coe, R. & Rhodes, R. J. Evaluation of alcoholism treatment: A comparison of three facilities. Archives of General Psychiatry, 1969, 20, 478-488.

Pattison, E. M., Headley, E. B., Gleser, G. C. & Gottschalk, L. A. Abstinence and normal drinking: An assessment of changes in drinking patterns in alcoholics after treatment. Quarterly Journal of Studies on Alcohol, 1968, 29, 610-633.

Pattison, E. M., Sobell, M. B. & Sobell, L. C. (Eds.), Emerging concepts of alcohol dependence. New York: Springer Publishing Co., 1977.

Pittman, D. J. & Tate, R. L. A comparison of two treatment programs for alcoholics. Quarterly Journal of Studies on Alcohol,
1969, 30, 888-889.

Pittman, D. J. & Tate, R. L. A comparison of two treatment programs
for alcoholics. The International Journal of Social Psychiatry
1972, 18, 183-193.

Popham, R. E. & Schmidt, W. Some factors affecting the likelihood
of moderate drinking in treated alcoholics. Journal of Studies
on Alcohol, 1976, 37, 868-882.

Robinson, D. The alcohologist's addiction--some implications of
having lost control over the disease concept of alcoholism.
Quarterly Journal of Studies on Alcohol, 1972, 33, 1028-1042.

Sobell, L. C., Sobell, M. B. & Christelman, W. C. The myth of
"one drink." Behaviour Research and Therapy, 1972, 10, 119-
123.

Sobell, M. B. & Sobell, L. C. Individualized behavior therapy for
alcoholics: Rationale, procedures, preliminary results and
appendix. California Mental Health Research Monograph, No. 13.
Sacramento, CA, 1972.

Sobell, M. B. & Sobell, L. C. Individualized behavior therapy for
alcoholics. Behavior Therapy, 1973, 4, 49-72(a).

Sobell, M. B. & Sobell, L. C. Alcoholics treated by individualized
behavior therapy: One year treatment outcome. Behaviour Research and Therapy, 1973, 11, 599-618(b).

Sobell, M. B. & Sobell, L. C. The need for realism, relevance and
operational assumptions in the study of substance dependence.
In H. D. Cappell & A. E. LeBlanc (Eds.), Biological and behavioral approaches to drug dependence. Toronto: Addiction Research Foundation, 1975.

Sobell, M. B. & Sobell, L. C. Second year treatment outcome of alcoholics treated by individualized behavior therapy: Results.
Behaviour Research and Therapy, 1976, 14, 195-215.

Sobell, M. B. & Sobell, L. C. Behavioral treatment of alcohol
problems: Individualized therapy and controlled drinking.
New York: Plenum Press, in press.

Verden, P. & Shatterly, D. Alcoholism research and resistance to
understanding the compulsive drinker. Mental Hygiene, 1971,
55, 331-336.

Vogler, R. E., Compton, J. V. & Weissbach, T. A. Integrated behavior change techniques for alcoholics. Journal of Consulting and Clinical Psychology, 1975, 43, 233-243.

Vogler, R. E., Weissbach, T. A. & Compton, J. V. Learning techniques for alcohol abuse. Behaviour Research and Therapy, 1977, 14, 31-38.

Vogler, R. E., Weissbach, T. A., Compton, J. V. & Martin, G. T. Integrated behavior change techniques for problem drinkers in the community. Journal of Consulting and Clinical Psychology, 1977, 45, 267-279.

Weisman, M. Letter in News and Views. Alcoholism Council of Greater Los Angeles. Jan., 1975, p. 4.

RELATIONSHIPS BETWEEN DRINKING BEHAVIOR OF ALCOHOLICS IN A DRINKING-DECISIONS TREATMENT PROGRAM AND TREATMENT OUTCOME

Arthur I. Alterman, E. Gottheil, H. K. Gellens, and
C. C. Thornton

Veterans Administration Hospital, Coatesville, Pa. and
Jefferson Medical College

In a recent paper we described the variety of drinking behaviors exhibited by patients participating in a drinking-decisions treatment program for alcoholics and the relationship between some of these patterns of behavior and treatment outcome at six months (Alterman, Gottheil, and Thornton, 1977). To briefly summarize the findings, it was found that nondrinkers on the program fared better generally than program drinkers following treatment but that various subgroups of program drinkers who were able to exercise some degree of control over their drinking did nearly as well after treatment as program nondrinkers. Patterns of drinking on the program were therefore found to be correlated with drinking following treatment. These findings appeared to be of import, first, in providing evidence for the validity of a drinking-decisions research model for the study of alcoholism and, secondly, because they suggested that alcoholics vary in their ability to regulate their drinking.

The primary objective of the present paper is to extend the examination of treatment outcome of the previously described drinking subgroups over a longer time period in order to evaluate the validity of their differentiation. The basic questions asked here are whether program drinking continues to be correlated with treatment outcome when the latter is examined over a longer time period and to determine the extent to which the differentiation made at the six-month outcome period are upheld. This paper will therefore first review our findings on various patterns of drinking during treatment and their relation to treatment outcome at six months by way of placing the current study into perspective. Then treatment outcome results for the same subgroups at one and two years posttreatment will be examined and the course of treatment outcome of program patients over the two-year evaluation period overviewed.

211

METHOD

Subjects

The subjects in the study were 249 male alcoholic veterans between the ages of 23 and 58 who completed the Fixed Interval
Drinking-Decisions (FIDD) program of the VA Hospital, Coatesville,
PA, as members of one of 40 independent groups treated over a five-
year period. All had volunteered for a six-week intensive treatment program on a closed ward in which alcohol was available. No
other motivational criteria were applied. Patients were recruited
from other wards within the hospital, neighboring VA and State outpatient sources, alcoholism treatment agencies and other hospitals.
All were screened for psychological, medical, and neurological disorders that could be aggravated by alcohol.

All patients had been diagnosed as alcoholics before referral
to our program and would be described generally as gamma alcoholics
(Jellinek, 1960). The average patient reported 12.6 years of heavy
drinking, 8.9 years of problem drinking, considered himself to be an
alcoholic for 6.6 years, and had previously been hospitalized an
average of 2.4 times for alcoholism. Eighty-three percent of the
patients had experienced blackouts, 84% had experienced shakes, and
86% felt that their alcohol problem was worsening.

General Procedure

The FIDD program is a six-week clinical research program for
male alcoholic veterans initiated by Gottheil and his colleagues
over seven years ago (Gottheil, Corbett, Grasberger, et al., 1971;
Gottheil, Corbett, Grasberger, and Cornelison, 1972) which allows
these patients to make decisions concerning drinking while undergoing treatment on a closed ward. Specifically, patients can decide
whether or not to drink up to two ounces of ethyl alcohol 13 times
a day on the hour from 9 a.m. to 9 p.m. on weekdays of the drinking-
decisions (D-D) phase or middle four program weeks (see Figure 1).
Patients obtain drinks by coming up to the ward nursing window at
the appointed time and requesting either one or two ounces of alcohol.

The basic treatment format of the program is conventional in
nature. Patients are exposed to a variety of treatment modalities
including two hours a week of individual psychotherapy, three
hours of group therapy, daily physical exercise, educational and
religious seminars, Alcoholics Anonymous, art therapy, marital
casework, and recreational and occupational therapy. Since patients
are given the opportunity, however, to make drinking decisions during the program, treatment also focusses and reflects upon feelings,

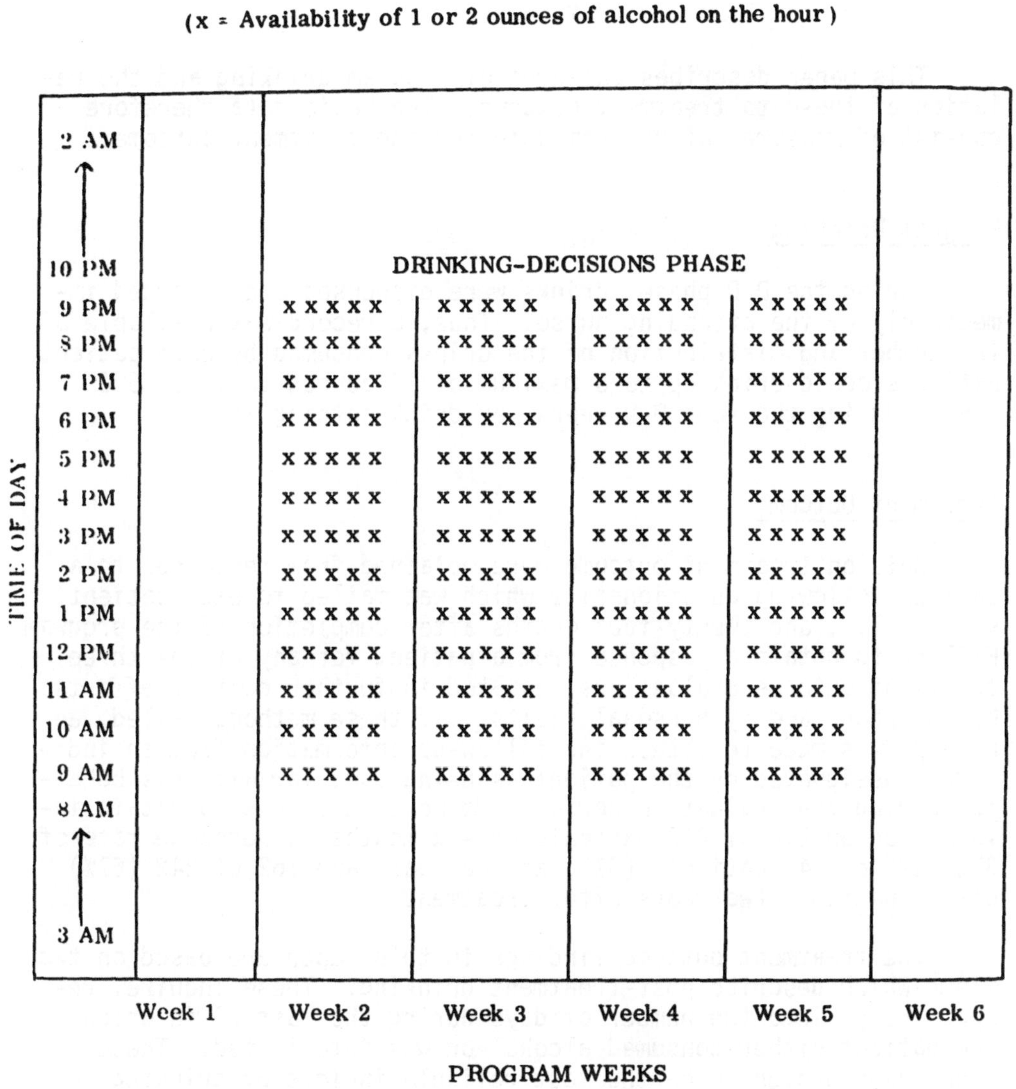

(x = Availability of 1 or 2 ounces of alcohol on the hour)

TIME OF DAY	Week 1	Week 2	Week 3	Week 4	Week 5	Week 6
2 AM						
↑						
10 PM		DRINKING-DECISIONS PHASE				
9 PM		x x x x x	x x x x x	x x x x x	x x x x x	
8 PM		x x x x x	x x x x x	x x x x x	x x x x x	
7 PM		x x x x x	x x x x x	x x x x x	x x x x x	
6 PM		x x x x x	x x x x x	x x x x x	x x x x x	
5 PM		x x x x x	x x x x x	x x x x x	x x x x x	
4 PM		x x x x x	x x x x x	x x x x x	x x x x x	
3 PM		x x x x x	x x x x x	x x x x x	x x x x x	
2 PM		x x x x x	x x x x x	x x x x x	x x x x x	
1 PM		x x x x x	x x x x x	x x x x x	x x x x x	
12 PM		x x x x x	x x x x x	x x x x x	x x x x x	
11 AM		x x x x x	x x x x x	x x x x x	x x x x x	
10 AM		x x x x x	x x x x x	x x x x x	x x x x x	
9 AM		x x x x x	x x x x x	x x x x x	x x x x x	
8 AM						
↑						
3 AM						

PROGRAM WEEKS

Figure 1: Schedule of alcohol availability over the six weeks of the FIDD program.

behavior, and beliefs associated with these decisions. While
therapeutically utilizing patients' drinking decisions, the staff
attempts neither to punish nor reward individual drinking decisions

Measures

This paper describes patterns of program drinking and the relation of these to treatment outcome. The basic data therefore
consist of measures of program drinking and treatment outcome.

Program Drinking

During the D-D phase, drinks were dispensed and recorded immediately by the attending nurse. Thus, a record was available of
the number and distribution of the drinks consumed by each patient.
Patients could drink up to a maximum of 26 ounces daily or 520
ounces during the four D-D weeks, or drink nothing at all.

Treatment Outcome

Data on treatment outcome were obtained from responses to a
14-item follow-up questionnaire which was mailed to each patient
six, twelve, and twenty-four months after completion of the program
Failure to obtain a response from a patient for any of the three
treatment outcome evaluations resulted in further contact efforts
by telephone and by personal visits. If these methods failed, an
attempt was made to obtain the follow-up information from an individual designated by the patient as being familiar with his behavior during the evaluation period. We were thus able to obtain information on 216 of 247 patients at six months, a response rate of
87%, 197 of 242 patients (81%) at one year, and 162 of 242 (67%)
of the patients two years after treatment.

The treatment outcome findings in this paper are based on two
items which describe post-treatment drinking. These inquire, respectively, into the number of days during the last 30 in which
the patient either consumed alcohol or was intoxicated. These
items have proven to be the most reliable indices of drinking behavior in our questionnaire; both items were found to be sensitive
to treatment outcome differences in the various groups of program
drinkers considered in this paper.

Since many of the analyses in this paper will be concerned
with improvement rates of various subgroups of patients at differen
follow-up periods and the number of responses obtained varied for
these periods, it was necessary to adopt a constant standard
throughout in order to allow meaningful comparisons to be made

across the different follow-up periods. The total number of patients completing treatment in each subgroup - excluding those de-
ceased - was therefore adopted as the baseline against which rate
of improvement could be assessed. This method naturally results in
a highly conservative estimate of rate of "improvement," since non-
responders are all treated as treatment failures.

RESULTS

Drinking Patterns

Our previous analyses of patients' drinking patterns revealed
that 120, or 48.2% of all patients remained entirely abstinent
throughout the program. The remaining 129 patients (51.8%) consumed alcohol in amounts ranging from a minimum of one ounce to the
maximum of 520 ounces. The drinking patterns of the 129 program
drinkers were examined from several perspectives. First, the amount
of alcohol consumed during the entire drinking-decisions phase was
examined. It was found that 16 out of 129, or about 1 out of 8 of
the program drinkers, drank 469 ounces or more (at least 90% of the
520 ounces available on the program). Approximately one-third of
the drinkers consumed 365 ounces or more, or at least 70% of the
available alcohol, while nearly 49% consumed at least half of the
available alcohol. One obvious conclusion that can be drawn from
this data is that the majority of the alcoholic patients who chose
to drink had considerable difficulty exercising control over the
amount of alcohol they consumed, even given the external control
imposed by the FIDD schedule of alcohol availability. On the
other hand, a small subgroup of 27 patients, or 20.9% of all program drinkers, were observed to consume no more than 10% of the
alcohol available to them, or a maximum of 52 ounces over a four-
week period. Thus, a small proportion of patients designated as
alcoholics were able to exercise some control over the amount of
their alcohol consumption.

The maximum amount of alcohol consumed on any day of drinking
has been used by some investigators to discriminate between alcoholic, problem, and normal drinking (Armor, Polich, and Stambul,
1976; Harris et al., 1974). Although it has not been possible
to arrive at total agreement on the amount of consumption distinguishing the categories of drinkers described, consumption of no
more than about three ounces of absolute alcohol on a given day is
generally taken to be indicative of "problem" drinking. It seems
reasonable, then, to use the maximum amount consumed by a patient
on any of the drinking-decisions days as another indication of his
ability to regulate his drinking behavior.

Examination of the data from this perspective indicated that

66, or 51.2%, of the program drinkers drank either 25 or 26 ounces, or nearly the maximum amount available, on one or more days. In addition, 20 (15.5%) patients consumed at least 23 or 24 ounces on at least one occasion. Thus, two thirds of the program drinkers drank at least 23 ounces of alcohol on one or more occasions. Depending upon whether patients were consuming a 40% or 43% alcohol solution, this amount was equivalent to between 9-11 ounces of absolute alcohol. The evidence, therefore, indicates that the majority of program drinkers drank in excessive amounts on one or more days and, accordingly, supports the conclusion that the majorit of alcoholics who chose to drink on the FIDD program were relatively unsuccessful in limiting the amount of their drinking. On the other hand, the data again revealed that a small proportion of the program drinkers exhibited considerable control over the amount of alcohol that they consumed daily. That is, 12 patients drank no more than two ounces on any day and a total of 19 patients (14.7%) never consumed more than eight ounces daily (about 3-3.5 absolute ounces of alcohol). These alcoholic patients, therefore, retained some control over their drinking behavior.

Drinking patterns were also described in terms of the relative ability of program drinkers to achieve abstinence on at least some of the D-D days. Since total abstinence represents the acme of control, we concluded that varying degrees of partial abstinence might represent steps in this direction. Along these lines, our clinical observations have indicated that the achievement of abstinence at either the outset or the termination of the D-D phase appears to have particular psychological significance for the patients. Indeed, based on this assumption we had earlier described a subgroup of "stoppers" who had decided to discontinue their drinking at least one week prior to the conclusion of the D-D phase and hypothesized that these patients might fare more favorably following treatment than other program drinkers. However, preliminary analyses failed to reveal any differences between this and other groups of program drinkers (Gottheil, Alterman, Skoloda, et al., 1973; Gottheil, 1973). For the purposes of the present analyses, all program drinkers were categorized on the ability to regulate their drinking to the extent of delaying the onset of drinking at least one day into the D-D phase or terminating drinking at least one day prior to the conclusion of that phase. Application of these relatively liberal criteria resulted in the differentiation of four subgroups of program drinkers as follows

   a. Patients abstaining entirely on the first D-D day, but not
       on the last day (first day abstainers - N = 18);

   b. Patients abstaining entirely on the last D-D day, but not
       on the first day (last day abstainers - N = 29);

    c. Patients abstaining on both the first and last D-D days
       (first + last day abstainers - N = 65);

    d. Patients neither abstaining on the first nor the last D-D
       days (first + last day drinkers - N = 17).

The data revealed that 65 patients, or fully half of the alcoholics who drank on the FIDD program, failed to achieve either of
the defined forms of abstinence, i.e., abstinence on either the
first or last D-D day; while about 3 out of 8 of the drinking alcoholics successfully accomplished at least one form of abstinence.
Finally, slightly more than one out of eight of the program drinkers
(17 or 13.2%) achieved both forms of abstinence. An analysis of
another program drinking characteristic of these four subgroups provided support for the validity of their categorization. The average
amount of alcohol consumed by the four subgroups on the days they
drank was found to vary in direct proportion to the number of defined forms of control over drinking that they had been able to
apply. Thus, patients who abstained on both first and last D-D days
were found to consume least on the days they drank, an average of
10.3 ounces. Patients who drank on both the first and last days,
by contrast, consumed the most when drinking (18.8 ounces); last
day abstainers (12.5 ounces) and first day abstainers (16.5 ounces)
were intermediate in this respect.

In summary, the results of an analysis of the relative ability
of drinking alcoholics to attain at least partial abstinence supported the conclusions drawn from the other analyses, namely, that
many of the alcoholics who drank on the FIDD program failed or were
unable to exercise much control over their drinking, but that some
of them did succeed in imposing some degree of control on their
drinking.

Findings on Treatment Outcome

Summary of Six-Month Treatment Outcome

Our analyses of relationships between program drinking and
six-month treatment outcome of program patients primarily focussed
on outcome differences between three categories of patients distinguished by the extent of their control over program drinking.
These three categories consisted of program abstainers, program
drinkers exhibiting little control over their drinking, and program
drinkers who exerted moderate control over their drinking. By
drawing from the preceding analyses of program drinking, it was
possible to define several subgroups of drinkers who exhibited
little control over their consumption; for example, those patients
consuming more than eight ounces on at least one day (high-maximum

group) or program drinkers who drank on both the initial and final
D-D days. More "moderate" program drinkers were taken to be either
patients who consumed no more than a total of 52 ounces throughout
the D-D phase, or one-tenth of the available alcohol, those who
never consumed more than eight ounces on any D-D day (low-maximum
group) or drinkers who abstained on both the initial and final D-D
days.

These three general categories of program drinkers, composed of
different subgroups depending upon the criterion employed to define
regulation of drinking, were compared on a number of treatment outcome variables--including both frequency of drinking and frequency
of intoxication during the final 30 days of the six-month post-
treatment evaluation period. The results, on the whole, indicated
that the best treatment outcome occurred for those patients who abstained throughout the FIDD program, particularly when more stringer
standards of "improvement" were employed. However, each of the thre
subgroups of program drinkers described above who were able to requ-
late their drinking to some extent were shown to fare considerably
better after treatment than the various subgroups of program drinker
who were less able to regulate their drinking. When less restrictiv
standards of "improvement" were employed, "moderate" drinkers were
found to have improved as much following treatment as abstainers.

Representative findings are illustrated by the data shown in
Figure 2 which describes the relative frequency of drinking of three
subgroups of patients defined in terms of maximum daily consumption.
It shows that the abstainers surpassed both groups of program
drinkers when improvement was taken to be abstinence during the previous 30 days. However, the more moderate group of program drinkers
(low-maximum group) were found to fare at least as well as abstainer
when a more liberal standard of "improvement," a drinking frequency
of no more than twice weekly, was adopted. Analyses indicated that
the differences in outcome between the various subgroups could not b
attributed to differences in background characteristics.

The remaining sections on treatment outcome attempt to determin
the extent to which the findings obtained at six months were maintained over a longer time period. The first two sections examine
the results for abstainers, moderate, and non-moderate program
drinkers at one and at two years post-treatment. The final two sections examine the course of treatment outcome for the various subgroups over the entire two-year follow-up period.

One-Year Post-Treatment Outcome

As indicated, 12-month treatment outcome was evaluated in
terms of frequency of drinking and frequency of intoxication during
the preceding 30-day period. Two parallel series of analyses were

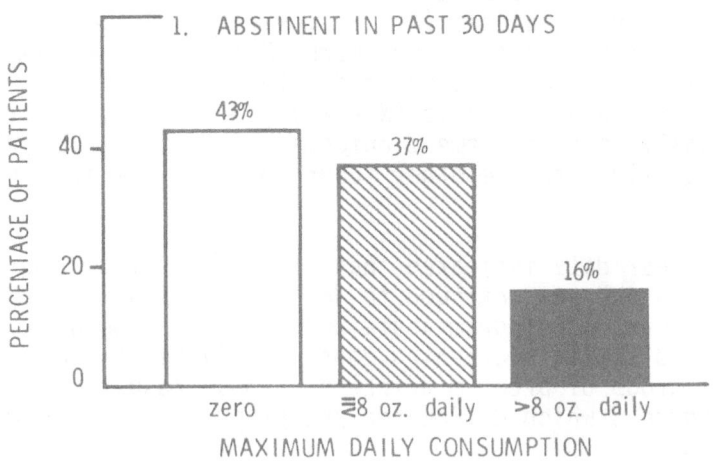

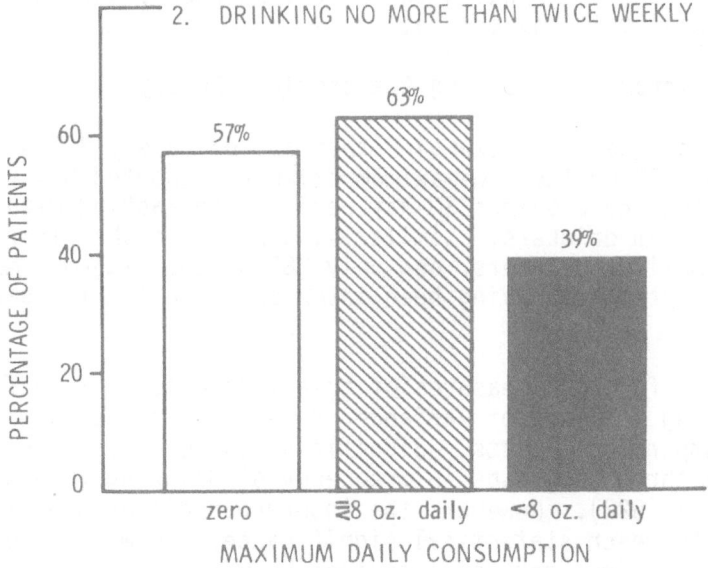

Figure 2: Frequency of drinking at six month follow-up of program drinking subgroups defined by maximum daily intake.

performed. The first was concerned with the subgroups defined in terms of maximum daily consumption and these therefore consisted of program abstainers (N = 120), drinkers who never consumed more than a maximum of eight ounces daily (low-maximum group - N = 19), and drinkers who exceeded an eight-ounce daily maximum (high-maximum group - N = 110). The second analyses series compared program abstainers (N = 120) with the four subgroups based on various combinations of first/last day abstinence consisting of first day abstainer (N = 18), last day abstainers (N = 29), first + last day abstainers (N = 17), and first + last day drinkers (N = 65). The treatment outcome data for the subgroup of "moderate" program drinkers who consumed no more than 52 ounces (N = 27) were not included in the subsequent analyses of treatment outcome, since this subgroup overlaps substantially with the other subgroups of "moderate" program drinkers.

The one-year data indicated that the low-maximum group of moderate program drinkers continued to do as well or nearly as well at one-year follow-up as program abstainers, and that both of these groups appeared to be functioning considerably better than the high-maximum group of program drinkers. These results are illustrated in Figure 3 which delineates the proportion of patients in each subgroup:

a. Abstinent or drinking twice or less weekly during the preceding 30-day period;

b. Not intoxicated during the previous 30 days.

It can be seen, for example, that approximately 34% of the abstainers and 32% of the low-maximum drinkers reported being abstinent during the previous 30-day period, contrasting with 20% of the high-maximum drinkers. Similarly, 50% of the abstainers, 53% of the low-maximum drinkers, and only 28% of the high-maximum drinkers reported not having been intoxicated during the past month.

The data for both measures of treatment outcome were formally analyzed using analysis of variance procedures and Newman-Keuls multiple comparison t tests. Significant differences were found between the three subgroups in frequency of drinking (F = 3.88, 2/191 df, p < .022). However, the intergroup differences were not sufficient to reach statistical significance. Between-group differences in the frequency of intoxication were also found to be statistically significant (F = 9.49, 2/190 df, p < .001). Both the low-maximum and abstainer groups reported being intoxicated less often than the high-maximum group. The first two groups were not found to differ from each other.

The results at one year post-treatment, therefore, supported

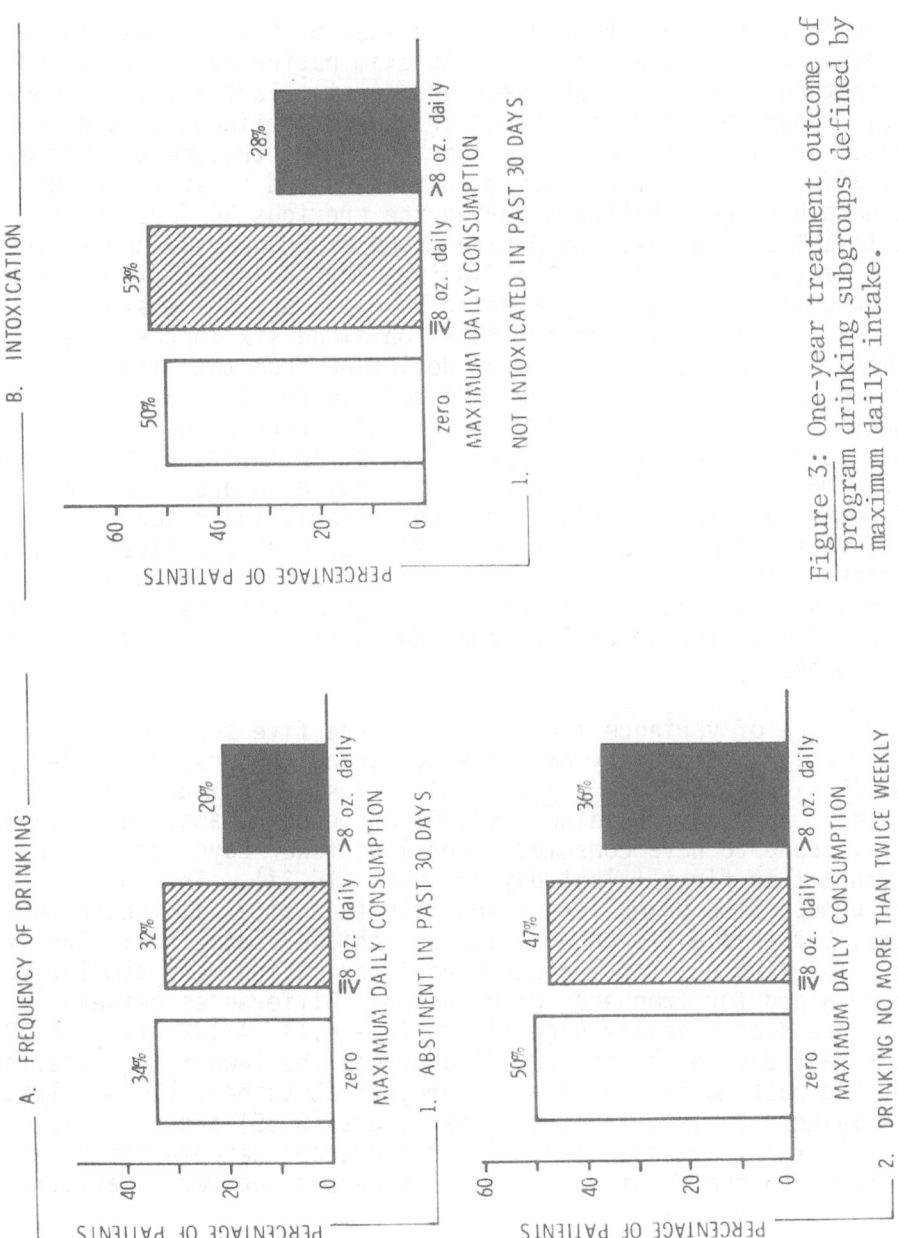

Figure 3: One-year treatment outcome of program drinking subgroups defined by maximum daily intake.

the findings obtained six months earlier in revealing that both
program abstainers and low-maximum program drinkers continued to
exercise considerably more control over their drinking behavior
following treatment than the high-maximum drinkers.

The program drinking subgroups defined by form of abstinence
were compared in similar fashion. An examination of Figure 4 indicates that both program abstainers and first + last day abstainers,
the most moderate group of program drinkers, continued to achieve
the most successful outcomes one year following treatment. For example, 34% and 29% of the program abstainers and first + last day
abstainers had been abstinent during the previous 30 days, in contrast to 15% of the most immoderate group of program drinkers, the
first + last day drinkers. The findings for the relative occurrence
of intoxication during the preceding 30 days were quite similar.
These results basically confirm those obtained six months after
treatment. However, one source of deviation from the six-month
follow-up can be seen in the outcome results for the last day abstainers. The level of adjustment for this group of program
drinkers had been previously reported as being intermediate between
the most moderate and least moderate program drinkers. However,
the findings at one year post-treatment clearly indicated the outcome of this group to be superior to that of both the first + last
day drinkers and the first day abstainers and nearly as good as
program abstainers and the first + last day abstainers. This somewhat surprising finding will be considered in a subsequent section
of this paper.

Analyses of variance indicated that the five subgroups differed
significantly in frequency of drinking during the preceding 30-day
period (F = 4.15, 4/193 $\underline{df}$, $\underline{p}$ < .003). First + last day abstainers (5.46), last day abstainers (7.80) and program abstainers (8.88)
were all found to have consumed alcohol on fewer days during the
past month than first + last day drinkers (15.37). The three
former groups also drank less frequently than first day abstainers
(12.14), but this difference failed to achieve statistical significance. The findings for frequency of intoxication were similar to
those obtained for frequency of drinking. Differences between
groups were statistically significant ($\underline{F}$ = 6.14, 4/192 $\underline{df}$, $\underline{p}$ <.001)
First + last day abstainers (2.62) averaged the fewest intoxications
during the past month, followed by program abstainers (3.44), last
day abstainers (5.15), first day abstainers (8.29) and, finally,
first + last day drinkers (10.02). Differences between the first
two groups and the first + last day drinkers group were statistically significant.

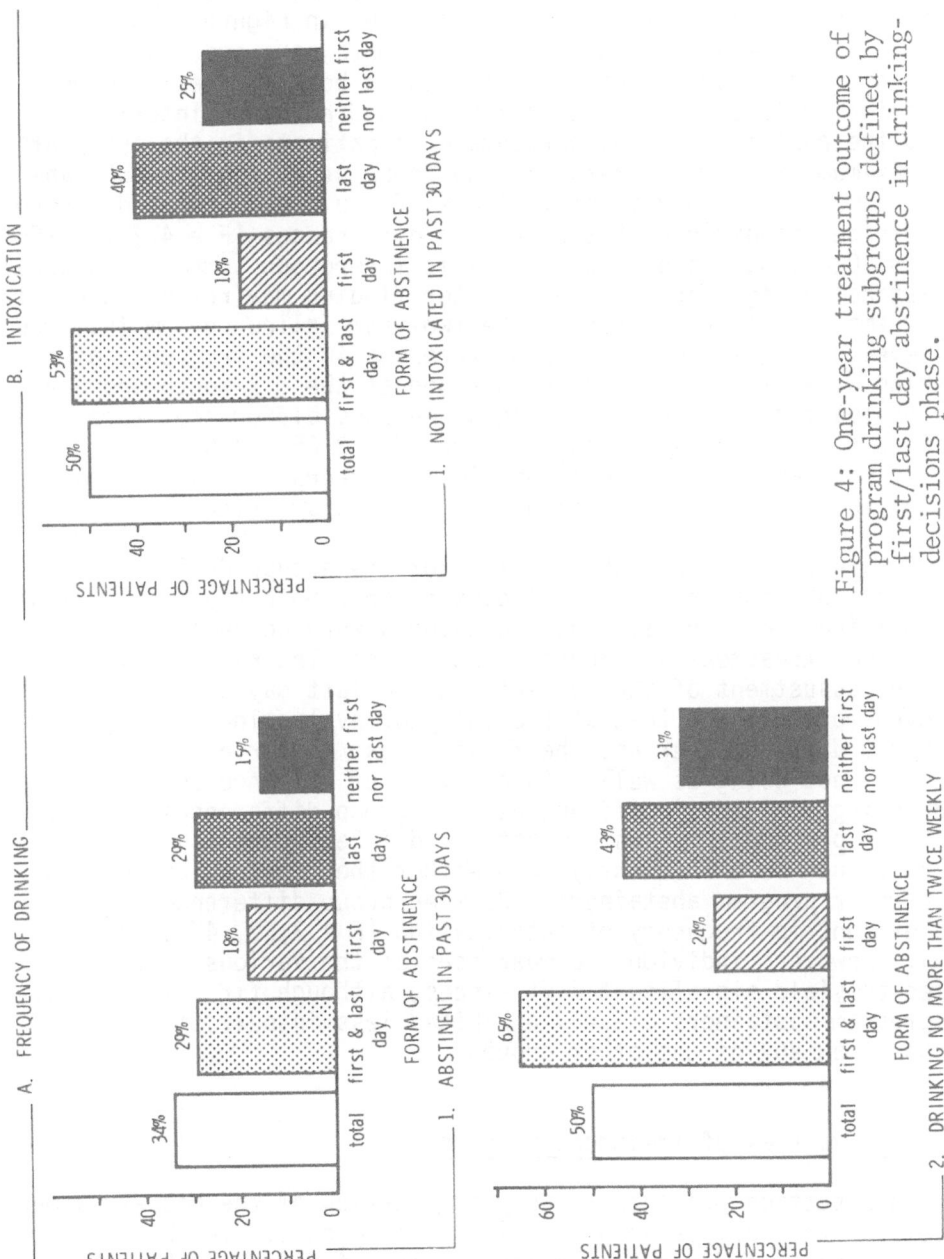

Figure 4: One-year treatment outcome of program drinking subgroups defined by first/last day abstinence in drinking-decisions phase.

Two-Year Post-Treatment Outcome

The two-year treatment outcome results for former FIDD program patients were quite similar to those obtained at the one-year post-treatment evaluation. The results, as shown in Figure 5, basically revealed no differences between program abstainers and low-maximum drinkers. More than 40% of the patients in both of these groups were drinking less than twice weekly and had not been intoxicated during the past month. The percentage of patients in the group of high-maximum program drinkers meeting these standards was substantially lower. An analysis of variance of the frequency of drinking data yielded significant between-group differences (F = 4.27, 2/156 df, p < .016); but none of the specific intergroup comparisons were statistically significant. This latter finding may reflect the rela tively smaller n's available in the two-year follow-up sample, since the level of magnitude of group differences in the frequency of drinking were of the same order obtained at the six-month and one-year evaluations. Significant between-group differences were also obtained for frequency of intoxication (F = 3.65, 2/157 df, p < .028). Both abstainers and low-maximum drinkers were significantly less often intoxicated than high-maximum program drinkers.

Figure 6 summarizes the results for the groups defined by first/last day abstinence. It indicates that both the first + last day abstainers and the last day abstainers were doing as well two years after treatment as program abstainers. The relatively high level of adjustment of the patients in the last day abstainer group was even more evident than at the one-year evaluation. Patients in the first day abstainer and the first + last day drinker groups were generally not doing as well. An analysis of variance on frequency of drinking yielded significant between-group differences (F = 5.09, 4/156 df, p < .001). First + last day drinkers were found to be drinking much more frequently than either the first + last day abstainers or program abstainers. Between-group differences were also evident in frequency of intoxication (F = 3.53, 4/157 df, p < .009). However, individual comparisons of the various groups failed to yield significant differences, although first + last day and program abstainers clearly exhibited less intoxication on the average than any of the other groups.

Two-Year Overview of Treatment Outcome

This section presents a two-year overview of the post-treatment outcome of the various program drinking subgroups. In this connection, Figures 7 and 8 graph the proportion of patients who were drinking twice or less weekly and who were not intoxicated during the 30-day period prior to each of the three outcome evaluations obtained over the two-year follow-up period. The figures indicate some decline from the six-month to two-year post-treatment evalua-

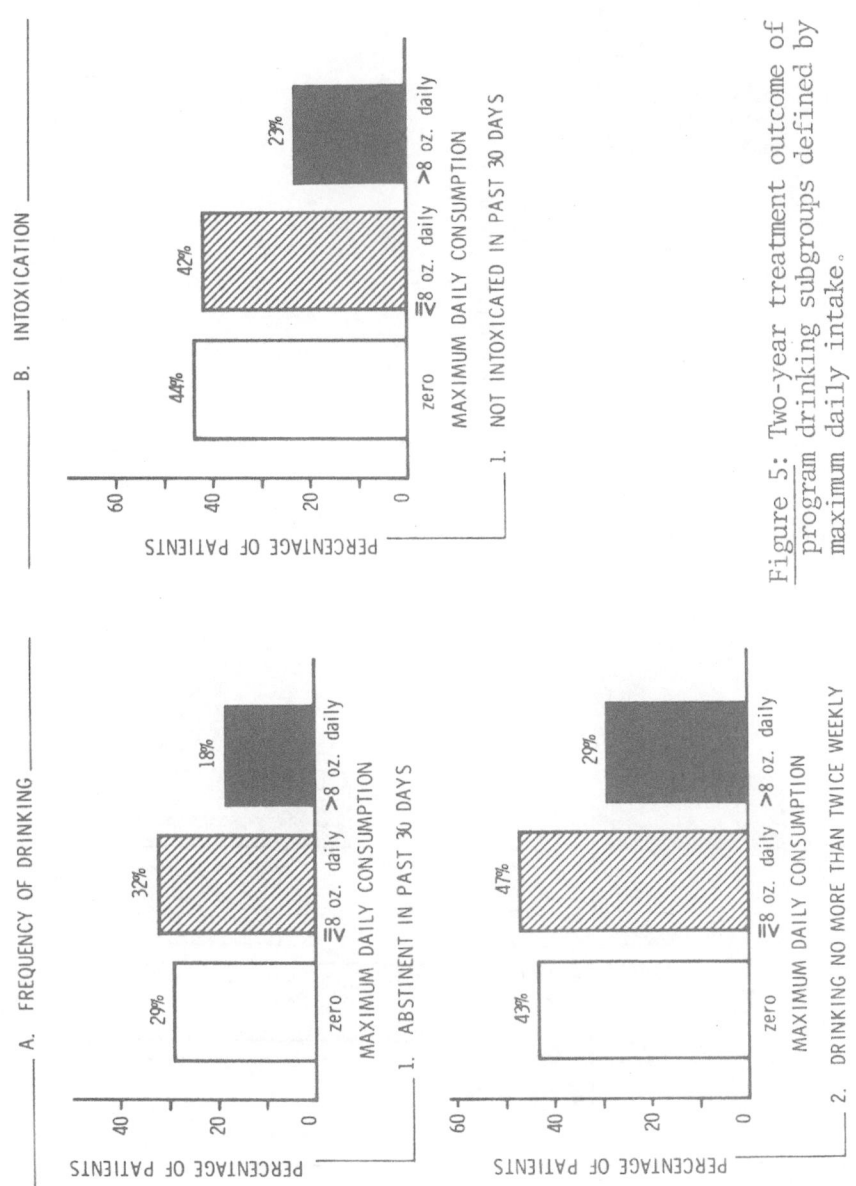

Figure 5: Two-year treatment outcome of program drinking subgroups defined by maximum daily intake.

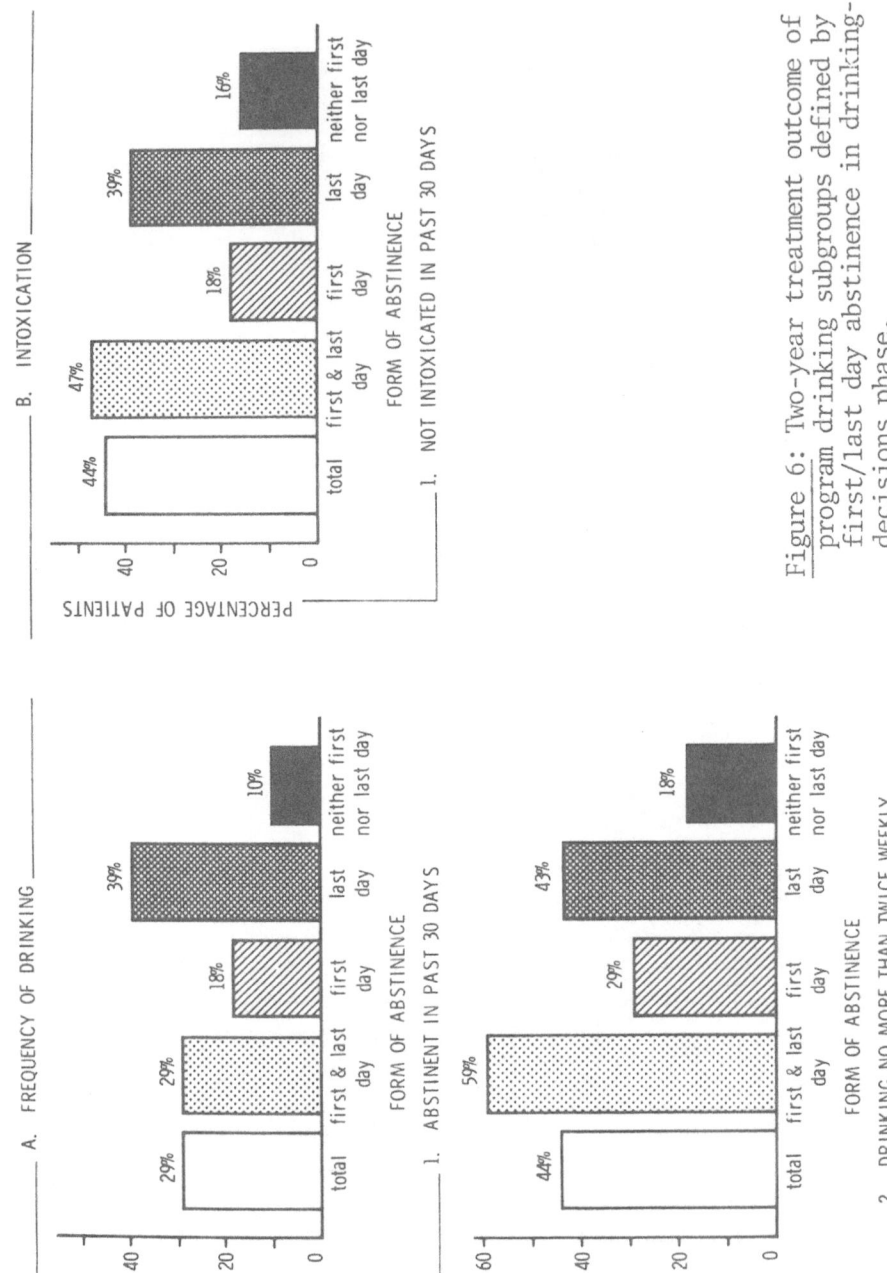

Figure 6: Two-year treatment outcome of program drinking subgroups defined by first/last day abstinence in drinking-decisions phase.

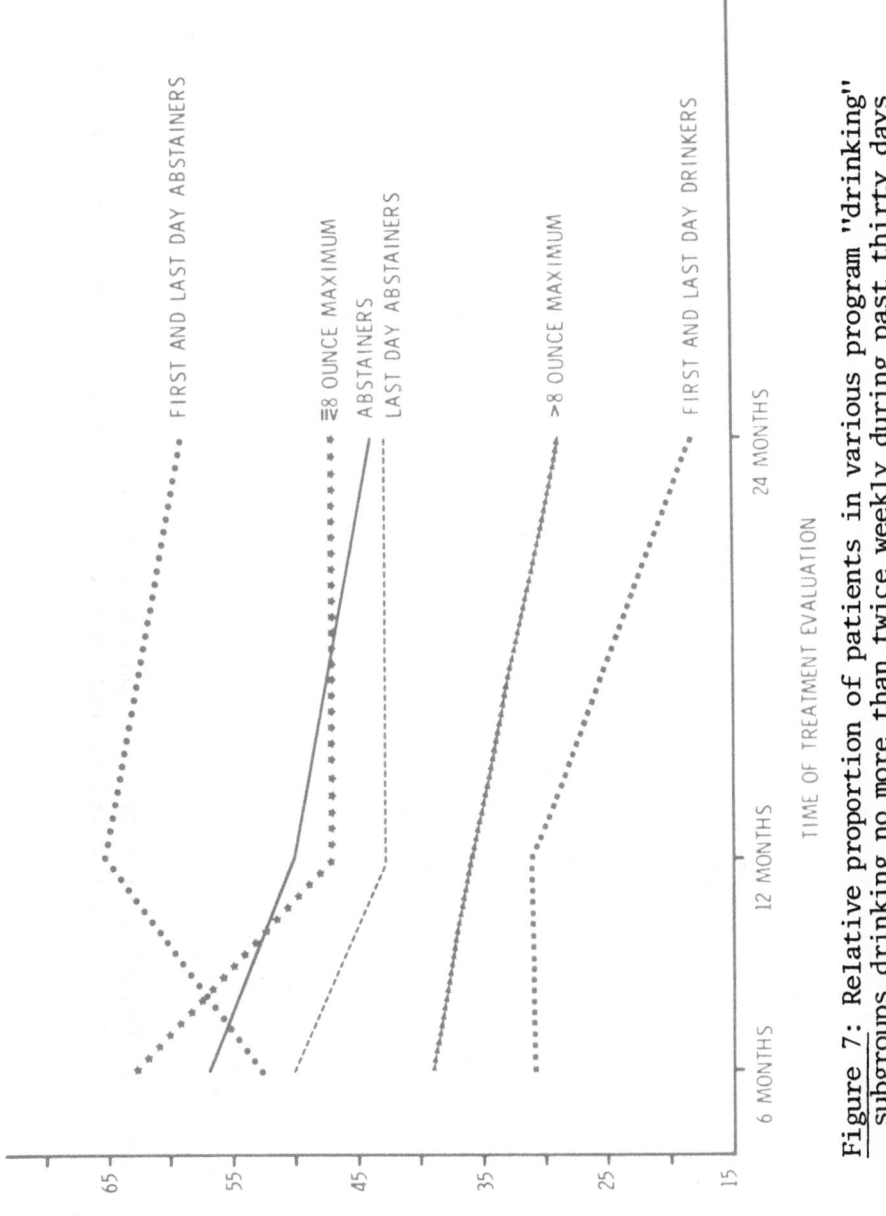

Figure 7: Relative proportion of patients in various program "drinking" subgroups drinking no more than twice weekly during past thirty days of treatment evaluation period.

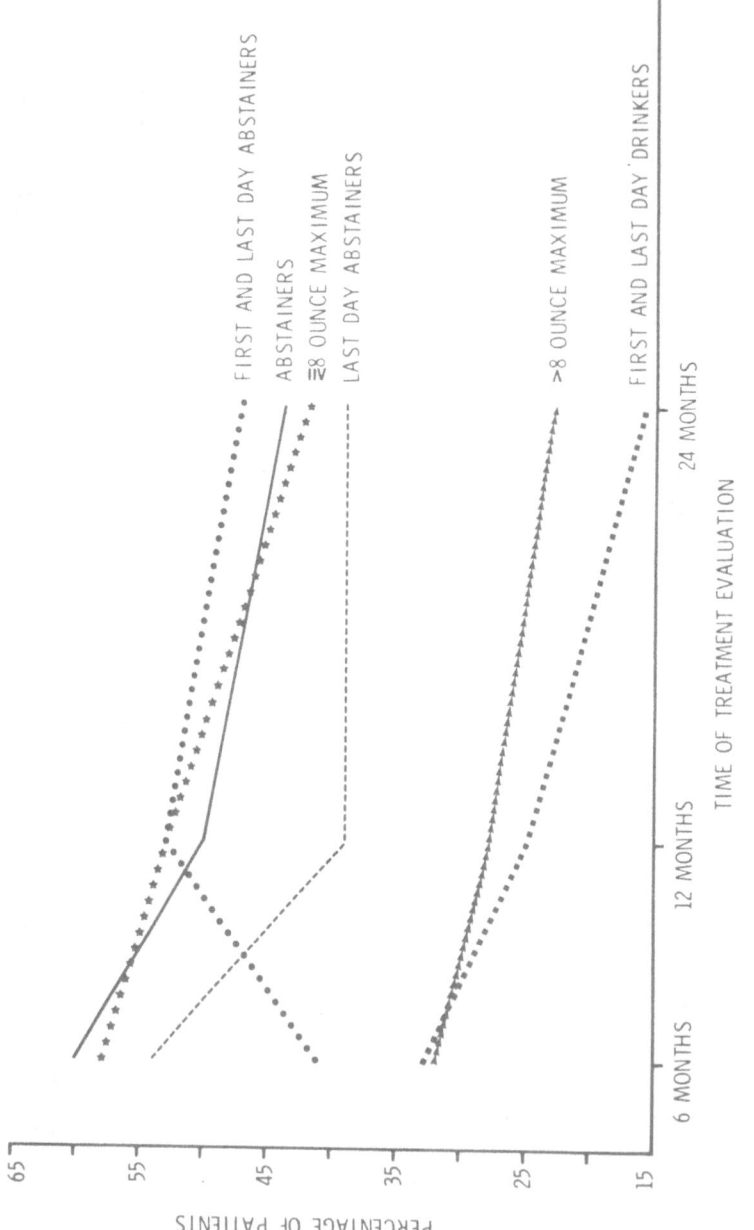

<u>Figure 8</u>: Relative proportion of patients in various program "drinking" subgroups not intoxicated during last thirty days of treatment evaluation period.

tions in the relative proportion of patients drinking at what could
be considered "acceptable" levels. For example, whereas 57% of
program abstainers were found to be drinking twice or less weekly
six months after treatment, only 44% of the same patients fell into
this category 18 months later. Similar trends were apparent for
most of the other groups. An interesting contrast was found, however, in the results for first + last day abstainers who generally
improved or at least declined relatively less over time than the
other groups.

An examination of these data pointed to several other interesting trends. The results indicated that the various subgroups could
basically be placed into one of two categories based on their general level of functioning over the two-year follow-up period. It
would appear that the program abstainer, first + last day abstainer,
low-maximum drinker and last day abstainer groups should be placed
in a common category characterized by moderately satisfactory levels
of success. The remaining subgroups of patients appear to belong in
a second category that clearly showed lower overall adjustment levels over the two-year follow-up period.

These results differ in several ways from those obtained in the
six-month follow-up evaluation. The evidence indicated at that time
that abstainers fared somewhat better than moderate program drinkers.
Although these findings based on two years of follow-up continued to
suggest a relationship between program drinking and treatment outcome, the distinction between abstainers and moderate drinkers no
longer appears to be tenable. The inclusion of last day abstainers
among the groups doing reasonably well after treatment also represents a noteworthy change from the previous findings. The patients
in this group were not originally categorized as moderate program
drinkers. Nevertheless, the additional information obtained over
two years of follow-up evaluation indicated that the level of post-
treatment adjustment of this group was more like that of abstainers
and moderate program drinkers than that of immoderate drinkers.
That is, both the first + last day abstainers and the last day abstainer groups of program drinkers did reasonably well after treatment. A total of 45 patients, or 35% of the program drinkers, are
included in one or the other of these two groups.

Longitudinal Analysis of Treatment Outcome

The treatment outcome data thus far described were based on
evaluations undertaken at three points in time. Since the persons
responding were not entirely the same in each of the evaluations,
the method employed was cross-sectional. Use of such an approach
implies a continuity in the individual and group that may not in
reality exist. An analysis was therefore undertaken in order to
provide, as well, a longitudinal analysis of treatment outcome. To

this end, only the data of patients who responded to all three follow-up evaluations were considered in assessing the relative success
rate of the various program drinking groups.

Frequency of drinking data was first analyzed to determine the
relative number of patients in each program drinking group who reported being abstinent at all three outcome evaluations. Number of
patients, exclusive of the preceding category, who drank no more
than twice weekly on any of the three follow-up evaluations was
also determined. The patients in these two categories were then
combined to yield an overall rate of "satisfactory adjustment" for
each of the subgroups considered in this paper. Finally, the relative number of patients reporting no intoxication at all three follow-up periods was determined from the intoxication item. These
data are shown in Table 1. It is evident for both frequency of
drinking and intoxication that the proportion of program abstainers
achieving satisfactory adjustment was not higher, and in some cases
was lower, than that of many of the groups of program drinkers. Of
even more interest than the overall results, however, were the
findings on the relative proportions of program abstainers and moderate program drinkers found to be either abstinent on all three
evaluations or drinking in moderate frequencies. We might assume
that program abstainers would be more likely than moderate program
drinkers to be found in the abstaining group of patients at followup
and that moderate program drinkers would be more likely to fall,
conversely, into the category of patients reporting moderate drinking frequencies after treatment. This did not prove to be the case;
neither program abstainers nor moderate program drinkers were more
likely to be in one category than in the other. For example, 12
(10.1%) of the program abstainers were abstinent on all three follow
up evaluations, but 15 (12.6%) were also drinking at moderate frequencies. Twenty-one percent of the low-maximum program drinkers
reported moderate frequencies of drinking over the two-year follow-
up period, as might be expected, but an additional 21.0% reported
being abstinent on all three follow-up evaluations. Thus, some of
the alcoholics treated in the FIDD program apparently succeeded in
attaining a large degree of abstinence for two years after treatment, while others appeared to be able to limit the extent of their
drinking; but either outcome was just as likely to be found in program abstainers as in moderate program drinkers.

DISCUSSION

The evidence presented here should, of course, be considered
to be suggestive, rather than conclusive, because of a lack of
definitiveness in the evaluation procedures employed. The evaluations were not intensive and provided only indirect information on
the amount of alcohol consumed by a respondent. Furthermore, information was obtained on drinking behavior at only three discrete

Table 1: Treatment outcome results of various program drinking subgroups limited to patients who responded to all three follow-up evaluations.

	Abstainers	≤8 oz. Maximum	First & Last Day Abstainers	Last Day Abstainers	First Day Abstainers	>8 oz. Maximum	First & Last Day Drinkers
A. Frequency of Drinking							
1. Number Responding to All Three Evaluations	(74/119) 62.2%	(10/19) 52.6%	(8/17) 47.1%	(18/28) 64.3%	(9/17) 52.9%	(60/104) 57.7%	(33/61) 54.1%
2. Abstinent Past 30 Days in All Three Evaluations	(12/119) 10.1%	(4/19) 21.0%	(1/17) 5.9%	(5/28) 17.9%	(2/17) 11.8%	(7/104) 6.7%	(2/61) 3.3%
3. Drinking No More Than Twice Weekly on All Three Evaluations	(15/119) 12.6%	(4/19) 21.0%	(4/17) 23.5%	(3/28) 10.7%	(3/17) 17.6%	(9/104) 8.7%	(2/61) 3.3%
4. Total Drinking "Within Limits" on All Three Evaluations (2 + 3)	(27/119) 22.7%	(8/19) 42.1%	(5/17) 29.4%	(8/28) 28.6%	(5/17) 29.4%	(16/104) 15.4%	(4/61) 6.6%
B. Intoxication							
1. Number Responding to All Three Evaluations	(74/119) 62.2%	(10/19) 52.6%	(10/17) 58.8%	(18/28) 64.3%	(9/17) 52.9%	(60/104) 57.7%	(31/61) 50.8%
2. Not Intoxicated on Any of the Three Evaluations	(32/119) 26.9%	(7/19) 29.6%	(5/17) 29.4%	(7/28) 25.0%	(2/17) 11.8%	(12/104) 11.5%	(5/61) 8.2%

points in time rather than more frequently which necessitated making
inferences about drinking behavior during intervening periods of
time which may not be justified. On the other hand, the follow-up
response rate obtained in this study was rather high for two years
of post-treatment evaluation with alcoholics and the results, whatever their limitations, applied equally to all of the subgroups
described. We should also realize that it becomes more difficult to
directly attribute effects to treatment as the post-treatment interval increases. The following discussion of the implications of the
findings should keep the above limitations in mind.

The findings raise a number of questions relevant to the treatment of alcoholics; for instance, should total abstinence be the
primary goal of treatment? Given the fact that more program abstainers did well after treatment than all other patients combined,
it would seem wise to continue to consider total abstinence the
preferred treatment goal for the majority of alcoholics. On the
other hand, the findings indicated that only a relatively small
proportion of the patients who achieved abstinence in our drinking-
decisions program were entirely successful in maintaining abstinence
over two years. A number of program abstainers drank occasionally
and apparently within limits after treatment. The results also
demonstrated, somewhat surprisingly, that some program drinkers were
just as likely as program abstainers to abstain or drink within
limits after release from treatment. Thus, although abstinence
during treatment may have been correlated with successful treatment
outcome, this does not necessarily signify that abstinence is basic
to or essential to treatment success. Rather, our findings would
seem to recommend the acceptance of degrees of partial abstinence
and limited drinking as realistic treatment options and outcomes for
some alcoholics.

The program drinking and the treatment outcome results indicated that the drinking behavior of many alcoholics may vary over
time. It is not clear, however, whether this applies to some alcoholics only, thereby suggesting the existence of types or subgroups
of alcoholics, or that such variations are possible in the majority
of alcoholics. This suggests that there is much that still has to
be learned about the long-term drinking behavior of these persons
designated as alcoholics (Pattison, 1974). That is, it would seem
important to know to what extent various alcoholics are able to
regulate their drinking over time and to what extent total abstinenc
is attainable in the larger body of alcoholics? The answers to such
questions should be of value in establishing treatment goals for
various alcoholic patients.

ACKNOWLEDGMENTS

Supported by U. S. Veterans Administration Project No. 0356-03 and NIAAA Grant No. R01AA02449-02. The authors would like to acknowledge their appreciation to Mrs. Donna Forbes and Mrs. Rita Gracia for their assistance in implementing the data analyses in this paper.

REFERENCES

Alterman, A. I., Gottheil, E. and Thornton, C. C. Variations in patterns of drinking of alcoholics in a drinking-decisions program. Presented at the 8th Annual NCA/AMSA Medical-Scientific Meeting, San Diego, CA, May 2, 1977.

Armor, D. J., Polich, J. M. and Stambul, H. B. Alcoholism and Treatment. Santa Monica, CA: Rand, June, 1976. (R-1739-NIAAA.)

Gottheil, E. Research on fixed interval drinking decisions in an alcoholism treatment program. Proceedings of 2nd Annual Alcoholism Conference of the National Institute on Alcohol Abuse and Alcoholism (June 1-2, 1972), DHEW Publication No. (NIH) 74-676, 139-157, 1973.

Gottheil, E., Alterman, A. I., Skoloda, T. E. et al. Alcoholics' patterns of controlled drinking. American Journal of Psychiatry, 1973, 130, 418-422.

Gottheil, E., Corbett, L. O., Grasberger, J. C. et al. Treating the alcoholic in the presence of alcohol. American Journal of Psychiatry, 1971, 128, 475-480.

Gottheil,, E., Corbett, L. O., Grasberger, J. C. and Cornelison, F. S., Jr. Fixed interval drinking decisions. I. A research and treatment model. Quarterly Journal of Studies on Alcohol, 1972, 33, 311-324.

Harris, L. and Associates, Inc. Public Awareness of the NIAAA Advertising Campaign and Public Attitudes toward Drinking and Alcohol Abuse, Feb., 1974.

Jellinek, E. M. The Disease Concept of Alcoholism. Highland Park, NJ: Hillhouse Press, 1960.

Pattison, E. M. Rehabilitation of the chronic alcoholic. In B. Kissin and H. Begleiter (Eds.), The Biology of Alcoholism. New York: Plenum Press, 1974, pp. 587-658.

SUBCULTURAL DIFFERENCES IN DRINKING BEHAVIOR IN U. S. NATIONAL SURVEYS AND SELECTED EUROPEAN STUDIES

Don Cahalan

Social Research Group, School of Public Health

University of California, Berkeley

It is particularly fortunate that the international conference on which this volume reports is being held in Scandinavia, in view of the many contributions to alcoholism made by researchers in all the Scandinavian countries during the last 20 years. This Scandinavian research has been an inspiration and material benefit to alcohol researchers in the U. S. for many years. We are looking forward to an expansion of collaborative research efforts with Scandinavian and other European colleagues in the immediate future.

Nobody has been impolite enough to ask me to explain what my topic of subcultural differences in drinking behavior has to do with the conference's central theme of experimental and behavioral approaches to alcoholism. Therefore I will merely say, in justification, that we have found such very substantial subcultural differences in the manifestation of drinking behavior and problems in our U. S. studies--and we have seen similar differences in studies elsewhere--that we know that clinical and experimental studies must take account of such subcultural differences in carrying out their work.

I will not be presenting a definitive review of cross-cultural comparisons of surveys of drinking behavior and problems in this paper because that task remains yet to be completed through cooperative international research efforts. Primarily, this paper is a presentation of some of our findings on subcultural differences from the drinking behavior surveys we have been conducting throughout the U. S. during the last 17 years, with some commentary on my perception of the general state of survey research on alcohol behavior and problems in Western countries. Sub-group differences in our U. S. surveys may be of special interest to our European colleagues because we have so many different European backgrounds represented in the

U. S. that the rather considerable differences in drinking behavior
that we find suggest that more detailed studies in European countries will be likely to find similar differences in drinking behavior between and within individual countries.

SUMMARY OF U. S. DRINKING BEHAVIOR SURVEYS

The series of national surveys on which I am reporting began
in the California Department of Public Health in 1960 under Wendell
Lipscomb, Ira H. Cisin, and Genevieve Knupfer in 1960, were continued through national surveys conducted by The George Washington
University under Professor Cisin in the mid-1960s, and have been
continued by our Social Research Group in Berkeley. In all, we
have conducted a dozen community surveys and five national surveys.
I will focus my discussion primarily on the findings from the na-
tional surveys, particularly those reported in our three books,
American Drinking Practices (Cahalan, Cisin and Crossley, 1969),
Problem Drinkers (Cahalan, 1970) and Problem Drinking Among American
Men (Cahalan and Room, 1974).

Our general-population surveys have been supported by the National Institute on Alcohol Abuse and Alcoholism and its predecessors because they agreed that we have to study the general popula-
tion--rather than merely the clinical alcoholic--if we are really
going to understand the processes by which people get into, and out
of, the drinking problems included in "alcoholism." Our program of
community and national surveys have had three general phases. For
the first few years, there was an emphasis upon description of
drinking practices and attitudes about drinking (Cahalan, Cisin,
and Crossley, 1969; Cisin, 1963; Knupfer et al., 1963). Next there
were a number of studies concentrating on specific problems related
to drinking (Clark, 1966; Cahalan, 1970; Cahalan and Room, 1974;
Knupfer, 1967). Our current final stage in this series concentrates
upon analysis of longitudinal data on drinking behavior and problems
within samples of individuals interviewed two or more times over a
span of years. All of these surveys have been strictly controlled
(scientifically randomized) probability samplings of the general
population aged 21 or older. And now to proceed to summarize the
findings from the first phase: drinking practices and attitudes in
the general population.

Drinking Behavior and Attitudes Toward Drinking

Our early national surveys, as well as those of other U. S.
researchers (Bacon, 1962; Gusfield, 1962; Haberman and Scheinberg,
1969; Harris and Associates, 1971; Keller, 1971; and Mulford and
Miller, 1961), consistently reflect the conflict in values between
U. S. groups (and within individuals) concerning drinking. The

generally uninformed state of knowledge and ambivalent attitudes
about drinking in our national surveys fit in very well with our
national heritage, especially the classical conflict between the
Protestant Ethic values and the hard-drinking behavior of a large
number of our citizenry, from the time of the frontiersman to the
age of the suburban cocktail circuit.

In our first national survey of 1964-65, we found the highest
proportions of heavier drinkers to be among those about age 40,
those of lower social status, those living in larger cities and in
the Middle Atlantic, New England, and Pacific areas; and those of
Irish, British, or Latin-American extraction (Cahalan, Cisin, and
Crossley, 1969). However, the groups with the highest proportions
of drinkers were not the groups with the highest proportions of heavy
drinkers. For example, while Jews and Episcopalians had the lowest
proportions of abstainers, they also had extremely low ratios of
heavy or problem drinkers.

Abstainers were found more likely to be older persons, or below-
average income, from the South or rural areas, of native-born parentage, and from conservative Protestant denominations. Alcohol appeared to be a greater threat to those of lower economic and social
status, possibly because these people are more vulnerable and economically insecure than those of higher status.

National Surveys of Drinking Problems

We have now completed four nationwide surveys on the prevalence
of specific drinking problems. The first was a followup in 1967 of
a subsample of the respondents in the 1964-5 survey of drinking practices just described; these findings have been reported in Problem
Drinkers (Cahalan, 1970). The second national survey on drinking
problems was conducted in 1969 with a sample of 978 men in the high-
risk group aged 21-59 and was reported in Problem Drinking Among
American Men (Cahalan and Room, 1974). In the third national study
on drinking problems, most of the men aged 21-59 were reinterviewed
in 1973 to measure changes in drinking behavior and problems over a
four-year span (Cahalan and Roizen, 1974; Roizen, Cahalan and
Shanks, in press). An additional national followup was completed
in 1974 among 900 men and women who were first interviewed in the
1964-5 survey and again in the 1967 survey; the data from this three-
wave survey are still being analyzed.

All these studies of drinking problems emphasize that what we
want to measure is the prevalence of specific types of problems (or
potential problems) associated with alcohol, rather than "alcoholism" however defined. As in Knupfer's 1964 San Francisco survey
(Knupfer, 1967), our national surveys have covered a number of actual, objective problems and also several potential problems (such

as drinking enough to constitute a potential health problem to the individual); we tabulate the various problems separately, as well as in certain combinations. Our national surveys take into account three types of variables in scoring responses on a dozen specific types of problems: 1) The severity of problems in terms of frequency and the presumed gravity of the problem; 2) The certainty or reliability of measurement (in terms of the number of items used in assessing the problem); 3) The timing of the problem in terms of whether it is a current problem (that is, within three years) or happened more than three years ago. The types of problems or potential problems in the 1967 survey were the following:

1) Frequent intoxication

2) Binge drinking

3) Symptomatic drinking: This potential problem refers to signs of Jellinek's "Gamma alcoholism" (1960, p. 37), including signs of physical dependence and loss of control (e.g., drinking to get rid of a hangover, having difficulty in stopping drinking, having blackouts or lapses of memory, skipping meals while on drinking bouts, drinking very rapidly for quicker effect, and sneaking drinks).

4) Psychological dependence

5) Problems with spouse or relatives

6) Problems with friends or neighbors

7) Job problems: Having lost a job or nearly lost one because of drinking, having people at work suggest the person cut down on drinking, and rating drinking as harmful to one's work and employment opportunities.

8) Problems with law, police, or accidents

9) Health: That drinking had been harmful to the respondent's health and that a physician had advised him to cut down.

10) Financial problems

11) Belligerence associated with drinking: Feeling aggressive or cross after drinking, or getting into a fight or heated argument. (This item was included as a potential problem in order, later, to test the hypothesis that belligerence may often be followed by later increase in interpersonal problems.)

Table 1 summarizes the prevalence of each of the eleven types of actual or potential problems in terms of occurrence during the prior three years. At the bottom of the table is an overall "Cur-

TABLE 1[a]

PREVALENCE OF DRINKING-RELATED PROBLEMS OVER THREE YEARS, BY SEX

	MEN (N = 751)			WOMEN (N = 608)		
	None %	Moderate %	High %	None %	Moderate %	High %
1. Index of Frequent Intoxication[b]	83	3	14	97	1	2
2. Binge drinking	97	–	3	100	–	*
3. Symptomatic drinking[c]	84	8	8	93	4	3
4. Psychological dependence	61	31	8	85	12	3
5. Problems with current spouse or relatives	84	8	8	96	3	1
6. Problems with friends or neighbors	93	5	2	97	3	*
7. Job problems	94	3	3	98	2	1
8. Problems with police or accidents	99	–	1	99	–	1
9. Health[d]	88	6	6	93	4	4
10. Financial problems	91	6	3	98	2	1
11. Belligerence associated with drinking	88	8	4	93	5	3
Current Problems Score, combining results for the 11 specific problems, (High score: 7+ of a maximum of 58)	57	28	15	79	17	4

[a]From Table 1, Problem Drinkers, Cahalan, 1970.
[b]Components: frequency and amount-per-occasion, and frequency of getting high or tight.
[c]Positive responses to such items as difficulty stopping drinking, blackouts, sneaking drinks.
[d]Physician told respondent to cut down drinking.

Note: All percentages in the tables in this paper represent weighted findings projectable against the household population. Numbers (e.g., N = 751) are the actual number of interviews in the group specified. Tables add horizontally for each sex. Totals may vary from 100 percent because of rounding each figure to nearest whole percent.

*Less than one-half of one percent.

rent Problems Score," in which each problem is given equal weight
(except for health, financial, and belligerence problems, which were
given a lesser weight because these had very few items). A "high"
overall current problems score could be attained only by having
problems in two or more areas, of which at least one was rated as
being in severe form; or problems in three or more areas, of which
at least two were at least moderate in severity; or moderate to
severe problems in five or more areas; or slight problems in seven
or more areas.

The most common problems for men were frequent intoxication,
symptomatic drinking, psychological dependence, and problems with
spouse or other relative. As for women, the only problem which attained a level as high as four percent was health problems.

Fifteen percent of the men and four percent of the women--nine
percent of the total--were rather arbitrarily classified in the
"high" problems category. But as the table shows, where one places
the cutting point has an enormous effect on the rate of problem
drinking: One can picture the adult population as having a high
rate of problem drinking since 43 percent of the men, 21 percent of
the women, and 31 percent of the total had one or more of the problems in the preceding three years; or one can put emphasis on the
relatively low rate of problems by noting that "only" fifteen percent of the men and four percent of the women were classified as
being in the more-severe-or-frequent-problems category.

Next, we concentrated on measuring drinking problems among men
aged 21-59, because we found this group to have the highest rate of
drinking problems overall. In analyzing this group, we combined
data from the 1967 and 1969 surveys to get a larger total in this
high-risk group. We will now discuss several tables presenting
findings from this group.

Note, in Table 2, that the general rate of problems is almost
invariably much higher among those 21-24 years old than among
older men, and is about twice as high in Current Overall Problems
Score for the 21-24 age group than even for the next oldest (25-29)
group. This concentration of drinking problems among men aged 21-
24 is dramatically at variance with data on the age of the average
clinical alcoholic, which is around 42. However, this pronounced
age trend is borne out in all our studies of the general population.
It is also borne out in studies Cisin and I (1972, 1975) have directed within the armed forces.

Concerning socio-economic status, Table 3 supports our earlier
findings that persons of lower status have a higher proportion of
abstainers, but that those of lower status who do drink tend to
have a higher-than-average rate of heavier drinkers than among
higher-status persons. This five-category typology of drinking

TABLE 2[a]

HIGHER SEVERITY LEVEL FOR SPECIFIC CURRENT PROBLEMS BY EIGHT AGE GROUPS (combined 1967-69 surveys)

	21-24 (147) %	25-29 (204) %	30-34 (186) %	35-39 (216) %	40-44 (226) %	45-49 (201) %	50-54 (199) %	55-59 (182) %	Total (1561) %
1. Heavy Intake	7	7	5	7	6	3	5	6	6
2. Binge Drinking	10	3	3	3	1	2	4	2	3
3. Psychological Dependence	5	4	4	4	4	5	4	3	4
4. Loss of Control	12	5	4	5	7	5	4	4	6
5. Symptomatic Drinking	26	11	8	7	6	10	9	3	9
6. Belligerence	15	12	10	8	7	8	6	2	8
7. Problems with Wife	19	17	15	10	9	9	11	6	12
8. Problems with friends or neighbors	15	5	7	5	4	4	6	4	6
9. Problems on job	10	4	3	5	5	5	6	2	5
10. Police Problems	10	4	2	2	2	2	4	1	3
11. Health or injuries from drinking	8	4	5	6	4	6	8	6	6
12. Finances	11	4	6	4	3	2	4	3	4
13. Current overall Problems Score 7+	40	22	20	21	17	17	17	11	20

[a]From Cahalan and Room, Problem Drinking Among American Men, 1974.

TABLE 3[a]

DRINKING PROBLEMS TYPOLOGY FOR FOUR

SOCIAL POSITION GROUPS, MEN AGED 21-59

	(n)	Non-drinker[f] %	Drank, no problems[e] %	Potential problems only[d] %	Heavy Intake or Binge, No Consequences[c] %	High Consequence[b] Score 3+ %
Lowest Social Position	(281)	19	22	19	14	26
Lower Middle	(411)	17	38	19	10	17
Upper Middle	(401)	14	39	24	15	8
Highest Social Position	(468)	11	47	22	11	9
Total	(1561)	15	38	21	12	14

[a]From Cahalan and Room, 1974, Table 22.

[b]High Consequences Score of 3+: Tangible Consequences, i.e., social consequences, health or injury problems associated with drinking, or financial problems.

[c]Intake or Binge: Not in above, but at least minimal severity Intake or Binge problems.

[d]Potential Problems Only: Not in the above groups, but a problem of at least minimal severity in any problem area.

[e]Drank, No Problems: Has been a drinker within the last three years, but not in any of the above groups.

[f]Non-drinkers: Did not drink during last three years.

problems shows that those of lowest status have a higher absolute rate of drinking problems than other persons, and that the ratio of Consequences of drinking to Heavy intake or binge drinking is also higher for those of lower status. This is another way of saying that more of the poor tend to get into trouble over their drinking, out of proportion to the number who drink heavily.

We classified regions of the country into "wetter" and "dryer," in terms of post voting or survey findings on prohibition sentiment. The "Dryer" areas are shown in Table 4 to have a higher proportion of men aged 21-59 with interpersonal problems related to drinking, relative to their rate of heavy intake without interpersonal consequences, than is true for men in "Wetter" areas. In other words, it is harder to avoid drinking problems if one drinks in a traditionally "Dryer" area. (Similar patterns of "wet" vs. "dry" areas have been reported by Nils Christie, 1965, in his comparison of Finnish and Danish drinking.) One implication of these findings is that efforts to minimize drinking problems could be more efficient if they took into account local cultural traditions, another indication that no single national prevention program will fit all regions.

Table 5, concerned with ethnoreligious groups, shows that most Jewish men drink at least a little, but relatively few drank heavily or got into trouble over their drinking. Most Catholics and liberal Protestants ("Liberal" in terms of their churches' stands on drinking) do drink, and a higher-than-average proportion of them get into trouble over their drinking; Conservative Protestants--denominations favoring abstinence--show a higher proportion of abstainers but also a relatively high ratio of Consequences in relation to Heavy Intake or Binge Drinking. These findings agree with Gusfield's writings (1962) about the history of connections between religious affiliation and attitudes toward alcohol.

Changes in Drinking-Problem Status Over Several Years

The final stage in this 15-year series of surveys has been to assess the correlates of change in drinking practices and problems over time. Longitudinal studies are crucial to the study of cause/ effect relationships because only with studies of individuals at two or more points in time is it possible to determine which prior events and states of mind preceded which later events.

A San Francisco probability sample of men aged 21-59 initially interviewed in 1967 and reinterviewed in 1972 first made it possible for us to measure detailed changes over a period of several years (Clark and Cahalan, 1976). The traditional expectation, if one follows Jellinek's models (1960), is that alcoholics will accumulate increasing numbers of problems over time, with the "early symptoms"

TABLE 4[a]

CURRENT PROBLEMS TYPOLOGY[b] BY REGION, IN PERCENT

	N	Non-drinker %	Drank: No Problems %	Potential Problems Only %	Heavy Intake or Binge: No Consequences %	High Consequences %
Dryer Regions						
South Atlantic	(156)	28	30	19	12	12
East South Central	(113)	29	30	16	5	20
West South Central	(188)	23	40	17	5	14
West North Central	(134)	10	47	17	12	14
Mountain	(61)	21	33	26	3	16
Totals	(652)	22	37	18	8	15
Wetter Regions						
New England	(89)	7	43	33	11	7
Middle Atlantic	(288)	7	39	20	19	15
East North Central	(353)	13	41	22	13	12
Pacific	(179)	10	32	27	14	16
Totals	(909)	10	39	23	15	13

[a]From Cahalan and Room, 1974, Table 17.
[b]Defined in Table 3 footnote, above.

TABLE 5[a]
CURRENT PROBLEMS TYPOLOGY BY ETHNORELIGIOUS CATEGORIES
(combined 1967-69 surveys)

	(n)	Non-drinker %	Drank, no problems %	Potential Problems only %	Heavy Intake or Binge, No Consequences %	High (3+) Consequences Score %
British:						
Catholic	(34)	12	35	27	18	9
Liberal Protestant	(48)	13	31	38	8	10
Conservative Prot.	(204)	20	39	22	3	16
Irish:						
Catholic	(77)	4	33	27	16	21
Conservative Prot.	(74)	27	46	15	3	10
German:						
Catholic	(76)	7	40	21	22	11
Liberal Protestant	(86)	12	47	24	9	8
Conservative Prot.	(120)	22	42	21	7	9
Italian:						
Catholic	(64)	5	52	14	23	6
Latin American:						
Catholic	(42)	10	10	21	17	43
Jewish:	(40)	8	60	25	5	3
Black:						
Conservative Prot.	(97)	18	23	16	13	31
Eastern European:	(71)	6	38	21	21	14
Other ethnicity:						
Catholic	(114)	7	42	26	11	13
Liberal Protestant	(56)	5	43	18	23	11
Conservative Prot.	(158)	34	35	15	8	8

[a]From Cahalan and Room, 1974, Table 27.
[b]Categories with small N's are omitted from this table. Ethnicity is defined by religion for Jews, by race for Blacks, by "country most ancestors come from" for the remainder.

of progressive alcoholism not being replaced by later symptoms, but
being merely added to problems already existing (Room, 1970). However, the San Francisco findings were at variance with the conventional expectation, because there was not found to be much "snow-
balling" of problems in more and more areas of the person's life
with the passage of time. While the proportion who no longer had
the same problem at Time 2 was quite high, ranging from 50 to 96
percent, people with any specific drinking problem at Time 1 tended
to have some type of problem at Time 2, although not necessarily the
same problem. Another finding was that those with specific problems
at Time 1 tended to have high rates of heavy intake or binge drinking at Time 2.

We conclude from this San Francisco study that continuity of
specific problems over time is rather low, but that the probability
of future involvement in some alcohol problems--but not necessarily
the same ones--is increased if one develops alcohol-related problems. Thus, the "progressive disease" concept of problem drinking
is open to question. However, the fact that those who have drinking problems at one time tend to have drinking problems of varying
kinds at a subsequent time may imply that environmental factors may
play a considerable part in determining the type of alcohol-related
problems that may occur.

A national probability sample measuring change over four years
was also conducted as part of this same series. In this survey,
725 men aged 21-59 were first interviewed in 1969 and again four
years later. The findings bear out the San Francisco findings of a
high turnover or change in drinking problems: About as many shifted
into, or out of, a drinking-problem status as stayed in a drinking-
problem status.

Our analysis of our longitudinal data is still in progress.
We completed another national followup survey two years ago which
will provide sufficient numbers to permit us to study changes in
drinking problems in even greater detail. However, we already have
sufficient evidence to conclude that the "symptom" or "prodromal"
status of specific drinking problems as predictors of later and more
serious problems has been considerably overrated, since our general-population change data suggest that an "early warning" prediction based on such "symptoms" will yield a substantial number of
false positives who will never get into serious trouble. We believe that the high rates of turnover in drinking problems will be
found to be highly associated with specific environmental circumstances and life events; and we are now conducting an intensive
analysis of the impact of life events or environmental circumstances
upon changes in drinking problems. We will have more to report on
this within the near future.

COMMENTS ON THE STATE OF SUBCULTURAL COMPARISONS THROUGH SURVEYS

And now, a few comments about the status of cross-cultural comparisons of drinking behavior through surveys. Again, I am not in a position to present a definitive review, because that task remains to be done--I hope, within the next few years--through international collaborative efforts of a number of researchers. However, I did want to draw your attention to a couple of excellent interim critiques of cross-cultural survey comparisons, and then to close by venturing a few opinions on what needs to be done in comparative cross-cultural alcohol survey reviews.

First, I hope all of you have read or will get a copy of Salme Ahlström-Laakso's excellent paper on "European Drinking Habits: A Review of Research and Some Suggestions for Conceptual Integration of Findings" which she presented at an anthropology conference in 1973. In her paper, she pointed out that while Europeans account for a disproportionately high share of the world's alcohol consumption, there are considerable variations between European countries. For example, a larger proportion of Danes and Swedes drink at least occasionally than is true for Finns and Norwegians. She also discusses how it is that while there is a high proportion of teetotalers in Finland, the Finns as a whole tend to drink to get drunk more often than is true for the Italians, more of whom both drink and tend to consider alcohol as part of everyday life insofar as having wine as an accompaniment to meals is concerned. I think it is obvious that such differences should be studied intensively, both between and within countries, to establish not only their extent but also what implications the social perceptions of the use of alcohol have on future preventive or treatment programs. Only through comparative studies is it possible to transcend one's own narrow cultural influences to get a fresh perspective on what preventive or remedial programs are most likely to be effective for what types of subcultures.

Dr. Ahlström-Laakso also recommends three foci of research which she believes will be fruitful in future cross-cultural alcohol research. One is to study the integration or segregation of alcohol consumption (for example, whether alcohol is used in conjunction with meals or is preferred on an empty stomach). A second area of interest she mentions is to study the interplay between formal and informal social controls (for example, Helsinki has higher rates of arrests for drunkenness than Copenhagen, even though the frequency of drunkenness is higher in the Danish capital than in the Finnish). She also points to the importance of studying changes in drinking habits due to cultural diffusion (for example, in Finland, more people are adding imported wine to their consumption of beer and distilled liquors; and all of us can think of anecdotal examples of cultural diffusion of drinking habits, such as the adoption of whiskey-drinking among Parisians and in the Tokyo expense-account community). I might add that one of my students recently completed

a dissertation bearing on cultural diffusion. Ben Simboli's study
of our national sample data on recent vs. multi-generational imi-
grants to the U. S. from Italy found that those who had been in the
States for three or more generations did keep a good deal of the
wine-drinking habits of their forefathers--BUT they also added a
substantial amount of spirits and beer to their drinking; and they
also showed a trend toward acculturating to the average American
rates for drinking problems (Simboli, 1976).

Dr. Ahlström-Laakso's paper also contains many useful suggestions on how the gathering of international alcohol-relevant statistics might be improved from their generally unreliable and un-
standardized state--on which I will have additional comments in a
moment. She ends her review article by saying,

> Epidemiological alcohol studies carried out from time to
> time have the important function of acting as a gauge
> for the alcohol or temperature situation. Parallel stu-
> dies are also needed, however, to bring to light the
> substantive content of the changes taking place and the
> processes by which they occur.

Another recent paper with a great deal of information and useful suggestions on what is available for cross-cultural comparisons
through alcohol surveys and how the content and utilization of such
surveys may be improved in the future is the article written by my
colleague, Robin Room, for the World Health Organization report on
Alcohol-Related Disabilities (1977). Robin's paper, entitled
"Measurement and Distribution of Drinking Patterns and Problems in
General Populations," provides a general account of what has been
happening in the increase in the number and sophistication of alcohol surveys in many Western countries during the last 30 years. He
makes the point (as does Dr. Ahlström-Laakso) that most of these
surveys are descriptive in nature and tend to have been done with
limited administrative or policy considerations in mind. He discusses demographic variations in drinking patterns which show differences by sex; differences by age groups (with abstention tending
to be declining among the young across a range of countries); by
social class (where, in the U. S. and England and Finland, abstention is linked with lower status in areas where drinking is traditionally heavy, but with higher status in areas where drinking is
less heavy); by urbanization (where, generally, drinking is heavier
in the more urbanized areas, but not so in Poland, as reported by
Swiecicki, 1972); and by regions of the country (where differences
were found for France and the U. S., where explosive and intermittent drinking in the traditionally alcohol "dry" areas appears to
play much the same role as in Finland.)

Robin also makes the point that surveys about drinking problems are everywhere much less available than they should be, that

there needs to be much more attention to dis-aggregating problematic drinking behavior and drinking problems, so that we can deal with specific drinking problems more directly and realistically, rather than putting so much emphasis upon trying to find scales to identify or predict a single classical type of clinical "alcoholic." He also makes a very strong case for more longitudinal studies of drinking practices to assess the etiology of drinking problems, as well as making a plea for the international alcohol research community's concentrating more effort toward achieving comparability in various countries' surveys insofar as measures of quantity, frequency, and variability of drinking are concerned. (An excellent illustration of early cooperation toward achieving comparability in cross-cultural alcohol studies was the work of the Nordic Committee for Alcohol Research which began working in 1959 toward that goal; see Bruun and Hauge, 1963. Dr. S. Brun-Gulbrandsen of the National Institute for Alcohol Research also was an early advocate of comparability in alcohol studies, 1973.) I heartily agree with Robin that we don't want to see an imperialistic approach in trying to bring about more comparability of measurements from country to country; but at least we should try harder to make sure that accidental differences in survey questions that can be avoided don't make it impossible to compare survey results between countries or subcultures.

Two additional papers, as yet unpublished, that are available from our Social Research Group upon request will be of interest to those interested in cross-cultural comparisons of drinking behavior. These papers have been prepared through grants or contracts from the NIAAA. One of these is a staff paper by Richard Bunce and Pamela Street entitled "Cross-National File of Surveys on Drinking Practices (Non-U. S.)"; it summarizes the content and sample sizes and characteristics of sample surveys from 53 different countries. The other is a new draft paper by Robin Room and Richard Bunce, entitled "International Alcohol Statistical Indicators," which compares many of the tratitional types of alcohol-related aggregate statistics and discusses problems of reliability and validity in the collection and analysis of these types of data.

CONCLUSION

In conclusion, I would like to summarize my perspective on the present state of the art in surveys of drinking practices and problems:

1) Substantial progress has been made in several European countries and the U. S. and Canada during the last 20 years in conducting surveys which meet sufficiently high technical standards to be useful in policy planning.

2) The available data from individual surveys of complex societies with many subcultures (such as the U. S. surveys I discussed earlier) and from comparative analyses of differences in drinking behavior in various European countries (as provided by Ahlström-Laakso's paper) clearly reveal important sub-group differences which should lend encouragement to our intensifying efforts to utilize cross-cultural comparisons to a greater extent. Only through such cross-cultural studies can we achieve the depth of understanding of how drinking behavior and problems develop that we need in order to provide research findings that are highly useful to the policy planners of the future.

3) There is, I believe, an urgent need to conduct more parallel surveys which permit comparisons of various countries and subcultures as regards drinking behavior and problems. Particularly needed are more longitudinal studies which permit charting relative changes over time in various subcultures, so we can do a better job of understanding the reasons behind change in drinking behavior and problems.

4) We should work with WHO and informal ad hoc associations of alcohol researchers to bring about better standardization of at least a minimum number of variables in surveys (such as frequency, quantity, and variability of drinking) and in the gathering of aggregate official alcohol-relevant statistics (such as cirrhosis mortality, accident rates, and alcohol-related crime and dependency). We need not only to have more opportunity to make comparisons among surveys of drinking behavior that are comparably conducted and among cross-cultural sets of official statistics that are comparably administered, but also to permit us to do a better job of interrelating survey findings and aggregate alcohol-indicator statistics within and between countries and subcultures, so as to achieve a better understanding of the reasons for trends in drinking behavior.

ACKNOWLEDGMENTS

The U. S. studies reported in this paper have been conducted through the support of the National Institute on Alcohol Abuse and Alcoholism and its predecessors during the last 15 years. The author wishes to express appreciation for the suggestions of his colleagues at the Social Research Group, especially Robin Room and Dick Bunce.

REFERENCES

Ahlström-Laakso, S. European Drinking Habits: A Review of Re-
    search and Some Suggestions for Conceptual Integration of
    Findings. In M. W. Everett, J. O. Waddell, & D. B. Heath
    (Eds.), Cross-Cultural Approaches to the Study of Alcohol.
    The Hague & Paris, Mouton: 1976.

Bacon, S. D. Alcohol and complex society. In D. J. Pittman and C.
    R. Snyder (Eds.), Society, Culture and Drinking Patterns.
    New York: Wiley, 1976.

Brun-Gulbrandsen, S. Analysis of consumption data. In The Epidem-
    iology of Drug Dependence. Copenhagen: World Health Organi-
    zation, 1973.

Bruun, K. & Hague, R. Drinking habits among Northern youth. Hel-
    sinki: Finnish Foundation for Alcohol Studies, 1963.

Bunce, R. & Street, P. B. Cross-national file of surveys on drink-
    ing practices (non-U. S.). Unpublished staff paper, Social
    Research Group, School of Public Health, Berkeley, 1977.

Cahalan, D. Problem drinkers. San Francisco: Jossey-Bass, 1970.

Cahalan, D. & Cisin, I. H. Final report on a service-wide survey
    of attitudes and behavior of naval personnel concerning alcohol
    and problem drinking. Washington, D. C.: Bureau of Social
    Science Research, Inc., 1975.

Cahalan, D., Cisin, I. H., & Crossley, H. American drinking prac-
    tices. New Brunswick, N. J. Rutgers Center of Alcohol Studies,
    1969.

Cahalan, D., Cisin, I. H., Gardner, G. L., & Smith, G. C. Drinking
    practices and problems in the U. S. Army, 1972. Report No.
    73-6, Information Concepts Incorporated, Arlington, Virginia,
    December 29, 1972.

Cahalan, D. & Roizen, R. Changes in drinking problems in a national
    sample of men. Paper presented at the North American Congress
    on Alcohol and Drug Problems, San Francisco, December, 1974.

Cahalan, D. & Room, R. Problem drinking among American men aged
    21-59. American Journal of Public Health, 1972, 62, 1473-1482.

Cahalan, D. & Room, R. Problem drinking among American men. New
    Brunswick, N. J.: Rutgers Center of Alcohol Studies, 1974.

Christie, N. Scandinavian experience in legislation and control. National Conference on Legal Issues in Alcoholism and Alcohol Usage, Boston University Law-Medicine Institute, 1965.

Cisin, I. H. Community studies of drinking behavior. Annals of New York Academy of Sciences, 1963, 107, 607-612.

Clark, W. Operational definitions of drinking problems and associated prevalence rates. Quarterly Journal of Studies on Alcohol, 1966, 27, 648-668.

Clark, W. & Cahalan, D. Changes in problem drinking over a four-year span. Addictive Behaviors, 1976, 1, 251-259.

Gusfield, J. R. Status conflicts and the changing ideologies of the American temperance movement. In D. J. Pittman and C. R. Snyder (Eds.), Society, culture, and drinking patterns. New York: Wiley, 1962.

Haberman, P. W. & Scheinberg, J. Public attitudes toward alcoholism as an illness. American Journal of Public Health, 1969, 59, 1209-1216.

Harris, L. & Associates, Inc. American attitudes toward alcohol and alcoholics. Report prepared for the National Institute on Alcohol Abuse and Alcoholism, 1971.

Jellinek, E. M. The disease concept of alcoholism. New Brunswick, N. J.: Hillhouse, 1960.

Keller, M. Ethanol: The basic substance in alcoholic beverages. Alcohol and Health. First special report to the U. S. Congress. National Institute on Alcohol Abuse and Alcoholism, Government Printing Office, Washington, D. C., 1971.

Knupfer, G. The epidemiology of problem drinking. American Journal of Public Health, 1967, 57, 973-986.

Knupfer, G., Fink, R., Clark, W., & Goffman, A. Factors related to amount of drinking in an urban community. Drinking practices Study, Report No. 6, Berkeley: State of California, Department of Public Health, 1963.

Mulford, H. A. & Miller, D. E. Drinking in Iowa. IV. Preoccupation with alcohol, heavy drinking, and trouble due to drinking. Quarterly Journal of Studies on Alcohol, 1960, 21, 279-291.

Mulford, H. A. & Miller, D. E. Public definitions of the alcoholic. Quarterly Journal of Studies on Alcohol, 1961, 22, 312-320.

Roizen, R., Cahalan, D. & Shanks, P. "Spontaneous remission" among untreated problem drinkers. In Kandel (Ed.), Longitudinal Studies in Drug Use: Substantive Findings and Methodological Issues, Washington, D. C.: Hemisphere Press - John Wiley, in press.

Room, R. Assumptions and implications of disease concepts of alcoholism. Paper delivered at the 29th International Congress on Alcoholism and Drug Dependence, Sydney, Australia, February, 1970.

Room, R. Measurement and distribution of drinking patterns and problems in general populations. In G. Edwards, M. M. Gross, M. Keller, J. Moser, & R. Room (Eds.), Alcohol-Related Disabilities. Geneva: World Health Organization, Offset Publication No. 32, 1977.

Room, R. & Bunce, R. International alcohol statistical indicators. Unpublished staff paper, Social Research Group, School of Public Health, Berkeley, 1977.

Simboli, B. Acculturated Drinking Practices and Problem Drinking Among Three Generations of Italians in America. Doctoral dissertation, University of California, Berkeley, School of Public Health, 1976.

Swiecicki, A. Drinking patterns in Poland. Drinking and Drug Practices Surveyor, 1972, 5, 1-7.

ALCOHOL TREATMENT OUTCOME EVALUATION: CONTRIBUTIONS FROM BEHAVIORAL RESEARCH

Linda C. Sobell

Dede Wallace Center and Vanderbilt University

Few would argue with the statement that alcohol treatment outcome evaluation studies have been replete with methodological problems. Several reviews, spanning thirty-five years, have found serious deficiencies in nearly all aspects of treatment evaluation studies (Crawford and Chalupsky, 1977, Hill and Blane, 1967; Miller, Pokorny, Valles and Cleveland, 1970; Sobell, 1978; Voegtlin and Lemere, 1942). These reviews, and the literature upon which they are based, demonstrate that the execution of good alcohol treatment outcome studies is difficult. Recently, however, well-designed alcohol treatment outcome studies have started to appear in the literature. Reasons why such evaluation studies have become more frequent and suggestions for performing better treatment outcome evaluation studies will be enumerated later in this paper.

Although imposing in number, most studies of alcohol treatment effectiveness have been greatly influenced and guided by traditional concepts of alcoholism--concepts which recently have been demonstrated to lack a significant empirical base (Pattison, Sobell and Sobell, 1977). Research conducted over the past several years now stands in direct contradiction to many of the traditional concepts of alcoholism. These new data have compelled a reformulation of ideas about alcohol dependence, including a rethinking and reevaluation of many of our existing treatment approaches and ways of measuring treatment effectiveness. This paper will examine issues in alcohol treatment outcome evaluation from two perspectives: (1) within the context of their departure from traditional concepts of alcohol dependence, and (2) as recent advances which have derived from behaviorally-oriented treatment programs and approaches.

It has often been suggested that alcohol abuse is an unusually recalcitrant clinical problem, with a relatively low rate of recovery. In explanation of this supposed state of affairs, some have suggested that, in the case of a condition for which there is not a definitive etiology, such a lack of treatment effectiveness is hardly surprising. Lest we accept such an explanation too hastily, I offer an alternative: that our low rates of success may very well be a function of how we measure success. Despite our recognition that the population of alcohol abusers is heterogeneous, traditional concepts have insisted upon a singular treatment goal and, thus, a singular index of treatment success. That is, if alcohol treatment success is viewed as an all-or-none phenomenon rather than in terms of gradual levels of recovery (as is typical with other mental and medical illnesses, e.g., depression, diabetes, pneumonia), then might we not be evaluating the effectiveness of this complex problem unrealistically. Insisting on adherence to a binary nomenclature (drunk versus abstinent) to evaluate alcohol treatment outcome may unintentionally demand levels of clinical excellence beyond our current capabilities. The traditional dichotomous view of the alcoholic's drinking behavior--drunk or abstinent--simply cannot reflect varying degrees of control over drinking. Moreover, recent studies have demonstrated that drinking behavior outcomes typically include various combinations of periods of abstinence, excessive drinking and limited drinking (Armor, Polich and Stambul, 1976; Orford, Oppenheimer and Edwards, 1976; Popham and Schmidt, 1976; Sobell, and Sobell, 1973, 1976, in press; Vogler, Compton, and Weissbach, 1975; Vogler, Weissbach, and Compton, 1977; Vogler, Weissbach, Compton, and Martin, 1977; Pomerleau, Pertschuk, Adkins, and Brady, Note 1). While reports of drinking outcomes other than abstinent or drunk have appeared in the literature and are proliferating in number, perhaps such reports have been limited in number because we look for results which are consistent with our established notions. Put more simply, to report or even measure outcomes of limited drinking places one in the position of recognizing that such outcomes are in fact distinctly possible for some individuals with drinking problems. One way of avoiding such reports - and the resultant threat to traditional beliefs - is to report any limited drinking outcome as "improved."

As noted earlier, when alcohol treatment programs have been evaluated using traditional outcome measures, success rates have been abysmally poor. To date, no treatment modality or approach has been unequivocally demonstrated to be successful or effective. Continuing frustrations with the lack of effective traditional alcohol treatment programs, coupled with the changing climate of opinion in the alcohol field, have paved the way for the emergence of behaviorally-oriented treatment programs and techniques for use with individuals who have alcohol problems.

BEHAVIORALLY ORIENTED ALCOHOL TREATMENT PROGRAMS

Most behaviorally-oriented alcohol treatment programs developed apart from - were not associated with - traditional alcohol treatment programs. Further, behaviorally-oriented treatment has origins in a scientific orientation which demands that premises be supported by data. The methods of evaluation and measurement associated with behaviorally-oriented treatment programs are, therefore, designed, whenever possible, to provide data which are complete and operationally defined. There are no intrinsic premises that particular types of outcomes are, by definition, impossible. For example, behaviorally-oriented alcohol treatment programs measure drinking behavior as they would any other behavior; they seek to define and quantify that behavior. Such an approach could easily have provided data to support traditional premises.

One of the studies which had the most influence on the emergence of new treatment evaluation measures and procedures is the Individualized Behavior Therapy (IBT) study by Sobell and Sobell (1972, 1973a, b, 1976, in press). This study, by virtue of its radical and controversial departure from traditional alcoholism treatment programs--an unorthodox treatment paradigm, including the goal of controlled drinking--made necessary the development of more sensitive and valid treatment outcome measures (e.g., daily drinking disposition) and follow-up procedures. Lest the reader be confused, I am not suggesting that the development of more adequate outcome evaluation measures could only have been accomplished vis-a-vis an unorthodox treatment paradigm, but rather that a behaviorally-oriented research-treatment study compelled the development of more precise and sophisticated evaluation methods and measures.

EVALUATION TECHNIQUES

Before considering treatment outcome measures, it is necessary to recognize the importance of efficacious data gathering procedures. The practical utility of treatment outcome evaluation data is highly dependent upon the follow-up procedures used to gather those data. Until recently, alcohol treatment studies have usually reported data for only about 30% to 75% of all subjects in a given study (reviewed by Sobell, 1978). The collective experience of several investigators (Bowen and Androes, 1968; Sobell and Sobell, 1976, in press; Barr, Rosen, Antes and Ottenberg, Note 2; Moos and Bliss, Note 3) suggests that subjects who are difficult to locate for follow-up are typically functioning less well than subjects who are more easily located. Since an adequate assessment of any alcohol treatment program requires that data be gathered for a sufficiently representative sample of treated clients, it is critical to locate and report data on as many subjects as possible.

Tracking alcoholics and their collaterals (significant others) to obtain adequate follow-up data has sometimes been viewed as a nearly impossible task. Reasons given for the difficulties experienced in conducting follow-up of alcoholics have included: (a) they are purported to lie about their drinking (Hill and Blane, 1967; Madsen, 1974; Sobell, 1976); (b) They are often difficult to locate for follow-up (Armor, Polich and Stambul, 1976; Crawford, Chalupsky and Hurley, 1973; Hill and Blane, 1967; Miller, Pokorny, Valles and Cleveland, 1970); and (c) there exist a paucity of appropriate follow-up procedures and treatment measures with which to evaluate alcoholism treatment programs (Sobell, 1978). Realistically, follow-up takes a good deal of time, expense and persistence.

Routine outcome evaluations have been rare in most alcohol treatment programs and most of the published evaluations of alcohol treatment studies were not planned in advance. In contrast, most behaviorally-oriented alcohol treatment outcome studies have included a planned evaluation component. The value in having a pre-planned evaluation component is obvious. First, difficulty in locating and tracking clients can be minimized by obtaining careful follow-up tracking data from clients prior to their discharge from treatment. Second, data useful in evaluating changes in clients' pretreatment to posttreatment adjustment can be gathered systematically. Third, clients can be informed in advance of the need for the ensuing evaluations.

In most alcohol studies, follow-up contacts have typically been impersonal, of limited duration, and conducted months or years after subjects were discharged from treatment programs. This kind of one-time follow-up usually covers periods ranging anywhere from six months to five years posttreatment. In such cases, when subjects are finally contacted, comprehensive information regarding the entire preceding interval must be obtained. Sometimes this includes questions about drinking behavior outcomes that cover periods of time ranging from one to five years posttreatment. Such an inordinate interval between discharge and follow-up has several obvious disadvantages, including (1) difficulty in locating subjects; (2) subjects' memory deficits due to the long time interval; (3) subjects' failure to report events which occurred early in the follow-up interval and are perceived as no longer relevant; and (4) selective reporting by subjects who want to represent themselves in a "good light" and, therefore, report information which is not fully representative of the entire follow-up interval.

One of the contributions of the IBT study was the development of more intensive follow-up tracking procedures. In order to adequately evaluate the treatment goal of controlled drinking, it was necessary to develop more sensitive quantifiable measures of drinking behavior outcomes and to gather follow-up data that would reflect changes in subjects' behavior over time. Thus, follow-up contacts were conducted monthly so as to minimize the difficulty of locating and tracking subjects and to collect more detailed and quantifiable data. As a result, complete two-year follow-up data were collected for 98.6% (69 of 70 subjects) of all subjects in this study.

Pretreatment Comparisons

In order to adequately measure and interpret treatment outcome data, it is necessary to collect measures of subjects' functioning prior to treatment. Pretreatment measures provide the basis for comparative evaluations of individual change. One prevalent problem in the presentation of treatment outcome evaluation data has been a failure to report pretreatment-posttreatment comparisons. It is unfortunate that most studies in the alcoholism literature have reported treatment outcome evaluation results only in terms of summarized group data. As a result, individual subject change cannot be evaluated. In order to portray changes in each subject's level of functioning, it is essential that treatment evaluation studies collect pretreatment as well as posttreatment data.

Obviously, some individuals have suffered more serious consequences than others as a result of their drinking. Similarly, some individuals will show no evidence of disruption in many areas of life health. Until recently, alcohol treatment programs attracted only chronic, debilitated alcoholics. In recent years, however, younger people and people with less chronic drinking problems have begun to seek treatment sooner than has been the case in the past. The significance of early problem drinkers and crisis-oriented drinkers for outcome evaluation is that many of them may only have problems in certain areas of their life. Consequently, improvement cannot be measured in some areas of life health because of a "ceiling effect." In cases where little, if any, disruption has occurred prior to treatment, it is necessary to present pretreatment-posttreatment comparisons to prevent these areas of life health from being evaluated as improved. Thus, in the end, cases must be evaluated in terms of specific individual changes resulting from treatment, in addition to any grouped data comparisons that are relevant.

Alcoholism is not a Unitary Phenomenon

For a long time, alcoholism has been viewed by many as a unitary phenomenon (Pattison et al., 1977). However, in reviewing the existing data, Pattison et al. (1977) suggest that there is no single entity which can meaningfully be defined as alcoholism. If so, this has serious implications for alcohol treatment outcome evaluations. For instance, outcome results are likely to vary as a function of the population under study. In support of this, Moos and Bliss (Note 3) recently reported that different populations of alcoholics not only show different patterns of outcome, but that different follow-up technologies were required to follow each population. Pattison, Coe and Doerr (1969) have also found different patterns of outcome for different populations of alcoholics. In a related regard, it has been shown that even the validity of self-reports varies slightly across different populations of alcoholics (Sobell and Sobell, Note 4).

Patterns of treatment outcome may also be very significant. While there has been much concern in the alcohol field over how many years follow-up data must be collected before an investigator can feel reasonably sure that outcome results are stable and conclusions valid, it might also be of great value to investigate various temporal patterns of outcome. That is, we should begin to evaluate drinking outcomes within the context of total life health and environment by looking at the interrelatedness of various outcome measures. It is possible, for instance, that reciprocal changes occur in various of the treatment outcome dimensions. There is now adequate evidence that grouped outcome results are relatively stable after a period of 18 months (see reviews by Gerard and Saenger, 1966; Sobell, Note 5). An equally important consideration, however, is what kinds of changes occur for what kinds of clients and the temporal patterning of these changes. Once this information is available, we can investigate why these changes occurred and if they, in fact, were related to the treatment received. The long range implications of these suggestions is that someday we may be able to make treatment goal assignments on an empirical basis, rather than relying on clinical judgment.

Measuring Drinking and other Areas of Life Health

Unlike their traditional counterparts, most behaviorally-oriented alcohol treatment programs have used a broad array of outcome measures to assess functioning in a variety of areas of life health. One of the most unique measures that emerged from the IBT study was an alternative dependent variable measure for evaluating drinking behavior outcomes--daily drinking dispositions. In brief, this measure attempted to operationally specify the amount of drinking or type of drinking-related behaviors (alcohol-related incarcerations) that occurred on each day of the follow-up interval. For an in-depth consideration of this measure, the reader is referred to the original source material (Sobell and Sobell, 1972, 1973a, b, 1976, in press). While the drinking behavior of alcohol abusers had not been assessed on a daily basis prior to the IBT study, several investigators have since begun to use similar measures to evaluate drinking behavior (Vogler, Compton, and Weissbach, 1975; Vogler, Weissbach and Compton, 1977; Vogler, Weissbach, Compton, and Martin, 1977; Caddy, Note 6; Pomerleau et al., Note 1; Taub, Note 7).

Irrespective of the outcome measures used, it is essential that they be operationally defined, quantifiable, and if possible, continuous. Two good examples of such outcome measures for other areas of life health were reported by Hunt and Azrin (1973)--"percent of time employed" and "percent of time away from home." Such measures are far more sensitive to inter-subject differences than categorical measures, i.e., "employed" or "not employed."

Drinking is usually not the only behavior affected when an individual abuses alcohol. While traditional treatment approaches have not been adverse to looking at and evaluating other aspects of alcoholics' life health which might be affected by drinking, few have actually employed multiple treatment outcome evaluation measures. Pattison, Headly, Gleser and Gottschalk (1968) and Gerard and Saenger (1966) were among the earliest to suggest that alcoholism treatment evaluations should employ multiple treatment outcome measures. Since that time, several other investigators (Belasco, 1971; Emrick, 1974; Lowe and Thomas, 1976; Sobell, 1978; Sobell and Sobell, 1976, in press) have further emphasized the need for a multivariate evaluation of alcohol treatment effectiveness. The cumulative findings of these investigators suggest that improvement in one area of life functioning does not necessarily imply or predict improvement in other areas of life functioning. Thus, when evaluating alcohol treatment outcome, it is suggested that the following areas of life health be considered: (1) daily drinking behavior, (2) alcohol-related incarcerations, (3) vocational assessment, (4) physical health, (5) use of therapeutic supports after treatment, (6) residential status and stability, (7) social and familial adjustment, (8) legal problems resulting from the use of alcohol, and (9) financial difficulties related to drinking.

Despite the intensity and depth of the outcome evaluation measures and procedures developed by behaviorally-oriented alcohol treatment programs, there is still a need for some standardization of measures and methods. Presently, because there are no standardized outcome measures, it is virtually impossible to compare treatment effects and programs adequately. Another concern regarding multiple measures of life health relates to the meaning and interpretation of various measures. Earlier, it was stated that a recurring issue in the alcohol field has been defining the criteria for recovery from alcohol problems. Even though most recent studies have not dichotomously reported drinking behavior as drunk or abstinent, we are still faced with deciding how to interpret gradual levels of recovery. For example, suppose that, over a one-year follow-up, a client is found to have reduced the number of days of drinking more than a pint of whiskey per day from 100 to 5 and was incarcerated only one time as opposed to five times during the 12 months preceding treatment. The question, of course, is whether this change is sufficient for this individual to be considered a treatment success. Unfortunately, at the present time, such

questions cannot be answered unequivocally, but such considerations
are necessary in shaping the future technology and assessment of
alcohol treatment programs.

Other, similarly difficult problems must also be addressed in
the future. For example, should divorces categorically be viewed
as detrimental or beneficial, is a subject's use of therapeutic
supports after hospitalization indicative of lesser or better functioning as compared to subjects who do not make use of such supports, and how do increases or decreases in social-recreational
activities relate to reduced drinking? These are but a few of the
problems that must be considered once we have standardized our
evaluation measures and adequately developed our assessment technology.

CAN WE BELIEVE WHAT THEY TELL US?

This paper's final topic is the reliability and validity of
alcoholics' self-reports. Most descriptive, comparative and evaluative data published about alcoholics are derived from self-reports.
Furthermore, most pretreatment and outcome data, including those
from behaviorally-oriented alcohol treatment studies, have been obtained from subjects' and/or collaterals' self-reports. Unfortunately, despite the frequent and extensive use of self-reports in
the alcoholism field, only a few studies have actually investigated
the reliability and validity of such reports (see review by Sobell,
1976).

At the same time that behaviorally-oriented alcohol treatment
outcome studies began measuring and reporting drinking behavior by
criteria which could be interpreted as indicating varying degrees
of control over drinking, traditionalists began to seriously question the reliability and validity of these self-reports of drinking
behavior. Since the findings that some alcoholics can drink in a
non-problem manner directly contradict the traditional concept that
no alcoholic can ever safely drink again, it is not surprising that
the validity of the outcome results of the behaviorally-oriented
programs were called into question. It is somewhat ironic, however, that questions regarding the veracity of self-reported data
had seldom been raised with regard to literally hundreds of treatment outcome reports from traditionally-oriented programs previously.

Interestingly, the existing limited evidence on the reliability
and validity of self-reports suggests that self-reports by alcoholics, interviewed when they were sober, are surprisingly accurate and
acceptable for use as outcome data, at least as concerns alcohol-
related events which can be verified by either official record information or collateral interviews (Sobell, 1976; Sobell and Sobell,

1975; Sobell, Sobell and Samuels, 1974; Maisto, Sobell, and Sobell, Note 8; Sobell, Sobell and VanderSpek, Note 9). In like fashion, it has long been thought that subjects would represent their drinking behavior in a more favorable light (e.g., fewer drunk days; more abstinent days) than collaterals. However, a recent study (Maisto, Sobell and Sobell, Note 8) comparing the self-reports of collaterals and subjects over a six-month posttreatment interval found that there were few discrepancies in reported drinking behavior outcomes between subjects and their respective collaterals. When differences occurred, it was usually because subjects reported a more negative drinking history than their collaterals. Similar findings have been reported by Guze, Tuason, Stuart, and Picken (1963). Finally, in one of the first studies to investigate the reliability of alcoholics' self-reports of daily drinking disposition (Maisto, Sobell and Cooper, Note 10), it was found using a six-week test-retest interval that outpatient alcoholics gave highly reliable self-reports of daily drinking disposition occurring over the 12 months prior to entering treatment (test reliabilities ranged from .78 to .98, with nearly all in excess of .85).

Regardless of the outcome measures used, however, self-reports by subjects and their collaterals have been and probably will continue to be the primary source of data describing drinking behavior and adjunctive measures of life health. The major question relevant to treatment outcome evaluation, therefore, concerns whether or not follow-up data are valid. In the case of alcohol-related arrests, hospitalizations and numbers of days employed, it is possible to assess the validity of these self-reported data using external criteria (records). Moreover, whenever possible, external criteria (records and collateral reports) should be used to validate self-reported treatment outcome data. Recently, a few behaviorally-oriented alcohol treatment programs (Miller, 1976; Sobell, Sobell and VanderSpek, Note 9; Sobell, Sobell, Ward, Cooper, Cooper and Maisto, Note 11; Taub, Note 6) have started to verify daily drinking reports on a "probe day" basis by obtaining unannounced in-field breath tests of subjects' blood alcohol levels. Pragmatically, it is not possible to obtain or verify daily drinking behavior more than on a few select probe days using these kinds of procedures. Thus, at the present time there are no ways of easily validating daily drinking disposition with external criteria. In a related regard, some investigators (Miller, 1976; Sobell, 1978; Pomerleau et al., Note 1; Sobell et al., Note 11) have suggested or begun to use periodic liver function tests to assess recent (within three to four weeks of the interview) episodes of heavy drinking. Again, liver function tests can neither assess nor verify drinking behavior that has occurred over long periods of time. Still, even with probe breath tests and liver tests, at the present time, there is no one specific measure which, in and of itself, is totally adequate for validly assessing a subject's daily drinking behavior.

When we question the reliability or validity of self-reports, what we are really asking is whether or not the results obtained in the interview reflect behavior that is occurring in the natural environment. To deal with this issue in a new way, it is proposed that questions about the reliability and validity of self-reports can best be answered by a convergence of outcome indicators. Treatment outcome information can and should be derived from a variety of sources: (1) subjects' self-reports; (2) multiple collateral sources of information--friends, relatives, employers, neighbors; (3) in-field probe breath samples of subjects' blood alcohol levels; (4) official records to verify subjects' self-reports of incarcerations, employment, driving histories, disabilities, etc.; and (5) periodic liver function tests (SGOT, ICD, etc.) to assess recent episodes of heavy drinking. Applying this convergent validity approach, to the extent that a number of independent measures of outcome (e.g., breath tests, subject self-reports, liver function tests, collateral reports) are mutually corroborative, we can have increased confidence in the validity of each individual measure and of the overall outcome conclusions.

CONCLUDING COMMENTS

Throughout the course of this paper, I have emphasized that the emergence of new treatment outcome measures and techniques has been related to the development of behaviorally-oriented alcohol treatment programs. Such treatment programs have forced several departures from traditional ways of measuring and evaluating alcohol treatment effectiveness.

1) Treatment outcome should be evaluated in terms of degrees of improvement or recovery rather than in terms of the traditional dichotomous classification of abstinence vs. drunk. The available evidence suggests that it is more realistic to examine various aspects of change in life functioning over time and that immediate and absolute changes in functioning (e.g., total abstinence) rarely occur.

2) In order to adequately measure and interpret treatment outcome data, measures of subjects' functioning prior to treatment should be obtained. Pre-posttreatment comparisons allow for the evaluation of individual change.

3) More sensitive and quantifiable measures of drinking behavior have been developed such as daily drinking dispositions. Quite obviously, varying degrees of control over drinking can be assessed using this type of measure.

4) In the alcoholism field, follow-up has usually been conducted six months to five years after subjects have left

treatment programs. In this respect, the development of more sensitive and quantifiable measures of drinking behavior raised serious questions about the accuracy of subjects' recall for drinking behavior which occurred over long time intervals. In an attempt to alleviate this problem, more frequent follow-up contacts with subjects were instituted. Contacting subjects more frequently might also contribute to the successful tracking of a higher percentage of subjects than using traditional one-time follow-up contacts.

5) As a direct result of the development of daily drinking disposition measures and the finding that alcoholics exhibited varying types of drinking behavior outcomes, questions were raised regarding the validity and reliability of alcoholics' self-reports. Concomitant with finding non-problem drinking by alcoholics, several procedures have been developed to aid in evaluating the reliability and validity of alcoholics' self-reports. In an effort to gain more confidence in the validity of various outcome measures, it has been suggested that the degree of convergence among outcome indicators is likely to reflect the probable validity of outcome conclusions. At this time, no single specific measure would seem totally adequate to verify reports of daily drinking behavior.

6) Behaviorally-oriented alcohol treatment programs have evaluated drinking outcome within the context of the subject's total life health. A broad range of treatment outcome measures have been used to assess various areas of life health. In addition to multiple measures of life health, it has been suggested that we also look at patterns of change in behaviors over time. That is, how are changes in drinking behavior related to other areas of life health, and vice versa?

7) Since there is substantial evidence contradicting the presumption of a unitary phenomenon known as alcoholism, we should closely examine differences among different populations of alcohol abusers, and generalize findings from one study to another very judiciously.

8) While there is a need for comparative evaluation of treatment outcome studies, treatment outcome measures and methodologies first need to be standardized.

In conclusion, while behavioral approaches to the treatment of individuals with alcohol problems have been regarded as very successful, one of the most important and often overlooked contributions of these approaches has been the development and implementation of more comprehensive and sophisticated treatment outcome measures and techniques.

REFERENCE NOTES

1. Pomerleau, O. F., Pertschuk, M., Adkins, D. & Brady, J. P.
    Comparison of behavioral and traditional treatment for problem
    drinking. Paper presented at the Annual Meeting of the American
    Association for Behavior Therapy, New York, December, 1976.

2. Barr, H. L., Rosen, A., Antes, D. E. & Ottenberg, D. J. Two
    year follow-up study of 724 drug and alcohol addicts treated
    together in an abstinence therapeutic community. Paper presen-
    ted at the 81st Annual Convention of the American Psychological
    Association, August, 1973, Montreal, Canada.

3. Moos, R. & Bliss, F. Difficulty of follow-up and alcoholism
    treatment outcome. Unpublished manuscript, Stanford University.

4. Sobell, L. C. & Sobell, M. B. Validity of self-reports in
    three populations of alcoholics. Paper presented at the 22nd
    Annual Meeting of the Southeastern Psychological Association,
    New Orleans, LA, March, 1976.

5. Sobell, M. B. Alternatives to abstinence: Evidence, issues
    and some proposals. Paper presented at the NATO International
    Conference on Behavioral Approaches to Alcoholism, Bergen,
    Norway, August 28 - September 1, 1977.

6. Caddy, G. R. Individualized behavior therapy for alcoholics'
    third year follow-up: Advantages and disadvantages of conduct-
    ing double blind treatment outcome evaluation studies. Paper
    presented at the 23rd Annual Meeting of the Southeastern Psy-
    chological Association, Hollywood, FL, May, 1977.

7. Taub, E., Steiner, S. S., Smith, R. B., Weingarten, E. & Camp-
    bell, J. Use of transcendental meditation, EMG, biofeedback,
    electronic biofeedback and electronic neurotherapy as therapies
    for alcoholism. Paper presented at the NATO International
    Conference on Behavioral Approaches to Alcoholism, Bergen,
    Norway, August 28 - September 1, 1977.

8. Maisto, S. A., Sobell, L. C. & Sobell, M. B. Comparison of
    alcoholics' self-reports of drinking behavior with reports of
    collateral informants. Unpublished manuscript, Vanderbilt
    University.

9. Sobell, L. C., Sobell, M. B. & VanderSpek, R. Three independent
    comparisons between clinical judgment, self-report, and physio-
    logical measures of blood alcohol concentrations. Paper pre-
    sented at the Annual Meeting of the Southeastern Psychological
    Association, New Orleans, March, 1976.

10. Maisto, S. A., Sobell, L. C. & Cooper, A. M. Reliability of alcoholics' self-reports of drinking and related behaviors one year prior to treatment in an outpatient alcohol program. Paper presented at the 23rd Annual Meeting of the Southeastern Psychological Association, Hollywood, FL, May, 1977.

11. Sobell, M. B., Sobell, L. C., Ward, W., Cooper, A. M., Cooper, T. & Maisto, S. A. Treatment outcome evaluation in a clinical setting. Paper presented at the conference on Alcohol and Drug Treatment Outcome Evaluation, Nashville, TN, September, 1977.

REFERENCES

Armor, D. J., Polich, J. M. & Stambul, H. B. Alcoholism and treatment. Report #R-1739-NIAAA. Santa Monica, CA: Rand Corporation, 1976.

Belasco, J. A. The criterion question revisited. British Journal of the Addictions, 1971, 66, 39-44.

Bowen, W. T. & Androes, L. A follow-up study of 79 alcoholic patients: 1963-1965. Bulletin of the Menninger Clinic, 1968, 32, 26-34.

Crawford, J. J. & Chalupsky, A. B. The reported evaluation of alcoholism treatments, 1968-1971: A methodological review. Addictive Behaviors, 1977, 2, 63-74.

Crawford, J. J., Chalupsky, A. B. & Hurley, M. M. The evaluation of psychological approaches to alcoholism treatments: A methodological review. Final Report. AIR-96502-3/73-FR, American Institutes for Research, Palo Alto, CA, 1973.

Emrick, C. D. A review of psychologically oriented treatment of alcoholism. Quarterly Journal of Studies on Alcohol, 1974, 35, 523-549.

Gerard, D. L. & Saenger, G. Out-patient treatment of alcoholism. Toronto: University of Toronto Press, 1966.

Guze, S. B., Tuason, V. B., Stewart, M. A. & Pickens, B. The drinking history: A comparison of reports by subjects and their relatives. Quarterly Journal of Studies on Alcohol, 1963, 24, 249-260.

Hill, M. J. & Blane, H. T. Evaluation of psychotherapy with alcoholics: A critical review. Quarterly Journal of Studies on Alcohol, 1967, 28, 76-104.

Hunt, G. M. & Azrin, N. H.   A community-reinforcement approach to alcoholism.   Behaviour Research and Therapy, 1973, 11, 91-104.

Lowe, W. C. & Thomas, S. D.   Assessing alcoholism treatment effectiveness: A comparison of three evaluative measures.   Journal of Studies on Alcohol, 1976, 37, 883-889.

Madsen, W.   The American alcoholic.   Springfield, IL: Charles C. Thomas, 1974.

Miller, B. A., Pokorny, A. D., Valles, J. & Cleveland, S. E.   Biased sampling in alcoholism treatment research.   Quarterly Journal of Studies on Alcohol, 1970, 31, 97-107.

Miller, P. M.   A comprehensive behavioral approach to the treatment of alcoholism.   In R. E. Tarter & A. A. Sugarman (Eds.), Alcoholism: Interdisciplinary approaches to an enduring problem.   Reading, MA: Addison-Wesley, 1976.

Orford, J., Oppenheimer, E. & Edwards, G.   Abstinence or control: The outcome for excessive drinkers two years after consultation Behaviour Research and Therapy, 1976, 14, 409-418.

Pattison, E. M., Coe, R. & Doerr, H. O.   Population variation among alcoholism treatment facilities.   International Journal of the Addictions, 1969, 8, 199-229.

Pattison, E. M., Headley, E. B., Gleser, G. C. & Gottschalk, L. A.   Abstinence and normal drinking: An assessment of changes in drinking patterns in alcoholics after treatment.   Quarterly Journal of Studies on Alcohol, 1968, 29, 610-633.

Popham, R. E. & Schmidt, W.   Some factors affecting the likelihood of moderate drinking in treated alcoholics.   Journal of Studies on Alcohol, 1976, 37, 868-882.

Sobell, L. C.   The validity of self-reports: Toward a predictive model.   Unpublished doctoral dissertation.   University of California, Irvine, 1976.

Sobell, L. C.   A critique of alcoholism treatment evaluation.   In G. A. Marlatt & P. E. Nathan (Eds.), Behavioral assessment and treatment of alcoholism.   New Brunswick, NJ: Rutgers Center of Alcohol Studies, 1978.

Sobell, M. B. & Sobell, L. C.   Individualized behavior therapy for alcoholics: Rationale, procedures, preliminary results and appendix.   California Mental Health Research Monograph, No. 13.   Sacramento, CA, 1972.

Sobell, M. B. & Sobell, L. C. Individualized behavior therapy for alcoholics. Behavior Therapy, 1973, 4, 49-72 (a).

Sobell, M. B. & Sobell, L. C. Alcoholics treated by individualized behavior therapy: One year treatment outcome. Behaviour Research and Therapy, 1973, 11, 599-618 (b).

Sobell, M. B. & Sobell, L. C. The need for realism, relevance and operational assumptions in the study of substance dependence. In H. D. Cappell & A. E. LeBlanc (Eds.), Biological and behavioral approaches to drug dependence. Toronto: Addiction Research Foundation, 1975.

Sobell, M. B. & Sobell, L. C. Second year treatment outcome of alcoholics treated by individualized behavior therapy: Results. Behaviour Research and Therapy, 1976, 14, 195-215.

Sobell, M. B. & Sobell, L. C. Behavioral treatment of alcohol problems: Individualized therapy and controlled drinking. New York: Plenum Press, in press.

Sobell, M. B., Sobell, L. C. & Samuels, F. H. The validity of self-reports of prior alcohol-related arrests by alcoholics. Quarterly Journal of Studies on Alcohol, 1974, 35, 276-280.

Voegtlin, W. L. & Lemere, F. The treatment of alcohol addiction: A review of the literature. Quarterly Journal of Studies on Alcohol, 1942, 2, 717-803.

Vogler, R. E., Compton, J. V. & Weissbach, T. A. Integrated behavior change techniques for alcoholics. Journal of Consulting and Clinical Psychology, 1975, 43, 233-243.

Vogler, R. E., Weissbach, T. A. & Compton, J. V. Learning techniques for alcohol abuse. Behaviour Research and Therapy, 1977, 15, 31-38.

Vogler, R. E., Weissbach, T. A., Compton, J. V. & Martin, G. T. Integrated behavior change techniques for problem drinkers in the community. Journal of Consulting and Clinical Psychology, 1977, 45, 267-279.

CRAVING FOR ALCOHOL, LOSS OF CONTROL, AND RELAPSE: A COGNITIVE-BEHAVIORAL ANALYSIS

G. Alan Marlatt

University of Washington

The purpose of this paper is to provide a critical review of
the relapse process as traditionally defined within the medical or
"disease" model of alcoholism. In the traditional approach, alcoholism is viewed as an addiction, and relapse is defined as the sequence of events leading to readdiction following a period of
abstinence from alcohol use. This is the common usage of the term
in medical parlance, and is reflected in the following definition
of relapse taken from Webster's New Collegiate Dictionary: "A
recurrence of symptoms of a disease after a period of improvement."
This same dictionary defines addiction as "compulsive physiological
need for a habit-forming drug." Consistent with this emphasis on
addiction as a physiological need, proponents of the medical model
frequently attribute an alcoholic's relapse to internal symptoms
such as physical craving for alcohol or an involuntary, compulsive
loss of control over drinking. Falling off the wagon after a period
of abstinence is thus taken as a pathognomonic symptom of alcoholism.

By its very nature, the designation of alcoholism as a disease
downplays the importance of environmental and psychological factors.
Many questions go unanswered because they are rarely even asked by
those who subscribe to the disease model, with its exclusive emphasis on internal, physiological determinants. Are there specific
environmental events which serve as triggers for relapse? Do the
alcoholic's emotions, moods and feelings affect the probability
that he or she will resume drinking following a period of sobriety?
Are the determinants of the first drink the same as those assumed
to govern subsequent drinking in the relapse process? How does the
alcoholic react to and conceptualize the events leading to and
following a relapse, and how do these reactions affect the individual's later drinking behavior? Is it possible to prepare alcoholics

271

in treatment to anticipate the likelihood of a relapse, so that
they may engage in preventive alternative behaviors? To borrow a
term from the medical model, can we develop treatment procedures
which would "inoculate" the alcoholic against the inevitability of
relapse? And the most controversial question of all: Can we teach
alcoholics to control or moderate their drinking, thus preventing
the readdiction usually associated with relapse?

I hope to provide answers to some of these questions and to
raise other related issues throughout this chapter. The plan of
the chapter is as follows. I will first present a case study of
a relapse seen through the eyes of one of our alcoholic clients.
The sequence of events in this case study will serve as a convenient example to illustrate the theoretical and experimental literature to be reviewed in later sections. Highlights of the medical
model of relapse will then be presented, with particular attention
devoted to the key concepts of craving and loss of control. Following a critical review of this traditional approach, an alternative
explanation of the relapse process will be offered, drawing upon
social learning theory and recent behavioral investigations of alcoholism and social drinking. Finally, the implications of this
alternative analysis for the prevention and treatment of the alcoholic relapse will be discussed. Until a better word becomes
available, I shall use the term "relapse" to refer to the resumption
of any drinking behavior (including a single drink or "slip") following a period of voluntary abstinence.

THE ALCOHOLIC RELAPSE: A CASE STUDY

The following case study concerns a young woman whom we shall
call Liane. Liane was a recent client in our addictive behaviors
treatment program at the Center for Psychological Services at the
University of Washington. She first came to my attention following
a lecture I had given on alcoholism. During our initial meeting,
Liane admitted that she was concerned about her own drinking, and
asked to be evaluated as a possible candidate for our treatment
program. A thirty-year old single woman who lived in a self-con-
tained apartment in her parent's home in Seattle, Liane was a senior
undergraduate student majoring in psychology at the university. In
the months prior to our first meeting, Liane's drinking began to
interfere with her studies, and she was doing poorly in her exams.
Rather than devoting time to her assignments, Liane would spend most
of her evenings alone, sipping vodka and watching television in her
apartment. She did not have a steady boyfriend, and reported that
most of her time was spent either alone or with her parents. She
claimed that she could not relate comfortably to most people unless
she had been drinking. A two-week daily self-monitoring report of
her alcohol intake revealed that she was consuming almost a full
quart of vodka each day (drinking mostly at night), along with a

variety of tranquilizers and barbiturates. Her rate of drinking had steadily increased over the past several years, and it was clear that most professionals would diagnose her as alcoholic. When I confronted her with the seriousness of her condition, she agreed to undergo a period of voluntary hospitalization to provide medical supervision for her detoxification from both alcohol and the other drugs she had been taking. Before she was released from the hospital, she committed herself to complete abstinence from alcohol for a period of at least one year. We agreed to provide a program of supportive therapy for her during this year. All went well during the first weeks of the program, and she maintained total abstinence during this initial period. On the 58th day following her discharge from the hospital, however, Liane experienced a "slip" and consumed one mixed drink during a luncheon date with a female friend. Although she did not take another drink on that day, and maintained abstinence again for the next twenty days, she consumed an additional single drink, again during a luncheon date with the same friend, on the 78th day. Finally, on the 81st day following her discharge, Liane drank to the point of intoxication during a weekend evening alone in her apartment. The course of her drinking during the first 150 days after her hospitalization is displayed in Figure 1. An examination of this figure reveals the overall pattern of a typical relapse. Following her initial intoxication experience on the 81st day, Liane again refrained from any drinking for about two weeks, followed by another occasion of "social drinking" (limiting herself to no more than two drinks), and shortly thereafter, another bout of intoxication. With the start of university classes in the fall, her drinking began to increase until it almost equalled her pretreatment rate of consumption. At the time of this writing, Liane has shown considerable improvement and is continuing treatment in our program.

For the present purpose, I would like to examine the events associated with her first two drinks (the "slips" on the 58th and 78th days), and her first intoxication (81st day). When these events occurred, I asked Liane to provide a detailed written account of everything she could remember: The events preceding the drinking, her feelings at the time, her reactions, and any other important details which stood out in her mind. As these reports were written within a few days of the actual events, I feel reasonably confident about their accuracy. The following excerpts are taken directly from her written account. The luncheon date reported for July 17 was the first social event of any importance Liane had participated in since her release from the hospital. Sharon, Liane's luncheon companion, was a woman who lived in the same neighborhood, although they were not close friends at the time.

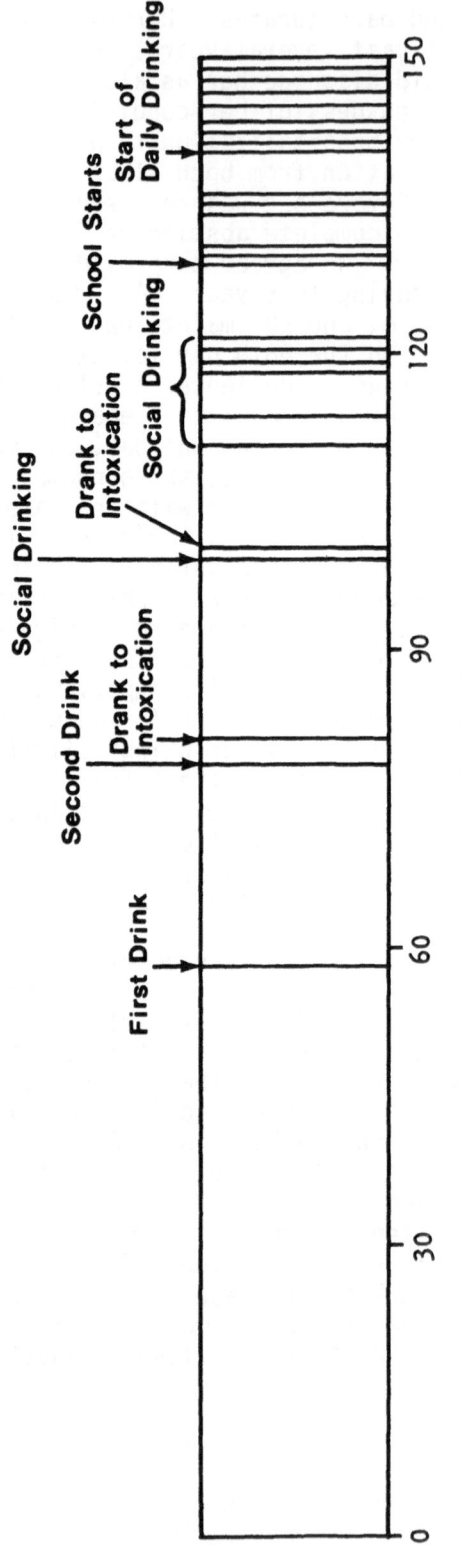

Figure 1: Course of Liane's drinking from the beginning of treatment.

Thursday, July 17, 58th day: The First Drink

Lunch with neighbor Sharon. Sharon's husband was in Detroit the week of July 17th because of the death of his nephew. I know Sharon and her husband casually. When I asked her to lunch, I told her that I thought it would be nice to wander through the shops in Edmonds, and then have lunch. I had no intention of having a drink, nor did I think about how I'd handle the problem, particularly since it was I who was paying for the lunch. This was, however, the first time I had invited anyone to dine with me as my guest since I had gotten out of the hospital, and more importantly, Sharon knew nothing about my problem drinking. I picked her up at 10:30 a.m., and we went to Edmonds. Both of us were dressed up, and I felt very at ease, cheerful and confident. I had tentatively planned to eat in Edmonds, but Sharon suggested that she enjoyed eating at a restaurant in Aurora Village, so I agreed that we go there. Throughout the entire time we were together, there was pleasant and interesting conversation. When we were seated, the waitress came by and asked if we wanted cocktails. Since I was taking her to lunch, I said, "Sharon, would you like a cocktail?" She ordered a scotch and soda, and then the waitress looked at me and said, "And you?" "Gin and tonic, please," just came out. I didn't plan it and I wasn't prepared for it.

The first sip was so strong, it literally shocked me. It was difficult for me to finish the drink. I am not entirely sure whether or not I imagined some of the physical effects, perhaps because of guilt, or if it was real. Sharon finished her drink before the lunch even came. I was just finishing when we were ready to leave, which was approximately 45 minutes from start to finish. I got a headache, my nose got stuffy, my thoughts were tangled and my voice and actions were abrasive...There were several things that came from this that did help me. One thing was that committing myself to one year of abstinence from alcohol, after an inebriated state of some duration, was a difficult reality to keep after a month or so because there is no reality in inebriation. Therefore, I was especially vulnerable to the unreality of alcoholic romanticism used in advertising, television, and movies. This sounds trivial, but this romanticism is a stimulus for envy and anxiety in me. Alcohol consumption is so intermeshed in the structure of society that it is very difficult not to feel the influence. For me, what was going through my mind was that a social drink at parties, lunches, dinners, etc., is condoned and exploited, making me feel extremely self-conscious for not participating. But, when one gets into the predicament of drug and alcohol abuse as myself, society doesn't fail to ostracize and condemn.

Twenty days later, on August 6, Liane again invited Sharon out for lunch, and once again she consumed a single mixed drink. Unlike the first occasion, however, Liane considered the possibility of having a drink in advance of the luncheon. The experience of July 17 had brought about a shift in her attitudes toward drinking, as she observes in the following passage:

I can't deny that the first situation was spontaneously reinforcing, greater than I had realized. I would only be kidding myself if I were to say that repeating the same scene, i.e., having one drink at lunch, was not planned; at least I had no real intentions of not having that drink... Having this drink was more of a pain than a pleasure. I didn't feel good about having that drink. It lowered my self-esteem and did create some guilty and remorseful feelings.

Three days later, Liane consumed enough alcohol to become in-toxicated. It was a Saturday, and she had no plans for the evening. Her parents were going out that night.

Saturday, August 9, 81st day: The First Intoxication

I got up early, and I felt content and at ease. I went down to the den and played the organ for several hours. I then put some records on which I hadn't done since I came home from the hospital. Unfortunately, as I recollect now, the records started my change in mood. It never occurred to me that the music conditioned specific moods in me without my being aware of it. So, as I played some records and sang along, my mood became melancholy and I began to feel sorry for myself and ill at ease. The point at which I noticed my "free floating anxiety" came around 4:00 in the afternoon. I have my little fantasies. This is not easy to write, and I don't know how to write it without sounding like I belong in a mental institution. My fantasy is to take a real person (male) whom I don't know personally, but through TV, etc., and, well, to imagine that we are together. So, suddenly, about 4:00, his face just popped into my mind. This really affected my moods, because, well, I was alone in the house, it was Saturday, and I knew I was going to be alone for the evening. I had no plans for doing anything that night, and I began to feel sad about being alone. After a while, I began to feel more and more anxious about this. At first, I thought I would feel better if I could take a tranquilizer, so I searched all over the house for the pills my mother had hidden. I couldn't find them, which made me all the more upset. What I did find, though, was two bottles of airline scotch, miniature bottles like you buy on the plane, hidden in a teapot in the kitchen. I also found three bottles of beer in the fridge upstairs. At

first, I tried to ignore them, but I couldn't. Finally, I
took out the two bottles of scotch and put them on the table.
I just looked at them for a long time, without even touching
them. After about half an hour, I finally opened one of the
bottles and poured it into a glass with some ice. I sipped
the drink but didn't feel anything at first, so I poured in
the other one and drank that too. After that, I drank the
bottles of beer in the fridge and then I don't remember what
happened. I woke up later that night with a hangover, and I
felt terrible about what I had done.

I would like to discuss the implications of this case in terms of
several different theoretical models of the relapse process. To
begin, let us examine the major assumptions of the traditional
medical approach.

CRAVING AND LOSS OF CONTROL AS DETERMINANTS OF RELAPSE:

THE MEDICAL MODEL

In 1954, the World Health Organization sponsored a joint meeting of their Expert Committees on Mental Health and on Alcohol to
discuss the physical basis of dependence on alcohol. Craving for
alcohol was one of the main topics of discussion. In one of the
position papers presented at this meeting, Isbell (1955) distinguished between two kinds of craving:

> Two kinds of "craving" are postulated. The first is a
> "physical" or nonsymbolic craving which occurs in persons who
> have been drinking excessive amounts of alcohol for long per-
> iods of time, and is manifested by symptoms on withdrawal of
> alcohol. This type of craving is believed to be due to physio-
> logical alterations, the mechanisms of which is not yet under-
> stood...The second kind (or kinds) of craving is thought to
> account for initial abuse of alcohol, and for relapse after
> abstinence. It is postulated that this second kind of craving
> is chiefly psychological in origin. (Isbell, 1955, p. 42;
> italics added.)

In another paper in the W. H. O. report, Mardones (1955) defines
craving for alcohol as "an urgent and overpowering desire to drink
alcoholic beverages...The urgency of the desire is a passive condition that is perceived by the individual, and 'overpowering' means
inducing an active attitude directed to surmount the obstacles op-
posing the desire" (pp. 51-52).

As Isbell noted, physical craving is most often defined as the
desire for alcohol experienced by a drinker undergoing the symptoms
of withdrawal following a prolonged drinking bout. The assumption

here is that drinking alcohol will relieve the unpleasant and often
painful physical sensations brought about by withdrawal: Drinking
"the hair of the dog which bit me," the saying goes. Physical craving defined in this manner resembles the notion of craving for
heroin experienced by the addict undergoing withdrawal.

Many authorities, however, have extended the concept of craving
to account for the circumstances leading to relapse in the abstinent
alcoholic, as in Isbell's second type of craving described above.
Relapse is thus precipitated by a subjectively experienced intense
desire or need for alcohol, which somehow "overpowers" the volitiona
control of the addicted drinker. Although Isbell described this
second kind of craving as psychological in origin, the motivational
emphasis is on the internal, physical "pull" of craving, with little
if any attention paid to the environmental or cognitive factors that
might be involved.

What happens after craving is responded to, and the first drink
is consumed? According to Jellinek (1960, pp. 145 ff.), one of the
leading proponents of the disease concept of alcoholism, the first
drink acts as a signal activating "the metabolism of nervous tissue
cells," and triggers off a strong desire for additional alcohol.
This desire leads to the compulsive ingestion of more alcohol, usually referred to as <u>loss of control drinking</u>. As Jellinek himself
first put it:

> Loss of control means that as soon as a small quantity of alco-
> hol enters the organism, a demand for more alcohol is set up
> which is felt as a physical demand by the drinker...the "loss
> of control" is effective after the individual has started
> drinking, but it does not give rise to the beginning of a new
> drinking bout. (Jellinek, 1952, p. 679.)

This compulsive desire to drink overrides any cognitive or volitional control; it is the complete <u>loss</u> of voluntary control:

> Recovered alcoholics in Alcoholics Anonymous speak of "loss of
> control" to denote that stage in the development of their
> drinking history when the ingestion of one alcoholic drink
> sets up a chain reaction so that they are unable to adhere to
> their intention to "have one or two drinks only" but continue
> to ingest more and more--often with quite some difficulty and
> disgust--contrary to their volition. (Jellinek, 1960, p. 41.)

Jellinek held the position that craving and loss of control
were both key symptoms in "gamma" alcoholism, considered by him to
be the predominant form of alcoholism in North America. This view
has been widely accepted by other proponents of the medical model
(see review by Maisto and Schefft, in press). Marconi, Fink, and
Moya (1967) define loss of control as a pathognomonic symptom, "the

most important clinical manifestation which usually accompanies intermittent (gamma) alcoholism and also the differential diagnosis between intermittent alcoholism and the unpathological forms of alcohol ingestion" (p. 544).

Mark Keller, Editor-Emeritus of the Journal of Studies on Alcohol, has taken issue with Jellinek's notion that loss of control drinking occurs only after the alcoholic has consumed his first drink. In his 1972 paper, Keller described a second form of loss of control which he believes is the essence of alcohol addiction. Acknowledging that an alcoholic may on occasion be able to refrain from further drinking after having the first drink or two, Keller goes on to say that, sooner or later, he will lose control over his drinking: "...If an alcoholic takes a drink, he can never be sure he will be able to stop before he loses control and starts on a bout" (p. 160). Furthermore, the alcoholic, according to Keller's reformulation of the loss of control construct, has little or no control over whether he or she will take the first drink: "...An alcoholic does not always have the choice of whether he will take the first drink or not, or whether he will start to drink, be it with or without the rationalization that it is to be just one drink or two. For that is precisely the nature and the essence of the addiction" (1972, p. 160).

What, then, are the factors determining the consumption of the first drink? Keller believes that certain environmental cues or stimuli precipitate a relapse:

> An addict may sometimes go about rationally debating the question, to drink or not to drink--sometimes. But not always, not consistently. His disease consists precisely in this, that at some time, under the impulsion of some cue or stimulus which may well be outside his conscious awareness, he will drink...The essence of the addiction is that, when the significant cue or signal impinges upon him, though he is unconscious of it, or conscious of it but unaware of its significance, he will reach out to drink. I don't want to be mystical about these fatal stimuli or signals that enforce drinking on alcohol addicts...it would go too far afield to discuss that in detail. I will therefore say here only this much: Addiction, in this conception, is thought of as a form of learned or conditioned response. (1972, p. 160.)

From this perspective, it would appear that loss of control is a kind of conditioned or "automatic" response, elicited by conditioned stimuli in the environment. Through the process of stimulus generalization, Keller believes further that almost any stimulus, including alcohol itself, can serve as a stimulus for drinking:

> For any alcoholic there may be several or a whole battery of critical cues or signals. By the rule of generalization, any

critical cue can spread like the tentacles of a vine over a
whole range of analogs, and this may account for the growing
frequency of bouts, or for the development of a pattern of
continuous inebriation. An exaggerated example is the man who
goes out and gets drunk every time his mother-in-law gives him
a certain wall-eyed look. After a while he has to get drunk
whenever any woman gives him that look. And after a while,
whenever any woman gives him any kind of look. With a bow to
Woman's Lib., the sex roles may just as well be reversed.
But perhaps the most important element in the spread of the
addictive conditioning is that alcohol itself may become the
significant cue; or it may be a particular blood alcohol level
(1972, p. 161.)

Keller is careful to note that, in accordance with his definition,
loss of control drinking is an inconsistent symptom. Sometimes it
happens, and sometimes it fails to occur:

...sometimes, the loss-of-control symptom happens to be in remission. Sometimes it is in remission as a response to treatment...So I have to make a special point of insisting that the
loss-of-control symptom in alcoholism, like other specific
symptoms in other diseases, appears only inconsistently.
Sometimes it is not operating, and at such times an alcoholic
may drink without tumbling into a bout. Now that I mention
it--haven't we all known it all the time? (1972, p. 163.)

While Keller does not evoke the concept of craving as a determinant of loss of control, his conditioning model emphasizes the
involuntary, almost automatic, nature of uncontrolled drinking as a
response to a wide variety of conditioned stimuli. This classical
conditioning approach is extended and elaborated by Ludwig and his
colleagues, who reintroduce craving as a mediating determinant of
relapse. Because of the complexity of Ludwig's approach, it is
worth examining his position in some detail.

Craving as a Determinant of Relapse

Ludwig's early studies were based on questionnaire data. The
first paper (Ludwig, 1972) reported the results of a questionnaire
which was administered to 176 male patients who had completed an
inpatient treatment program for alcoholism. The 161 patients who
had relapsed were asked to give their own reasons why they took
their first drink after discharge from the hospital. A majority of
the sample gave a variety of environmental and emotional reasons
whey they had "fallen off the wagon," including such factors as
psychological distress (25%), family problems (13%), sociability
(10%), and employment problems (5%). Ludwig notes that, "Surprisingly, only the smallest percentage of patients (1%) offer reasons

akin to the subjective feeling of craving as the cause of their relapse" (p. 94). Although it is not specified in the paper, it appears that the patients were asked to provide their own reasons for
relapse, in their own words, and that these responses were later
classified into categories by the investigator.

In a second study, Ludwig and Stark (1974) administered a
Drinking and Craving Questionnaire (DCQ) to 60 male alcoholics in
an inpatient treatment setting. This time the questionnaire contained structured questions which the subjects were asked to respond to by filling out a rating scale. Among the items on the DCQ,
the following questions pertained directly to craving: "When you
are drinking steadily, to what extent does it feel like your body
'craves' or needs alcohol?" Phrased in this form, 70% checked the
rating "very much" and 15% checked "a fair amount." In another section of the DCQ it was found that, "Although only 78% of the alcoholics reported having experienced 'craving,' 98.4% (all but one)
responded when asked to 'define craving and what it means to you'"
(p. 902.) When the subjects were asked to define craving in their
own words, 57.5% described it in terms of a need or desire for alcohol (e.g., "Have to have it"), and 42.5% defined craving as a desire for the effects of alcohol (e.g., "Don't crave the taste but
the feeling"). Although there were no items on the DCQ which specifically asked if the subjects experienced craving prior to the
initiation of drinking (as opposed to craving during a drinking
bout or session), the authors conclude that "regardless of the particular symbolism alcoholics employ to describe the experience of
craving, the common denominator of this experience is that it provides them with an acceptable excuse to resume or continue drinking" (p. 904). The data presented, however, fail to support the
notion that craving may be a determinant of relapse, in contrast
to craving as a desire for more alcohol once drinking has begun.
In comparison with the results of the first paper (Ludwig, 1972),
the second study shows that 78% of the sample will report having
experienced craving--if one asks the right questions. Curiously,
there is no reference to the earlier paper in Ludwig and Stark's
1974 publication.

In 1974, Ludwig and Wikler published a paper entitled,
"'Craving' and Relapse to Drink." Although no supporting data are
offered, the authors describe a theoretical model which attempts
to show that "craving and loss of control represent crucial determinants of relapse and excessive drinking behavior" (p. 113).
As if in response to the data presented in the first study (Ludwig,
1972), the authors state conclusively that:

> Because alcoholics do not spontaneously report craving or be-
> cause they offer some other reason for drinking does not nec-
> essarily mean that they are not experiencing craving or that
> craving is not an important determinant in the initiation and

perpetuation of drinking...On further probing, experimenters
and clinicians will likely find an intense desire, compulsion
or drive for alcohol consumption beyond the veil of the glib
labels or reasons supplied by alcoholics for their drinking.
(1974, p. 120.)

In a direct analogy with narcotic craving, Ludwig and Wikler consider "craving for alcohol, comparable to craving for narcotics, as
representing the psychological or cognitive correlates of a 'subclinical' conditioned withdrawal syndrome" (p. 114). Based on
Wikler's earlier conditioning theory of the opiate withdrawal process, the authors claim that the craving for alcohol which is
assumed to occur during an alcoholic's withdrawal from the drug can
become classically conditioned to stimuli (e.g., "physical environment, drug-using or drug-dispensing associates, certain emotional
states," p. 114) which are contiguous with the withdrawal experience. If an abstinent alcoholic is exposed to these same conditioned stimuli at a later date, he will experience symptoms of
craving (conditioned responses) which may predispose him to relapse
In this sense, craving is similar to a "mini-withdrawal"--a "subclinical conditioned withdrawal syndrome" as they put it. "The
more frequent and severe the prior withdrawal experiences, the
greater the predisposition to conditioned withdrawal symptoms with
consequent desire for relief (i.e., 'craving') through drink" (p.
115).

The authors claim that craving can become conditioned to a
wide variety of interoceptive or exteroceptive stimuli. In this
stage of their theory, however, they go far beyond limiting these
conditioned stimuli to those directly associated with withdrawal.
Stimuli become capable of eliciting craving even if they are associated with drinking prior to withdrawal:

As for exteroceptive conditioning, this would pertain to a
variety of situations associated with prior heavy drinking or
with the uncomfortable psychological and physical effects of
prior withdrawal experiences. A conditioned withdrawal syndrome, generally subclinical in nature, with associated
craving might result, therefore, whenever the alcoholic passed
a bar or was in the presence of other people drinking or encountered cues relevant to previous drinking practices. (1974,
p. 116, italics added.)

As with Keller's discussion of the stimulus generalization effect
embracing more and more stimuli which may elicit drinking, the present analysis suggests that any cue, present during either drinking
or withdrawal, can become a conditioned stimulus for craving.

How does the experience of craving relate to actual drinking
behavior? Ludwig and Wikler state that the relation between craving

and drinking is not an invariant one. Rather, craving is viewed as a necessary but not sufficient determinant of relapse: "While the experience of craving provides an alcoholic with the necessary cognitive symbolism for goal-directed, appetitive behavior (i.e., the negative reinforcement provided by alcohol), there is no cogent reason (as with anger, hunger or sexual urge) why this subjective desire for alcohol should be directly acted upon or expressed in overt behavior, especially if there are competing drives or motivations" (p. 116). The authors describe a number of "modifiers" of craving, including the influence of setting and other situational factors, the availability of alcohol, and the ability of the alcoholic to correctly label his physiological state of arousal as craving for alcohol. Under the proper set of circumstances, however, craving will eventuate in drinking and associated loss of control. Loss of control is described as the alcoholic's relative inability to regulate the consumption of alcohol:

> Craving is the cognitive state of designating ethanol consumption as a source of relief or pleasure; it need not inevitably lead to drinking. Loss of control is the behavioral state initiated by craving and characterized by activities indicative of a relative inability to modulate ethanol consumption; it need not eventuate in gross intoxication or stupor. It may even take the form of total abstinence, when an alcoholic has incorporated the belief that he must remain abstinent because he cannot handle alcohol. (1974, p. 122.)

Consumption of the first drink acts to further increase craving for alcohol, thus increasing the probability that loss of control will occur: "This first drink, then, would act like an 'appetizer,' stimulating hunger (craving, as a conditioned withdrawal response) even further because it has become sequentially conditioned to the later consumption of an 'entree' (intoxication)" (p. 128).

In an empirical test of their theory of craving as it relates to relapse, Ludwig, Wikler and Stark (1974) describe a study in which 24 detoxified male alcoholics were administered either a high or low dose of alcohol, under different labeling conditions, to determine the effects on reported craving and alcohol acquisition behavior. Half the subjects received their drinks in an appropriate labeling context (subjects consumed their preferred drink in the presence of alcohol-related cues), and half were assigned to the "nonlabel" condition (subjects consumed ethyl alcohol with an artificially sweetened mixer in the absence of alcohol cues). Subjects were also assigned to one of three alcohol dose conditions: high dose (1.2 ml/kg of body weight), low dose (0.6 ml/kg), or a placebo group (a small amount of alcohol floating on top of the mixer). A variety of subjective, behavioral, and physiological measures was obtained at four time intervals following administration of the drinks (ranging from 20 to 200 minutes later). The

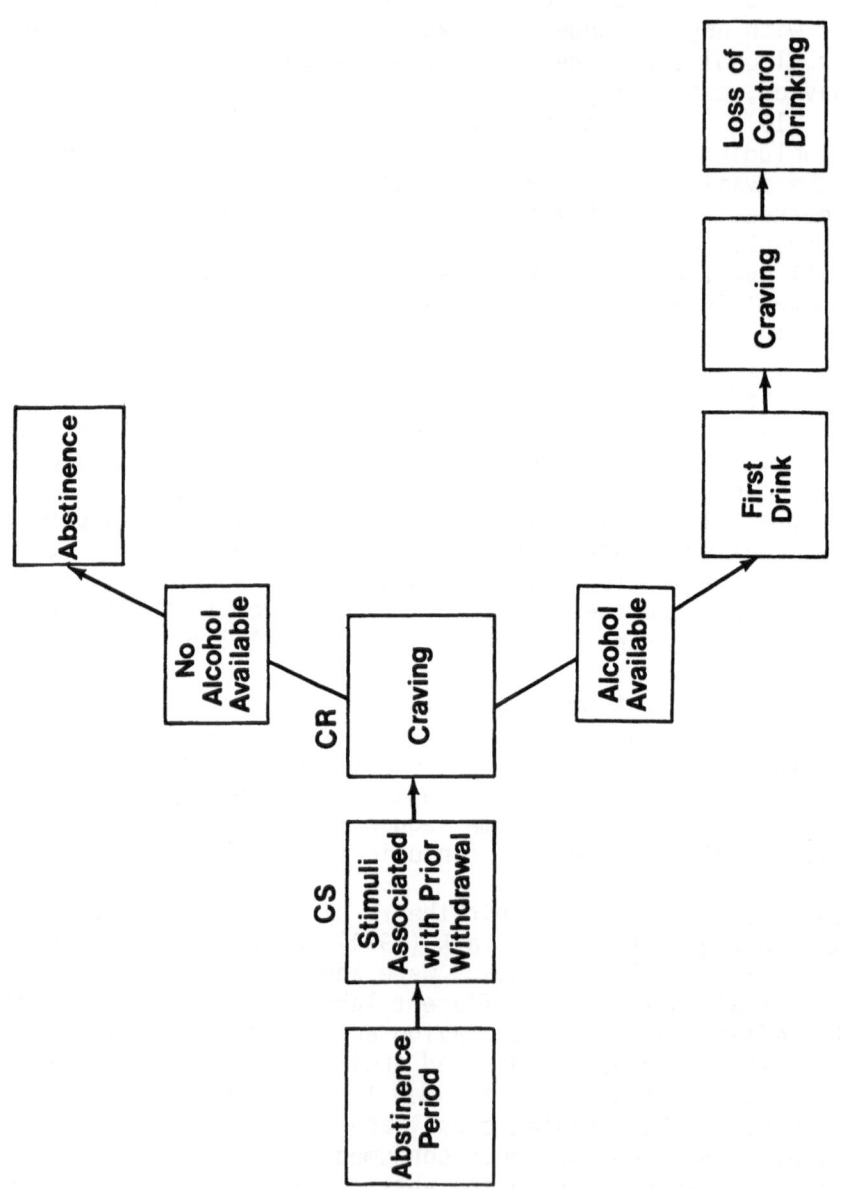

Figure 2: Ludwig and Wikler's analysis of relapse.

authors hypothesized that the low dose of alcohol should produce a greater craving for alcohol than the high dose, because of the "appetizer" effect associated with the lower dose. Presumably, the high dose would "satisfy" the craving, and the low dose would stimulate craving for more alcohol. It was also predicted that craving and associated behaviors would be highest in the appropriate labeling condition. Subjective reports of craving were assessed by means of a "craving meter," a device that permitted the subject to indicate his degree of craving on a scale ranging from 0 to 100. A behavioral measure of alcohol acquisition was also included, in which the subject could work for alcohol on a button-pressing device. The number of button presses for alcohol was taken as the main measure of loss of control (although the alcohol could not be consumed until the end of the experimental session).

The results of the study are extremely difficult to interpret because of numerous problems in the design and analysis of the data. The groups were not independent, as all subjects received all three alcohol doses in different sessions. Because of the small number of subjects and the "highly skewed distribution and wide range of responses for craving and work measures" (p. 542), nonparametric statistics were used to analyze the data and "$\underline{P}$ values of .10 or less represented the criteria for statistical significance" (p. 542). In the absence of specific instructions to the subjects, the use of the craving meter may have biased the results because of the experimental demand characteristics involved.

The results as reported fail to support the authors' hypotheses. At the first assessment of craving, 20 minutes following the administration of the drink, the highest level of craving was reported by the placebo group in the appropriate labeling condition. Although the low dose group (appropriate labeling) showed somewhat higher levels of craving during the later assessment periods (140 and 200 minutes following consumption of drinks), it did not differ significantly from the high dose group (appropriate labeling) at any time period. In all beverage groups, the appropriate labeling condition led to significantly higher levels of craving than the nonlabel condition--a finding which would seem to support the role of cognitive expectancy factors rather than the physically "appetizing" effects of a low dose of alcohol. A very similar pattern of results was obtained with the operant measure of alcohol acquisition on the button-pressing task. Again, there was no significant difference in responses for alcohol for the low and high alcohol dosage groups in the appropriate labeling conditions.

In an attempt to rescue their theoretical model from the morass of conflicting data reported in the paper, Ludwig et al. introduce additional hypothetical constructs into the hopper. First, they propose that the extent of craving is limited by a "hypothetical blood alcohol ceiling"--a kind of upper limit of intoxication

that differs from individual to individual. Second, the setting in
which the first drink is consumed and the mental set of the alcoholic may impose additional "cognitive ceilings" against the expression of craving:

> ...we propose that within a physical setting or exteroceptive
> situation conducive to natural cognitive labeling the degree
> of craving should be a direct function of the difference be-
> tween the actual blood alcohol level attained by the first
> drink and the hypothetical blood alcohol ceiling for each al-
> coholic. In the presence of inappropriate or incongruous set-
> tings, not conducive to natural cognitive labeling, the magni-
> tude of craving should be substantially less than under all
> appropriate exteroceptive conditions, regardless of actual
> alcohol dose. This is largely because the hypothetical ceiling
> for desired blood alcohol levels will more likely be determined
> by individualized mental constraints (i.e., fluctuating "cogni-
> tive ceilings") against the expression of craving and alcohol
> acquisition in incongruous settings. (pp. 545-546.)

According to Ludwig and his co-workers, the concept of craving
is paramount to the understanding of the relapse process. It would
appear to be a very global and all-encompassing phenomenon indeed,
since it can be elicited by a wide variety of conditioned stimuli,
ranging from cues associated with drinking to those associated with
withdrawal from alcohol. The conditioning can also occur with both
interoceptive and exteroceptive stimuli, including almost any state
of physiological arousal: "...since anxiety, nervousness, jitteri-
ness, and other types of emotional dysphoria may produce such
physiological responses as increased heart rate and respiration,
tremulousness, autonomic lability, increased sweating, insomnia,
all of which represent changes associated with the alcohol withdrawal syndrome, we should anticipate that these states, induced by
either arguments with spouses, employment difficulties or loneliness, may likewise evoke craving" (Ludwig, Wikler & Stark, 1974,
pp. 539-540). Craving is certainly not limited by the authors to
situations associated with acute withdrawal from alcohol--otherwise
we would expect the greatest craving to occur in such settings as
alcoholism detoxification and treatment centers. Unfortunately,
the experience of craving does not always lead to relapse or loss
of control drinking, because of the influence of numerous "modifiers," such as the setting, appropriate labeling conditions, the
"subclinical" intensity of the craving symptoms, and the restraints
imposed by both the hypothetical blood alcohol ceiling and the
fluctuating cognitive ceiling. And even if craving is elicited,
despite these various constraints, it may not be recognized or labeled correctly by the alcoholic because of its similarity to a
variety of other arousal states. The explanatory power of the
craving construct seems to far outweigh its ability to predict
behavior.

Craving: A Critique

Let us return to the case study presented earlier, to see how this theoretical model of craving might be used as an explanation of Liane's relapse. It is difficult to see how craving could have served as a determinant of her first drink on July 17. To the best of my knowledge, Liane had never experienced withdrawal symptoms in a restaurant setting, so there were no stimuli present in that situation to elicit a conditioned craving response. Her luncheon companion was not someone she drank with on a regular basis, and it is unlikely that Sharon alone could have served as a cue for craving. Did the consumption of the first drink serve as an "appetizer" for more alcohol? This seems most unlikely, as it took Liane almost an hour to finish her drink. No loss of control drinking occurred that day as a result of taking the first drink, and she was able to maintain total abstinence for an additional twenty days until she repeated the luncheon drinking experience on August 6. Once again, there was no report of craving, and no loss of control drinking occurred.

It was not until August 9, 23 days following her first drink, that she drank to the point of intoxication. Prior to drinking that day, Liane reported that she felt anxious and upset about being alone. It is possible, of course, that she experienced craving that afternoon (elicited by the cues present in her apartment, where she used to engage in heavy drinking), and that she mislabelled this feeling as loneliness and anxiety. But why would the cues associated with prior heavy drinking and/or withdrawal elicit a conditioned craving response on that <u>particular</u> day, the 81st day she had spent in the apartment since her discharge from the hospital? It seems more likely that Liane did label her feelings correctly, and then sought a remedy for her anxiety and tension: she tried to find some tranquilizers. Unfortunately, in her unsuccessful search for the tranquilizers, she accidentally discovered some alcohol. After considerable hesitation, she drank the first scotch, probably hoping that it would exert a tranquilizing effect. When she "didn't feel anything at first" (no reduction of anxiety, no craving), she took the second drink. It was as though she sought a certain effect from the alcohol (reduction of anxiety), and when she failed to obtain it after the first drink, she decided to increase the dosage. Rather than an involuntary response to the physical demand of craving, her drinking seemed to be determined more by the cognitive <u>expectation</u> of the effects of alcohol. We will return to elaborate this important difference in a later section of the paper.

It should be clear by now that, from a scientific perspective, craving is a superfluous construct, and the hypothesis that craving is a crucial determinant of relapse is a basically untestable tautology. As Mello (1975) has also noted in her review of craving and loss of control, to say that craving is a determinant of drinking is to engage in circular reasoning: craving is defined by the very behaviors it is evoked to explain. Because craving is described as a subjective, internal "subclinical" state, its presence can only be verified objectively in terms of the behavior it is supposed to determine--namely, loss of control drinking. Craving is a superfluous construct because it presumably can be elicited by almost any stimulus associated with prior drinking or withdrawal, subject to the restraint of a wide variety of modifiers. As such, it has almost no utility to predict behavior. Craving has become reified as an apparent explanation of relapse within the medical model of alcoholism; it has become a convenient catch-all construct to attribute as the "cause" of relapse, much as "will-power" has been reified as the apparent "cause" of continued abstinence.

To say that an alcoholic drinks because he craves alcohol is as circular as to say that an individual eats because he is hungry, or that a person trembles because he is anxious. In each case, the existence of the internal state is inferred from observation of the very behavior it is supposed to determine (see Sarbin, 1968, for a parallel analysis of anxiety as a "determinant" of behavior). Craving is not a "cause" of relapse or loss of control. Unless it can be defined operationally by specifying observable criteria (i.e., physiological or biochemical measures) which are independent from the behavior it is supposed to explain (relapse or loss of control), craving is best thought of as a hypothetical construct. From a scientific standpoint, the construct of craving serves only to distract the investigator in the search for independent variables which actually determine drinking behavior.

There are a number of undesirable consequences of teaching an alcoholic to believe that relapse or loss of control is somehow "caused" by craving. If the abstinent alcoholic "slips up" and takes a drink, he or she can then attribute the cause of the drinking to craving, an internal physical state over which the alcoholic has no control. Whether craving is construed as a "symptom" of the underlying addiction or disease or as an elicited "conditioned response," the emphasis is on the involuntary nature of the reaction. Loss of control drinking becomes an inevitable consequence of this self-fulfilling prophecy. The alcoholic is absolved of any personal responsibility in the relapse process--he becomes a helpless victim of the symptomatology of his disease, over which he has no control. It is easy to see how this belief could serve as a convenient excuse for drinking.

Should we do away altogether with the concept of craving? Despite the arguments advanced by Ludwig and his colleagues, the concept of craving seems to have little or no scientific utility as a predictor of relapse or loss of control drinking. Use of the term should be limited to phenomenological descriptions of the subjective desire for alcohol experienced by addicts undergoing withdrawal. In this context, it would be more accurate to speak of craving as a strong desire for the alleviation of unpleasant withdrawal symptoms: craving for the relieving effects of alcohol, rather than craving as some kind of internally-based cellular "need" for alcohol itself. Although there is no scientific evidence to support the assumption that craving is an important determinant of relapse, the fact that some alcoholics believe this to be the case may have important effects on their behavior. The role of personal beliefs and expectancies as they affect relapse will be discussed shortly.

In recent years, a number of experimental studies have been published which have challenged the validity of both craving and loss of control as determinants of relapse or uncontrolled drinking (i.e., "binge" or "spree" drinking). Many of these studies have involved the direct observation of the drinking behavior of alcoholics in controlled settings. Since these studies have recently been reviewed in detail by other authors, the interested reader may wish to consult these sources (e.g., Maisto, Lauerman and Adesso, 1977; Maisto and Schefft, 1977; Mello, 1975; Pomerleau, Pertschuk and Stinnett, 1976) for further information. As is clear from an examination of the recent experimental literature, the constructs of craving and loss of control as key concepts in the disease model of alcoholism have received little or no empirical support.

THE RELAPSE PROCESS: A COGNITIVE-BEHAVIORAL ANALYSIS

Situational Determinants of Relapse

According to the basic assumptions of the medical model of alcoholism, relapse is an all-or-none event. Abstinence is the indication of treatment success and relapse is an indication of treatment failure. From this perspective, the particular details associated with the relapse (the specific antecedents, time, situation, personal feelings and emotions involved, etc.) are of little importance; the emphasis is on the "emergence" of the key symptom of the "underlying" addiction or disease, loss of control drinking. To relapse implies that one has "lapsed" back into the disease, succumbing to the influence of internal, biological factors beyond the individual's control. As a symptom of an underlying disease, loss of control drinking is no different from other symptoms: it is treated much like a fever, a symptom which emerges from within, acting as a sign that the disease has "broken out" again.

A review of the literature dealing with the environmental or situational aspects of relapse reveals only a few studies to date. In a paper describing the characteristics of the relapse situation

of alcoholics treated with aversion therapy, Burt (1974) reported
that about 80% of his sample (30 men and 4 women) took their first
drink in a location which differed from their preferred location
prior to treatment, but no detailed descriptions of these locations
are given in the paper. Hore (1971) followed a group of 22 alcohol
patients receiving treatment in an outpatient clinic for periods
ranging up to six months. He attempted to relate relapses in this
group to significant life events reported to occur by the patients
in the same general time period. Hore analyzed the events associated with relapse (41 episodes of relapse and 52 significant
events), and assigned them to the following categories: (a) Personal interaction, or disturbances in an emotional relationship
(33%), (b) Work events, involving a change in the patient's working
life (33%); (c) Events involving a health change in the patient or
in members of the family (20%); and (d) Events involving a change
of residence (13%). Gregson and Taylor (1977), among others, have
attempted to isolate factors which are predictive of relapse, such
as degree of cognitive dysfunction, membership in AA, and selected
demographic data, but no data are reported on the situational or
intrapersonal factors immediately preceding the relapse.

I first became interested in looking at the situational factors
associated with relapse while I was engaged in a study to determine
the effectiveness of electrical aversion as a treatment procedure
for alcoholism (Marlatt, 1973). To increase the generalization of
the aversion effects, we conducted the treatment sessions in a simulated bar, assuming that the contextual cues of the drinking setting would also acquire aversive properties through the principal
of higher-order conditioning. To assess the effectiveness of this
procedure, I decided to gather as much information as I could about
the particulars of the relapse situation for those patients who
took at least one alcoholic drink during the follow-up phase. By
doing so, I hoped to determine whether the aversion procedures had
any generalization effects in terms of the setting in which relapse
occurred.

The patients in our sample consisted of 65 males, most of them
chronic alcoholics drawn from a rural catchment area. During the
first ninety days of the follow-up period, 48 subjects in the sample consumed at least one drink. Each of these patients was personally interviewed within a few weeks of the relapse episode as
part of our follow-up evaluation. We asked a variety of questions
about the events leading up to the consumption of the first drink,
soliciting information about the location, time of day, presence of
others, characteristics of the drinking situation, external or environmental events occurring in that general time period, and the
patient's reported feelings and emotions on the day of relapse. We
then attempted to sort the list of relapse situations into independent, operationally defined categories. Raters were trained to
assign specific episodes to categories until they achieved at least

80% interrater agreement. The results of this analysis are reported in Table 1.

It is clear from the results presented in Table 1 that all of the relapse cases could be assigned to relatively few categories. The first two categories, accounting for over half the cases, usually involved an interpersonal encounter. Most of the situations in the first category (29%) involved an episode in which the patient was frustrated in some goal-directed activity (typically by another person who criticized or negatively evaluated him), and reported feelings of anger. Rather than expressing this anger overtly, however, the patient ended up taking a drink. In the second category (23%), the patients reported being unable to resist either direct or indirect attempts by others to engage him in drinking (social pressure). The other two major categories, accounting for about a third of the cases, were basically intrapersonal in nature. Temptation situations were quite common (21%), although it is possible that the sudden "urge" to drink was determined by other environmental or emotional factors which were not identified in the assessment interview. Negative emotional states (10%), including feelings of depression, anxiety, and boredom, accounted for only relatively few relapse episodes.

How limited are these findings? If the results applied only to our first sample, then the analysis would provide little more than a list of idiosyncratic relapse situations, possibly masking a more central underlying determinant of relapse, such as craving for alcohol. A recent follow-up study with another sample of male alcoholics (n = 25), revealed that the same categories accounted for almost all of the relapse episodes, although the percentage of cases in each category differed from the first sample: frustration and anger, 16%; social pressure, 17%; intrapersonal temptation, 15%; negative emotional state, 43%; and miscellaneous other, 9% (Chaney, O'Leary and Marlatt, 1978).

In a related series of studies, we have also been investigating the extent to which the key features of the relapse situations act as determinants of alcohol consumption in nonalcoholic drinkers. Using heavy social drinkers as our subjects, we have manipulated situational and emotional variables drawn from our analysis of relapse episodes to determine how such factors affect drinking behavior in standardized alcohol consumption tasks (cf., Marlatt, 1978). These studies have shown that feelings of anger (Marlatt, Kosturn and Lang, 1975), anticipation of interpersonal evaluation (Higgins and Marlatt, 1975), and modeling as a form of "social pressure" to drink (Caudill and Marlatt, 1975) all serve as powerful elicitors of increased drinking behavior in nonalcoholic subjects.

Based on this accumulating evidence, it seems clear that environmental/situational factors, and the emotional and cognitive

TABLE 1

ANALYSIS OF SITUATIONS IN WHICH RELAPSE OCCURRED

Situation Category	Example	Number	Percent
Frustration and anger	Patient tried to call his wife (they were separated); she hung up on him; he became very angry and took a drink.	14	29%
Social pressure	Patient went with the "boys" to a bar after work. They put pressure on him to "join the crowd" and he was unable to resist.	11	23%
Intrapersonal temptation	Patient walked by a bar, and "just unconsciously walked in, no real reason"; could not resist the temptation to take a drink.	10	21%
Negative emotional state	Patient living alone, no job; complained of feeling bored and useless; could see no reason why he should not take a drink.	5	10%
Miscellaneous other situations	Patient reported that everything was going so well for him that he wanted to celebrate by having a drink.	5	10%
No situation given or unable to remember		3	7%

reactions which accompany them, are important determinants of relapse. While other factors may also play a role in the relapse
process, such as demographic and other individual differences, past
drinking history, degree of cognitive impairment, and the type of
treatment received, it is the effect of these high-risk situations
and the individual's response to them that is most closely associated with the occurrence of the relapse itself. If the alcoholic
does not know how to cope with these high-risk situations when they
occur in the natural environment, the probability of relapse will
increase.

Expectancies about Alcohol and its Effects

The evidence reviewed in the last section suggests that the
probability of relapse increases in high-risk situations, particularly for those individuals who lack the appropriate skills to cope
with the problem. In the latter instance, alcohol represents the
only "coping" alternative available. For most alcoholics, drinking
has become the predominant, overlearned, and habitual response to
stressful situations. Over years of heavy drinking, alcohol begins
to acquire the properties of a potent elixir, to turn "lows" into
"highs," an image which is constantly reinforced by the romantic
display of liquor and drinking presented in advertising and the
media.

What is there about alcohol that makes it such a powerful reinforcement? The experimental literature on the reinforcing effects
of alcohol in human subjects has yet to provide an agreed-upon
answer to this question. The original hypothesis, that alcohol is
reinforcing because it reduces tension and anxiety, has failed to
receive consistent support in recent studies (see reviews by
Cappell, 1975; Marlatt, 1976). This issue is complicated by the
fact that the physiological response to alcohol appears to be biphasic in nature, and that the affective consequences of drinking
depend on the dosage consumed:

> There is increasing evidence to support that alcohol, at
> least in small doses, has a stimulating effect on physiological
> arousal. Emotional responsiveness to alcohol may be biphasic
> in nature, with the depressant or sedative effects occurring
> only at higher doses. The increase in arousal occurring at
> relatively low blood-alcohol levels may be one of the motivat-
> ing factors in social drinking: people often say that they
> drink to get high or to experience a lift, not to experience a
> low emotional state. A person who returns home after a hard
> day's work and consumes a couple of cocktails before dinner
> may experience a physiological boost that temporarily relieves
> his state of fatigue or physical exhaustion. The drinker, how-
> ever, may label this effect as one of relaxation instead of

excitation. (Marlatt, 1976, p. 290.)

Whatever the physiological effects may be on the brain and nervous system, it is the drinker's subjective experience of these effects that determines his personal attitudes and beliefs about alcohol. Like many drugs which have variable and nonspecific effects on emotional and physical response systems, the consumption of alcohol produces a range of ambiguous effects which can be interpreted differentially by the drinker, depending on various personal and situational factors (cf., Schachter, 1964). To take but one example, Pliner and Cappell (1974) have shown that the presence or absence of social companions can influence subjects' evaluation of the effects of alcohol. Social drinkers who consumed alcohol in a group setting interpreted its effects in terms of changes in mood and emotional state, whereas solitary drinkers described the effects solely as changes in physical symptoms. The influence of set and setting will often determine whether a particular drinking experience is evaluated as pleasant or unpleasant by the individual drinker.

More research needs to be done to pin down the effects of alcohol on cognitive processes in human subjects. At the present time, we know almost nothing about the effects of varying doses of alcohol, consumed in qualitatively different settings by subjects who differ in terms of their drinking histories, upon subjective states of consciousness. In one of the few investigations in this area, McClelland and his associates conducted a series of studies dealing with the effects of alcohol on cognitive fantasies and ideation in male drinkers (McClelland, Davis, Kalin and Wanner, 1972). Personal fantasies were assessed after consumption of alcohol with use of the TAT, a projective story-telling test. Although the research suffers from a number of methodological flaws, the results are provocative enough to warrant further investigation. The major conclusion drawn from the studies was that alcohol facilitates or stimulates a sense of "power" in male social drinkers. With relatively low doses of alcohol, the fantasies centered around themes of "social power," or power for the good of others or for a cause. At higher dose levels, the fantasies were more associated with themes of "personal power"--less altruistic and more self-aggrandizing in nature. If alcohol acts to increase the drinker's sense of perceived power or control, at least at the cognitive level, it seems likely that the attractiveness of alcohol as a coping strategy would increase in situations where the individual feels powerless or otherwise lacking in personal control (Marlatt, 1976).

In addition to the reinforcing subjective effects of alcohol, a case can be made for the hypothesis that alcohol consumption provides a convenient excuse for the execution of otherwise unacceptable forms of behavior (cf., Sobell and Sobell, 1973). An individual can avoid personal responsibility for engaging in certain antisocial acts, for example, if he attributes the cause of the behavior to the disinhibiting effects of alcohol. "I was drunk, and I didn't know what I was doing," goes the typical excuse. Recent research has supported the idea that cognitive processes and expectation effects outweigh the physiological effects of alcohol in facilitating both aggressive behavior (Lang, Goeckner, Adesso and Marlatt, 1975), and sexual arousal (Wilson and Lawson, 1976) in male social drinkers.

The importance of cognitive expectancies about alcohol and its effects was highlighted for me by the results of a study we conducted on the determinants of loss of control drinking (Marlatt, Demming and Reid, 1973). In this experiment, both nonabstinent male alcoholics and matched social drinkers were asked to sample and compare the taste qualities of either alcoholic or nonalcoholic beverages. The taste-rating task served as an unobtrusive measure of consumption, as subjects were free to sample as much of each drink as they wished to in making their ratings (see Marlatt, 1978, for a full description of the taste-rating task procedure). In order to control for both the physiological effects of alcohol and the subject's expectancy or belief about the effects of alcohol, we employed a diacritical factorial design with subjects assigned to one of four independent conditions. Half the subjects were told they would be rating drinks containing alcohol (vodka and tonic) and half were told they would be rating nonalcoholic beverages (tonic and water only). At the same time, we independently varied the alcoholic content of the drinks in such a way that half the subjects in each of these two conditions actually received drinks containing vodka and half received drinks containing only tonic water. We had found on the basis of pilot studies that a drink containing one part of vodka to five parts of tonic could not be reliably identified as containing alcohol on a better than chance basis. Prior to the taste-rating task, all subjects were given a "primer" dose of the drink they would be sampling, to determine if the alcoholic subjects receiving alcohol would show loss of control or increased consumption in the drinking task.

The results of the study were clear cut. The only significant determinant of overall beverage consumption, and subjects' later estimates of the alcohol content of their respective drinks, was the expectancy factor. Regardless of the actual alcohol content of the drinks, both alcoholic and social drinker subjects consumed significantly more beverage if they believed they were sampling drinks containing vodka. Similarly, subjects assigned to the told tonic/receive alcohol condition consumed relatively little beverage, whether or not the drinks actually contained vodka. This experimental design is superior to the traditional two-group design (alcohol administration vs. placebo), because it controls for the expectancies of receiving both alcoholic and nonalcoholic beverages. The results of the study provided evidence that cognitive mediational factors must be taken into account in any investigation of the ef-

fects of alcohol on human behavior. Other investigations support
the finding that prior expectations exert a stronger influence than
the physiological effects of alcohol for a variety of behaviors in
both alcoholics and social drinkers (Engle and Williams, 1972; Lang
et al., 1975; Maisto et al., 1977; Wilson and Lawson, 1976).

If expectancies are based on past experiences with the effects
of alcohol, why doesn't the alcoholic maintain a negative expectancy
about drinking? Most of the long-range effects are predominantly
negative, including both the physical symptoms of excessive consumption (nausea, hangover, and withdrawal) and the personal and social
consequences of problem drinking (domestic strife, employment difficulties, and the like). Despite repeated exposure to these unpleasant consequences, the fact remains that these negative effects
are delayed in time, relative to the immediate positively reinforcing effects--the initial "high" of the biphasic response to alcohol.
In addition, the alcoholic's memory of the long-range effects may be
impaired because of blackouts and state-dependent learning effects
(cf., Overton, 1971). In a series of studies by Mendelson and his
colleagues (e.g., McGuire, Mendelson and Stein, 1966; Tamerin,
Weiner and Mendelson, 1970), alcoholics were asked to describe their
initial expectancies about the effects of alcohol on their feelings
and moods prior to a scheduled period of drinking in a research ward
setting. Almost all the subjects anticipated that alcohol would
make them feel more relaxed, more comfortable, and less depressed.
These expected effects were directly opposed to the actual dysphoric
effects of prolonged drinking reported by the subjects during the
intoxication phase of the study.

The expectation of positive effects, combined with the initial
reinforcing influence of alcohol itself (the immediate stimulation
of the "high"), act together as potent sources of motivation to take
the first drink. The desire for the anticipated reinforcing effects of drinking is so strong that it is understandable that some
alcoholics speak of "craving" for alcohol--much as someone else
might speak of an overwhelming craving for hot fudge sundaes or
pecan pie. In this sense, to crave is to desire the anticipated
pleasures of consumption; wanting to experience the pleasurable
effects of a food or drug is not the same thing as experiencing an
internal need for the substance itself.

The Abstinence Violation Effect

The combination of positive expectancies for the effects of
alcohol and the stress of a high-risk situation greatly heightens
the probability of relapse. What happens when the abstinent alcoholic responds to these forces by taking a drink? How does the
consumption of the first drink influence the probability of taking
the second drink or drinking to the point of intoxication? Why is
it more difficult to return to a position of complete abstinence

following the experience of a single slip? What is there about a "relapse" that makes an alcoholic "lose control" over subsequent drinking?

Our previous review of the assumptions of the disease model provided one answer to these questions: that the consumption of one or two drinks elicits an addictive craving for more alcohol, triggering an uncontrolled bout of excessive drinking. But the phenomena of "relapse" and "loss of control" extend far beyond the alcoholic and his reputed disease. In my clinical practice (and in my various attempts to control my own behavior), I have experienced the same chain of events associated with other behaviors, particularly smoking and eating. The following examples will illustrate the point. (a) A young man decides to quit smoking. After a month of abstinence, he smokes one cigarette at a party. Within two weeks, he is smoking at the same rate he smoked prior to quitting. (b) A young woman goes on a strict diet to lose weight. During one afternoon when she is feeling depressed, she goes out and treats herself to a banana split. Afterwards, she feels that because she "blew" her diet, she can eat anything she wants to until the next diet begins. In both examples, the behavioral sequence is the same: the individual commits himself or herself to an indefinite period of abstinence or strict control over patterns of consumption. Then, for whatever reason, the taboo behavior occurs. The probability of repeating the behavior increases dramatically, and a full-blown "relapse" ensues. Are we then entitled to conclude that the ex-smoker, the dieter, and the abstinent alcoholic are all victims of an underlying, latent addictive disease which can be triggered by one cigarette, one ice-cream sundae, or one drink into compulsive craving and loss of control consumption?

To account for the similarity of the relapse process across different consummatory behaviors, we have postulated a common cognitive denominator: the Abstinence Violation Effect (AVE). The AVE is postulated to occur under the following conditions: (a) The individual is personally committed to an extended or indefinite period of abstinence from engaging in a specific behavior; (b) The behavior occurs during this period of voluntary abstinence. The intensity of the AVE will vary as a function of several factors, including the degree of commitment or effort expended to maintain abstinence, the length of the abstinence period (the longer the period, the greater the effect), and the importance or value of the behavior to the individual concerned. We hypothesize that the AVE itself is characterized by two key cognitive elements:

(1) A cognitive dissonance effect (Festinger, 1957, 1964), wherein the occurrence of the previously restricted behavior is dissonant with the cognitive definition of oneself as abstinent. Cognitive dissonance is experienced as a conflict state, and underlies what most people would define as guilt for having "given in to temptation."

(2) A personal attribution effect (cf., Jones, Kanouse, Kelley, Nisbett, Valins and Weiner, 1972), wherein the individual attributes the occurrence of the taboo behavior to internal weakness or personal failure (e.g., to "lack of will power" or "insufficient personal control" over one's behavior), rather than to external situational or environmental factors.

Although the dissonance and attribution components of the AVE may overlap to a degree, the additive effects of both reactions will greatly increase the probability of repeating the restricted behavior, according to our theoretical model. The dissonance state will vary in intensity depending on the duration of the abstinence period and the degree of personal and public commitment to the role of an abstainer. A member of Alcoholics Anonymous, for example, who refers to him/herself as a "recovering alcoholic" and has made a public pledge never to take another drink, is particularly susceptible to an intense dissonance reaction upon consuming a single drink. According to classical cognitive dissonance theory, dissonance is experienced as a negative emotional drive state and can serve as a motivating force to engage in behaviors (or changes in cognition) which serve to reduce dissonance. Consumption of more alcohol, following the first drink, is a particularly effective behavior in this regard because (a) to the extent that alcohol reduces tension or anxiety, it should also be effective in reducing feelings of dissonance or guilt; and (b) the alcoholic can modify his self-image to achieve consonance with his drinking behavior (e.g., "I guess I still haven't fully recovered; the disease has got me again"). From the perspective of dissonance theory, it is very difficult to "undo" the act of taking the first drink: a single drink is enough to shatter the image of oneself as an abstainer.

Most abstinent alcoholics are justifiably proud of their continued sobriety. Although they may partially attribute their success to a specific experience, such as involvement in a particular treatment program or in A.A., it seems likely that they also feel personally responsible for having taken the "first step" towards a new life, totally free from alcohol. As each day of abstinence passes, "one day at a time," their self-confidence and feelings of personal control over the temptations to drink increase. How is this self-image affected by taking the first drink? To the extent that the individual feels personally responsible for "giving in" and consuming that drink, attribution theory would predict that he or she would attribute this "failure" to internal, personal causes (cf., Weiner, Frieze, Kukla, Reed, Rest and Rosenbaum, 1971). Even though an objective observer of the relapse might ascribe the causes to environmental stressors beyond the individual's control, the individual concerned will frequently attribute the event to personal weakness or other shortcomings. People often draw inferences about their own internal states through the observation of their behavior

(Bem, 1972). Thus, if a relapse occurs, the alcoholic is likely to infer a lack of will power or personal control as the determinant of the relapse. If the relapse is viewed as a personal failure in this sense, the individual's expectancy for continued failure will increase as a result (Weiner et al., 1971). An expectancy of failure is likely to mediate decrements in actual performance. If one feels weak-willed and powerless for giving in to the temptation of the first drink, the expectation of resisting the second or third drink is correspondingly lower: "Once a drunk, always a drunk."

From this perspective, loss of control drinking is determined in large part by the alcoholic's perception of having "lost control" when the relapse first occurred. By attributing the relapse to personal failure, the absence of will power or strength, the alcoholic relinquishes control to alcohol. Attributing the relapse to an underlying disease or addiction has much the same effect: the alcoholic still believes he is personally helpless in the face of internal forces beyond his control. The relapse is still "caused" by internal, intrapersonal factors--whether they are labeled as symptoms of a disease state, or as signs of personal weakness.

The conditions for a full-blown relapse are as follows: (a) The abstinent alcoholic feels "in control" until he encounters a high-risk situation which challenges his perception of control; (b) The individual lacks an appropriate method of coping with the high-risk situation, or fails to engage in a coping response; (c) He has positive expectancies about the effects of alcohol, and alcohol is available; (d) He takes the first drink; (e) He experiences one or both components of the Abstinence Violation Effect (cognitive dissonance and/or attribution of failure to internal weakness); (f) The probability of continued drinking markedly increases. Whether or not the individual will continue to drink immediately following the consumption of the first drink will depend on a number of factors, including the intensity of the AVE and the effect of situational constraints (presence of observers, time limitations, availability of alcohol, and other factors). Consumption of the first drink shifts the probability of subsequent drinking upwards, to a much higher level than it was during the abstinence period.

Loss of Control Drinking Revisited

If "loss of control" drinking does occur following consumption of the first drink, how is this behavior explained? We have already noted that consumption of the first drink increases the probability of subsequent drinking. The specific form of this behavior may vary, depending on a number of factors. It may be possible, for example, that alcoholics can be explicitly trained to exert control upon their subsequent drinking, as has been demonstrated recently by some behavioral investigators (esp. Sobell and

Sobell, 1973, 1976). Treatment interventions to modify the typical
response to relapse will be discussed in the final section of this
paper.

The precise form of behavior traditionally associated with loss
of control drinking has never been clearly specified in the literature. From descriptions which are available (e.g., Jellinek,
1960), usually drawn from retrospective reports given by alcoholics
or from the A.A. literature and lore, one gets the picture of a
crazed alcoholic compulsively driven by addictive craving, gulping
down drink after drink of whiskey until he falls on the floor in a
drunken stupor. Unfortunately, this behavior has never been reliably demonstrated in any observational studies of alcoholics' drinking behavior (Mello, 1975). Although these reports certainly do
show that alcoholics maintain a very high blood-alcohol level during
periods of drinking, they fail to provide support for the notion of
loss of control drinking as a compulsive, uncontrolled behavior.

One can only infer that control has been "lost" by observing
the drinking behavior itself--a tautology similar to the circular
definition of craving reviewed earlier. A strict operational definition would omit any reference to the construct of control, and
would focus on the rate of drinking itself (sip rate and amount
consumed) and the resultant blood-alcohol level. There are a number of factors which may account for the high rate of drinking
associated with the loss of control phenomenon:

(1) <u>Absorption of alcohol rate</u>. It takes a certain amount of
time for alcohol to be absorbed into the blood stream and distributed throughout the brain and nervous system before its effects can
be perceived by the drinker. This time period varies as a function
of the subject's sex and weight, amount of undigested food in the
stomach, dilution of the alcohol in the drink consumed, and other
such factors, but it is frequently estimated to take from 15 to 30
minutes before the effects are "felt" (cf., Kalant, 1971). Because
of the alcoholic's strong positive expectancies about the effects
of alcohol, the rapid rate of beverage intake may reflect his "im-
patience" to experience the desired effects during the absorption
delay period. From the standpoint of reinforcement theory, the
high response rate indicated by gulping down a number of drinks in
quick succession resembles the burst of responses typically observed in subjects exposed to a Variable Interval (VI) schedule of
reinforcement. In a VI schedule, reinforcement occurs at variable
time intervals, and is associated with a very high response rate
prior to reinforcement (e.g., Verhave, 1966). The alcoholic may
show a similar burst of responses during the absorption period until the alcohol begins to "hit" him.

(2) <u>Tolerance effects</u>. Alcoholics typically show an acquired
tolerance for alcohol, such that a progressively increased dosage

of alcohol is required to obtain the same degree of effect in a given individual over time (Kalant, Leblanc and Gibbins, 1971). While this tolerance effect may decrease over a prolonged period of abstinence, it still must be considered as a factor in determining the high rate of consumption associated with loss of control drinking.

(3) Overdose effects. Because of the combined influence of the absorption delay and tolerance effects, the alcoholic may frequently exceed a desired blood-alcohol level (associated with the optimal "high") by overdosing himself during the period of rapid intake. Although the literature contains little or no mention of the blood-alcohol levels considered optimum by various alcoholics, let us consider a hypothetical example. Suppose an alcoholic desires to experience the "high" effects associated with a blood-alcohol level in the .05% to .07% range. Under normal circumstances, this level would be achieved approximately 30 minutes or so after the consumption of about two or three drinks. If he tosses down the first drink within a minute or so, and nothing happens, he is likely to consume another drink in short order. Within the space of 20 minutes, a number of additional drinks could be consumed before the effects of the first drink or two are experienced. As a result of this overdose, the ascending limb of the blood-alcohol curve rises quickly past the desired level and may peak in the intoxication range (greater than .10%). Direct observation of the drinking behavior of alcoholics reveals that they frequently do drink in this rapid manner (Sobell, Schaefer and Mills, 1972). The high blood-alcohol level resulting from the overdose is likely to elicit the range of dysphoric effects associated with high dose levels. The reinforcing "high" experienced at lower blood-alcohol levels is superceded by the unpleasant "low" of an overdose (cf. the biphasic response to alcohol with increasing dosage). The alcoholic may harbor the expectation that consuming even more alcohol would alleviate these unpleasant effects, and a vicious circle of drinking may ensue.

Overview of the Relapse Process: Return to the Case Study

A schematic representation of the cognitive-behavioral analysis of relapse is presented in Figure 3. In reviewing the main points of this analysis, let us return to the case study of Liane's relapse presented at the beginning of this paper. Liane was abstinent for 57 days following her release from the hospital. During this period, she saw herself as "in control" and became increasingly confident of her ability to maintain abstinence as the days passed. She soon felt confident enough to establish a new social relationship and invited her neighbor Sharon out to lunch. It is significant that Liane played the role of hostess during this encounter: She took the initiative to ask Sharon out and she paid for the

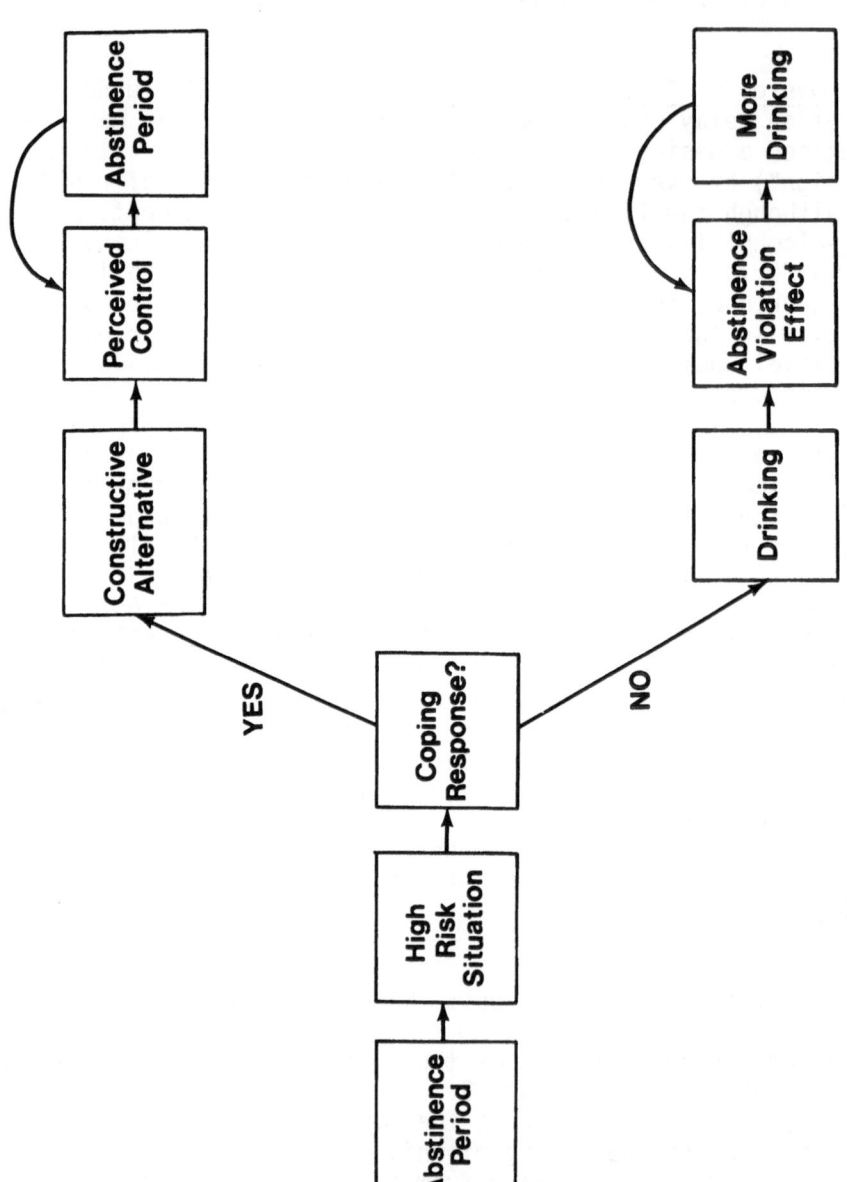

Figure 3: Cognitive-behavioral analysis of relapse.

lunch. Prior to arriving at the restaurant, Liane was in control of the situation ("Both of us were dressed up, and I felt very at ease, cheerful and confident"). Suddenly, a high-risk situation developed, and she ended up taking her first drink.

There are a number of factors which contributed to this high-risk situation: (a) This was Liane's first venture into the social realm, and she was with someone she did not know well. Although she wanted to become friends with Sharon, Liane did not yet feel comfortable in telling her about her past drinking problems. Had Liane disclosed this information previously, she would have had little difficulty in dealing with the cocktail waitress when she came to take the order; (b) The entire social occasion was a challenge to Liane's confidence, because of her general lack of social skills. Through the influence of liquor advertising and the romanticism associated with drinking as presented on television and the media, Liane had developed an image of herself as secure and poised in social settings--provided she had a drink in her hand ("I was especially vulnerable to the unreality of alcoholic romanticism used in advertising, television, and movies...This romanticism is a stimulus for envy and anxiety in me"); (c) Liane had not prepared herself for the encounter with the cocktail waitress. Had she anticipated this event, she could have avoided it altogether (by choosing a restaurant without a liquor license), or she could have come up with some excuse and ordered either nothing or a nonalcoholic beverage. As it turned out, Sharon ordered her drink first, thus providing a powerful modeling cue for Liane. In Liane's eyes, Sharon was a prestigious model: She was older, married, and could drink with impunity. What else could Liane do? After all, she was the hostess, and supposedly in control of the situation. "'Gin and tonic, please,' just came out. I didn't plan it and I wasn't prepared for it." The first drink was taken.

As indicated in Figure 3, if Liane had managed to cope with the high-risk situation by engaging in any of the constructive alternative behaviors described above, she would have reinforced her sense of perceived control and confidence in handling similar situations. The abstinence period would then continue, and her perception of her ability to cope successfully with future problematic situations would grow in strength. Unfortunately, she was unable to cope with the situation, and ordered the drink instead. Her reaction to the first drink appears consistent with the Abstinence Violation Effect described earlier. From the first sip, her response is characterized by conflict and guilt: "The first sip was so strong, it literally shocked me. It was difficult for me to finish the drink. I am not entirely sure whether or not I imagined some of the physical effects, perhaps because of guilt, or if it was real." It seems most unlikely that a single mixed drink, consumed slowly during the course of a 45-minute luncheon, would produce much of a physical effect under normal circumstances.

The intensity of Liane's AVE is experienced as a combination of physical and psychological effects ("I got a headache, my nose got stuffy, my thoughts were tangled and my voice and actions were abrasive"). Although she did not order a second drink, probably due to the constraints of the particular social situation and the intensity of the unpleasant reaction, she had nonetheless crossed over the line between abstinence and drinking.

Our model predicts that the probability of subsequent drinking will increase following the experience of the first drink. About three weeks later, Liane again took Sharon out to lunch, and again ordered a drink. On this second occasion, she admitted that she had considered having the drink in advance ("At least I had no real intentions of not having that drink"). Perhaps she felt an increased sense of confidence because she had not "lost control" after the experience of having the first drink. Nevertheless, she again felt the pangs of the AVE: "I didn't feel good about having that drink. It lowered my self-esteem and did create some guilty and remorseful feelings." The stage was now set for the full relapse to occur.

It happened just three days later. Again, Liane was faced with a high-risk situation. She faced a lonely Saturday evening at home; she had "no plans for doing anything that night." Her loneliness increased during the afternoon, facilitated by moods evoked by listening to her records. Late in the afternoon, her romantic fantasies increased and she began to feel increasingly sad and anxious. At this point, Liane felt that she might be able to cope effectively with her feelings by taking a tranquilizer. In her unsuccessful search for tranquilizers, she discovered the miniature bottles of scotch and the beer belonging to her parents.

Liane had failed to cope successfully with the situation. Knowing that she would be alone, she could have made alternative plans to go out for the evening, spending time with her parents or friends, or going to a movie. Another point of intervention arose when she began to feel anxious about being home alone. Had she been trained in a method of relaxation, such as muscle relaxation or meditation, she might have calmed herself down and considered other activities to occupy her time. Instead, Liane reverted to her old pattern of coping and sought out the tranquilizers. Not being able to find any, she became even more upset. At that point, finding the alcohol must have been like discovering a bottle of liquid tranquilizer.

After her test of will power had failed (she put the bottles on the table and "just looked at them for a long time, without even touching them"), she poured one of the bottles of scotch into a glass with some ice. At that moment, she probably harbored a strong expectation that the alcohol would provide a source of relief from her tension and anxiety. But, at first, nothing happened. "I sipped the drink but didn't feel anything at first, so I poured in the other one and drank that too." As this was by now her third drink since she broke her abstinence, the AVE was probably of minimal intensity. In addition, her expectation of relief was not immediately satisfied, because of the delay in absorption of alcohol. The conditions were ripe for an overdose: "After that, I drank the bottles of beer in the fridge and then I don't remember what happened."

In terms of the high-risk situations presented in Table 1, I would classify the circumstances leading to the consumption of the first drink as a "social pressure" situation. The components of this situation include the influence of Liane's romanticized image of herself as a drinker and the role that alcohol plays in social situations ("For me, what was going through my mind was that a social drink at parties, lunches, dinners, etc. is condoned and exploited, making me feel extremely self-conscious for not participating"), and the influence of Sharon who acted as a powerful model when she ordered the first drink. When she drank to the point of intoxication, the situation could be classified as falling into the "negative emotional state" category--attempting to cope with feelings of loneliness and anxiety.

IMPLICATIONS FOR TREATMENT AND PREVENTION OF RELAPSE

Many traditional alcoholism programs fail to contend with the components of the relapse process in their treatment procedures. Often the topic is avoided, as if the very mention of relapse would somehow increase the chances of its occurrence. Instead, the emphasis is on abstinence as the only "cure" for alcoholism. In my own experience as a clinical psychologist in two inpatient alcoholism programs, the total thrust of the treatment seemed to be to convince the patient that he was indeed an alcoholic and that he should never drink again. There seemed to be a double message implicit in this approach: You must accept the diagnosis that you are an alcoholic, that you have a disease which is beyond your own personal control; yet the only way to "control" the disease is to desist from drinking on the strength of your own will power or volitional control. From the disease model perspective, there is not much an alcoholic can do once the first drink has been consumed: He is a treatment "failure" and relapse is inevitable--drink, drank, drunk.

In recent years, there has been considerable controversy about the possibility of controlled drinking as a treatment goal for some alcoholics (cf., Hamburg, 1975; Lloyd and Salzberg, 1975; Pattison, 1976. The findings of the recent Rand report (Armor, Polich and Stambul, 1976), showing that a sizeable proportion of

alcoholics appear to be drinking in a normal or controlled fashion
after discharge from conventional, abstinence-oriented treatment
programs, have recently stimulated a barrage of criticism. Programs which have been geared specifically to train alcoholics to
engage in moderate or controlled drinking, using newly-developed
behavioral treatment techniques, have been quite successful (Sobell
and Sobell, 1973, 1976). But even in the traditional abstinence-
oriented program, there are a number of specific procedures which
may reduce the probability of relapse or lessen the likelihood
that a single drink will be followed by a bout of intoxication. I
will conclude this paper with a brief review of three intervention
strategies which appear promising in this regard.

Coping with High-Risk Situations: Specific Alternatives

In a recent treatment study with male alcoholics who participated in an abstinence-oriented program, Chaney and other members
of our research team (Chaney, 1976; Chaney, O'Leary and Marlatt,
1978) investigated the effectiveness of a social skill training
program designed to help alcoholic inpatients cope with high-risk
relapse situations. Using a variety of high-risk situations drawn
from our previous research and the self-reports of patients in the
program, the therapists conducted a series of skill-training group
meetings, in which each patient acted out his responses to individual situations. Using a training model previously developed by
other researchers (e.g., D'Zurilla and Goldfried, 1971; McFall,
1976), Chaney used a combination of techniques, including modeling,
therapist coaching, group feedback, behavioral rehearsal, and repeated practice, to teach each patient a variety of specific skills
to cope effectively with a full range of high-risk situations. Each
subject was assessed on several measures of social skill effectiveness, including standardized tests of assertive behavior and a
specially designed Situational Competency Test, administered both
prior to and following the training program. In this test, each
subject is asked to respond verbally to a high-risk situation presented by a narrator on an audiotape; the subject's response is
scored for a number of components related to the quality and overall
competency of the reply.

Subjects in the skill training condition were compared with
two control groups: A discussion group in which patients talked
about personal feelings associated with the same situations covered
in the skill training condition, but did not participate in any of
the behavioral techniques; and a third group that received only the
usual inpatient treatment procedures. All patients were followed
at regular intervals for a period of one year following discharge
from the hospital. At the one-year follow-up assessment, the subjects who had received skill training showed a significant decrease
in the duration and severity of relapse episodes compared to the two

control groups. In addition, performance on the posttreatment administration of the Situational Competency Test was found to be predictive of subjects' drinking behavior during the follow-up period. Chaney's findings are supported by the results of a laboratory study in which we found that providing heavy drinkers with a constructive alternative response in a high-risk situation (expression of aggression when provoked to anger in an interpersonal encounter) reduced their subsequent consumption of alcohol in an analogue drinking task (Marlatt, Kosturn and Lang, 1975).

Chaney's study demonstrated that training alcoholics to cope with a variety of standardized high-risk situations is an effective treatment procedure. The results of this program could be enhanced even more by training the alcoholic to be his own behavior therapist. By learning to self-monitor his own urges to drink (by keeping an ongoing daily record of his temptations) the alcoholic should be able to recognize the antecedent conditions leading up to a high-risk situation, and take appropriate preventive action. The client who anticipates these situations in advance should be able to execute self-control procedures to circumvent the problems or minimize their impact. A variety of self-control techniques have been described in the recent behavioral literature (e.g., Mahoney and Thoresen, 1974; Meichenbaum, 1975). Many of these procedures have been applied to the problem drinker in a recent self-help manual by Miller and Munoz (1976).

Increased Perception of Control: Global Alternatives

According to the model presented in Figure 3, each time the alcoholic engages in a constructive alternative to drinking in a particular high-risk situation, his perception of personal control should increase in strength. It may also be possible to influence the client's general sense of confidence and control by instructing him or her in a globally effective coping strategy. Techniques such as Jacobson's progressive muscle relaxation (Jacobson, 1938), or a form of meditation, may increase the alcoholic's overall ability to cope with stressful situations. Meditation may be a particularly useful adjunct procedure in the treatment of alcoholism, because it offers an alternative "high" to alcohol. It is also somewhat easier to experience and may be more intrinsically reinforcing than muscle relaxation techniques (cf., Glueck and Stroebel, 1975).

We recently found that if heavy social drinkers take time out each day to perform a simple relaxing exercise, whether it is meditation, muscle relaxation, or simply reading quietly, their daily consumption of alcohol can be reduced substantially compared to subjects who do not practice such a technique (Marlatt and Marques, 1977). In this study, the regular practice of a relaxation technique led to a significant increase in internal scores on Rotter's

(1966) test of locus of control. Thus, subjects who feel more "in control" of their own behavior, as defined by the locus of control measure, show a marked decrement in their day-to-day alcohol consumption. The effects of both skill training (to deal with specific high-risk situations) and a global relaxation technique would seem to be a particularly effective combination in reducing the risk of relapse for alcoholics.

The Programmed Relapse

There is one additional procedure that deserves attention as a possible treatment technique, although it has yet to be evaluated on an empirical basis. One of the strongest factors in the relapse process is the Abstinent Violation Effect, described earlier in this paper. Failure to cope adequately with the dissonance and attribution components of this reaction may turn the single slip into a full-blown relapse. It is possible, of course, to prepare the alcoholic to anticipate this reaction during the course of treatment. The therapist could explain the nature of the AVE, and how it may motivate one to continue drinking. The alcoholic in treatment could be provided with a list of things to remember and to do if a slip does occur. For example, he or she could be given a wallet-sized card to carry around after discharge from the hospital. Such a card might read as follows:

WHAT TO DO IF A SLIP OCCURS

1. Stop after you have finished the first drink. Take half an hour out to consider the following information.

2. A single slip is not all that unusual. It does not mean that you have failed or that you have now lost control over your drinking.

3. You are now probably feeling guilty about what you have done, and blame yourself for having taken the first drink. This feeling is to be expected; it is part of the Abstinence Violation Efect. There is no reason why you have to give in to these feelings and take another drink. The feeling will pass.

4. Look upon the slip as a learning experience. What were the elements of the high-risk situation which led up to the slip? What coping response could you have used to get around this situation? Can you use this coping respor now?

5. If you are still having trouble resisting the temptation to drink again, call your therapist. Telephone number: _____. Back-up number:_____.

6. Remember: ONE DRINK DOES NOT MEAN A DRUNK!

Dealing with the pressures of the Abstinence Violation Effect should be a key element in a relapse prevention program. The best way to learn to cope with the demands created by a single slip may be to experience the AVE directly, under supervised conditions. Why not schedule a "programmed relapse" before the alcoholic is discharged from the treatment program, or during a specified follow-up contact period? To experience the effects of a single drink, under a therapist's close supervision, may help the alcoholic to cope with the potential problems of a relapse should it occur in the outside world. A programmed relapse could serve as a "dry run" to iron out difficulties before they occur--much like a fire drill is used as a prevention measure for future emergencies. The experience may also help the alcoholic in treatment to overcome some traditional beliefs about the magical potency of alcohol. He may come to believe instead that one drink is not strong enough to push him off the wagon.

REFERENCES

Armor, D. J., Polich, J. M., & Stambul, H. B. Alcoholism and treatment. Report #R-1739-NIAA. Santa Monica, Calif.: Rand Corporation, 1976.

Bem, D. J. Self-perception theory. In L. Berkowitz (Ed.), Advances in Experimental Social Psychology, Vol. 6. New York: Academic Press, 1972.

Burt, D. W. Characteristics of the relapse situation of alcoholics treated with aversion conditioning. Behaviour Research and Therapy, 1974, 12, 121-123.

Cappell, H. An evaluation of tension models of alcohol consumption. In R. J. Gibbons, Y. Israel, H. Kalant, R. E. Popham, W. Schmidt & R. G. Smart (Eds.), Research Advances in Alcohol and Drug Problems. Vol. 2. New York: Wiley, 1975.

Caudill, B. D. & Marlatt, G. A. Modeling influences in social drinking: An experimental analogue. Journal of Consulting and Clinical Psychology, 1975, 43, 405-415.

Chaney, E. F. Skill Training with Alcoholics. Unpublished doctoral dissertation, University of Washington, 1976.

Chaney, E. F., O'Leary, M. R. & Marlatt, G. A. Skill training with alcoholics. Unpublished manuscript, 1978.

D'Zurilla, T. J. & Goldfried, M. R. Problem solving and behavior modification. Journal of Abnormal Psychology, 1971, 78, 107-126.

Engle, K. B. & Williams, T. K. Effect of an ounce of vodka on alcoholics' desire for alcohol. Quarterly Journal of Studies on Alcohol, 1972, 33, 1099-1105.

Festinger, L. A theory of cognitive dissonance. Stanford: Stanford University Press, 1957.

Festinger, L. Conflict, decision and dissonance. Stanford: Stanford University Press, 1964.

Glueck, B. C. & Stroebel, C. F. Biofeedback and meditation in the treatment of psychiatric illness. Comprehensive Psychiatry, 1975, 16, 303-321.

Gregson, R. A. M. & Taylor, G. M. The prediction of relapse in chronic alcoholics. Journal of Studies on Alcohol, 1977, in press.

Hamburg, S. Behavior therapy in alcoholism: A critical review of broad-spectrum approaches. Journal of Studies on Alcohol, 1975, 36, 69-87.

Higgins, R. L. & Marlatt, G. A. Fear of interpersonal evaluation as a determinant of alcohol consumption in male social drinkers. Journal of Abnormal Psychology, 1975, 84, 644-651.

Hore, B. D. Life events and alcoholic relapse. British Journal of Addiction, 1971, 66, 83-88.

Isbell, H. Craving for alcohol. Quarterly Journal of Studies on Alcohol, 1955, 16, 38-42.

Jacobson, E. Progressive Relaxation. Chicago: University of Chicago Press, 1938.

Jellinek, E. M. The phases of alcohol addiction. Quarterly Journal of Studies on Alcohol, 1952, 13, 673-684.

Jellinek, E. M. The disease concept of alcoholism. New Brunswick, N. J.: Hillhouse Press, 1960.

Jones, E. E., Kanouse, D. E., Kelley, H. H., Nisbett, R. E., Valins, S. & Weiner, B. (Eds.), Attribution: Perceiving the causes of behavior. Morristown, N. J.: General Learning Press, 1972.

Kalant, H. Absorption, distribution and elimination of alcohols: Effect on biological membranes. In B. Kissin & M. M. Begleiter (Eds.), The biology of alcoholism. Vol. 1. New York: Plenum, 1971.

Kalant, H., LeBlanc, A. E. & Gibbons, R. J. Tolerance to, and dependence on, ethanol. In Y. Israel & J. Mardones (Eds.), Biological basis of alcoholism. New York: Wiley-Interscience, 1971.

Keller, M. On the loss-of-control phenomenon in alcoholism. British Journal of Addiction, 1972, 67, 153-166.

Lang, A. R., Goeckner, D. J., Adesso, V. J. & Marlatt, G. A. The effects of alcohol on aggression in male social drinkers. Journal of Abnormal Psychology, 1975, 84, 508-518.

Lloyd, R. W., Jr., & Salzberg, H. C. Controlled social drinking: An alternative to abstinence as a treatment goal for some alcohol abusers. Psychological Bulletin, 1975, 82, 815-842.

Ludwig, A. M. On and off the wagon: Reasons for drinking and abstaining by alcoholics. Quarterly Journal of Studies on Alcohol, 1972, 33, 91-96.

Ludwig, A. M., & Stark, L. H. Alcohol craving: Subjective and situational aspects. Quarterly Journal of Studies on Alcohol, 1974, 35, 899-905.

Ludwig, A. M. & Wikler, A. "Craving" and relapse to drink. Quarterly Journal of Studies on Alcohol, 1974, 35, 108-130.

Ludwig, A. M., Wikler, A. & Stark, L. H. The first drink: Psychobiological aspects of craving. Archives of General Psychiatry, 1974, 30, 539-547.

Mahoney, M. J. & Thoresen, C. E. (Eds.). Self-control: Power to the person. Monterey, Calif.: Brooks/Cole, 1974.

Maisto, S. A., Lauerman, R. & Adesso, V. J. A comparison of two experimental studies investigating the role of cognitive factors in excessive drinking. Journal of Studies on Alcohol, 1977, 38, 145-149.

Maisto, S. A. & Schefft, B. K. The constructs of craving for alcohol and loss of control drinking: Help or hindrance to research. _Addictive Behaviors_, in press.

Marconi, J., Fink, K. & Moya, L. Experimental study on alcoholics with "Inability to Stop." _British Journal of Psychiatry_, 1967, _113_, 543-545.

Mardones, R. J. "Craving" for alcohol. _Quarterly Journal of Studies on Alcohol_, 1955, _16_, 51-53.

Marlatt, G. A. _A comparison of aversive conditioning procedures in the treatment of alcoholism._ Paper presented at the meeting of the Western Psychological Association, Anaheim, Calif., 1973.

Marlatt, G. A. Alcohol, stress, and cognitive control. In I. G. Sarason & C. D. Spielberger (Eds.), _Stress and anxiety._ Vol. 3, Washington, D. C.: Hemisphere Pub. Co., 1976.

Marlatt, G. A. Behavior assessment of social drinking and alcoholism. In G. A. Marlatt & P. E. Nathan (Eds.), _Behavioral approaches to the assessment and treatment of alcoholism._ New Brunswick, N. J.: Center of Alcohol Studies, 1978.

Marlatt, G. A., Demming, B. & Reid, J. B. Loss of control drinking in alcoholics: An experimental analogue. _Journal of Abnormal Psychology_, 1973, _81_, 233-241.

Marlatt, G. A., Kosturn, C. F. & Lang, A. R. Provocation to anger and opportunity for retaliation as determinants of alcohol consumption in social drinkers. _Journal of Abnormal Psychology_ 1975, _84_, 652-659.

Marlatt, G. A. & Marques, J. K. Meditation, self-control, and alcohol use. In R. B. Stuart (Ed.), _Self-management: Strategies, techniques and results._ New York: Brunner/Mazel, 1977.

McClelland, D. C., Davis, W. M., Kalin, R. & Wanner, E. _The drinking man._ New York: Free Press, 1972.

McFall, R. M. _Behavioral training: A skill-acquisition approach to clinical problems._ Morristown, N. J.: General Learning Press, 1976.

McGuire, M. T., Mendelson, J. H. & Stein, S. Comparative psychosocial studies of alcoholic and non-alcoholic subjects undergoing experimentally-induced ethanol intoxication. _Psychosomatic Medicine_, 1966, _28_, 13-25.

Meichenbaum, D. Self-instructional methods. In F. H. Kanfer & A. P. Goldstein (Eds.), Helping people change. New York: Pergamon, 1975.

Mello, N. A semantic aspect of alcoholism. In H. D. Cappell & A. E. LeBlanc (Eds.), Biological and behavioural approaches to drug dependence. Toronto: Addiction Research Foundation, 1975.

Miller, W. R., & Munoz, R. F. How to control your drinking. Englewood Cliffs, N. J.: Prentice-Hall, 1976.

Overton, D. A. State-dependent learning produced by alcohol and its relevance to alcoholism. In B. Kissin & H. Begleiter (Eds.), The biology of alcoholism. Vol. 2. New York: Plenum, 1971.

Pattison, E. M. Nonabstinent drinking goals in the treatment of alcoholics. In R. J. Gibbins, Y. Israel, H. Kalant, R. E. Popham, W. Schmidt, & R. G. Smart (Eds.), Research advances in alcohol and drug problems. Vol. 3. New York: Wiley, 1976.

Pliner, P., & Cappell, H. Modification of affective consequences of alcohol: A comparison of social and solitary drinking. Journal of Abnormal Psychology, 1974, 83, 418-425.

Pomerleau, O., Pertschuk, M. & Stinnet, J. A critical examination of some current assumptions in the treatment of alcoholism. Journal of Studies on Alcohol, 1976, 37, 849-867.

Rotter, J. B. Generalized expectancies for internal versus external control of reinforcement. Psychological Monographs, 1966, 80, No. 1 (Whole No. 609.

Sarbin, T. R. Ontology recapitulates philology: The mythic nature of anxiety. American Psychologist, 1968, 23, 411-418.

Schachter, S. The interaction of cognitive and physiological determinants of emotional state. In L. Berkowitz (Ed.), Advances in experimental social psychology, vol. 1. New York: Academic Press, 1964.

Sobell, M. B., Schaefer, H. H. & Mills, K. C. Differences in baseline drinking behavior between alcoholics and normal drinkers. Behaviour Research and Therapy, 1972, 10, 257-267.

Sobell, M. B. & Sobell, L. C. Individualized behavior therapy for alcoholics. Behavior Therapy, 1973, 4, 49-72.

Sobell, M. B. & Sobell, L. C. Second-year treatment outcome of alcoholics treated by individualized behavior therapy: Results. Behaviour Research and Therapy, 1976, 14, 195-215.

Tamerin, J. S., Weiner, S. & Mendelson, J. H. Alcoholics' expectancies and recall of experiences during intoxication. American Journal of Psychiatry, 1970, 126, 1697-1704.

Verhave, T. The experimental analysis of behavior. New York: Appleton-Century-Crofts, 1966.

Weiner, B., Frieze, I., Kukla, A., Reed, L., Rest, S. & Rosenbaum, R. M. Perceiving the causes of success and failure. Morristown, N. J.: General Learning Press, 1971.

Wilson, G. T. & Lawson, D. M. Expectancies, alcohol, and sexual arousal in male social drinkers. Journal of Abnormal Psychology, 1976, 85, 587-594.

BOOZE, BELIEFS, AND BEHAVIOR: COGNITIVE PROCESSES IN ALCOHOL USE

AND ABUSE

G. Terence Wilson

Rutgers University

New Brunswick, New Jersey

The study of alcohol use and abuse has for the most part been steeped in the biomedical model that emphasizes the overriding importance of the pharmacological properties and physiological consequences of alcohol. Of course, alcohol is a potent drug and biomedical analyses are vital to our fuller understanding of its effects. However, the inappropriate overextension of the biomedical model to psychological phenomena that cannot be reduced to the physical effects of alcohol has retarded the development of effective means for the assessment and treatment of alcohol abuse. It is for this reason that perhaps the most important feature of the behavioral approach lies in the alternative conceptual model it provides for understanding and modifying patterns of alcohol use and abuse. The details of the behavioral model are discussed elsewhere (Bandura, 1969). Suffice it to state here that this model entails a rejection of the quasi-disease or psychodynamic model of psychopathology and regards abnormal behavior that is not a function of specific brain disturbance or biochemical disorder as governed by the same principles that regulate normal behavior. Examples of alcohol-related phenomena that are customarily ascribed to the alleged physiological or psychodynamic effects of alcohol but which are more accurately explained in terms of a cognitive-behavioral analysis are discussed below.

It should be remembered that only two decades have passed since Wolpe (1958) wrote the landmark text that essentially ushered in the study and practice of behavior therapy. In that time the dual thrust of the adoption of a different model of abnormal behavior and the commitment - in principle if not always in practice - to scientific method, measurement, and evaluation (the defining characteristics of behavior therapy) has resulted in considerable progress in the assessment and treatment of diverse clinical disorders. Despite some wrong turns and conceptual cul-de-sacs along what, as Skinner (1975) has reminded us, is the "steep and thorny" road to scientific respectability, impressive and often unprecedented therapeutic accomplishments have been recorded (cf. Franks and Wilson, 1973, 1977; Kazdin and Wilson, in press). The systematic application of behavioral principles and procedures to alcohol use and abuse is a relatively recent development and as yet is not as advanced as other areas of behavioral research and therapy. However, early indications are that it promises to be as rewarding as the behavioral treatment of other disorders (cf. Marlatt and Nathan, 1978).

A SOCIAL LEARNING ANALYSIS

Thus far, the behavioral treatment of alcoholism has relied almost exclusively on the principles of classical and operant conditioning (Franks, 1970; Nathan and Briddell, 1977). However, just as behavior therapy in general has become increasingly more complex and sophisticated with the added emphases on self-regulatory functions and cognitive mediating processes, it is now time that behavioral approaches to alcohol use and abuse follow suite in embracing a broader cognitive social learning orientation (cf. Bandura, 1969; Mahoney, 1974; O'Leary and Wilson, 1975). This paper seeks to illustrate the importance of hitherto largely neglected cognitive processes in the understanding and modification of alcohol use and abuse.

Emphasis on cognitive factors in alcoholism is not new. McClelland's (1972) thesis, for example, that non-alcoholics drink to increase thoughts of socialized power where alcoholics are addicted to alcohol because of exaggerated needs for personalized power thoughts is a cognitive theory of alcohol use and abuse. A complete critique of this theory is beyond the scope of this paper, although some of the methodological flaws in the type of study on which McClelland (1972) bases much of his theorizing are described below. The advantages of a social learning analysis of the role of cognitive processes in clinical behavior change are discussed by Bandura (1977a, 1977b) and Wilson (in press a). Social learning theory accomodates existing empirical findings on behavior change; integrates the three psychological determinants of behavior of antecedant (classical conditioning), consequent (operant conditioning) and mediational (cognitive) influences in a consistent yet testable framework; and is heuristic in generating novel conceptual analyses and treatment methods. Unlike other cognitive theories of behavior change, cognitive mediating mechanisms in social learning theory are deliberately tied to overt action. This reciprocal determinism between cognitions and behavior is highlighted in the final section of this paper on the maintenance of treatment-produced sobriety or controlled drinking.

EXPECTANCIES AND THE BEHAVIORAL EFFECTS OF ALCOHOL

Available evidence now clearly shows that there is no simple, direct relation between the pharmacologic effect of alcohol and its behavioral consequences. For example, the expectations a person has about the nature and effects of alcohol can contribute importantly to its influence on human behavior. The influential role of expectation in determining alcohol's effects is illustrated in the following examples drawn from two ongoing research programs at the Rutgers Alcohol Behavior Research Laboratory.

Alcohol and Human Sexual Behavior

In the first study of the separate and interactive effects of alcohol and expectancy on sexual responsiveness, male social drinkers received one of four dose levels of alcohol (0.08, 0.4, 0.8, and 1.2 g/kg respectively) and one of two instructional sets designed to influence expectancies regarding changes in sexual arousal (Briddell and Wilson, 1976). Half the subjects were led to believe that alcohol increases sexual arousal, the other half, that it decreases sexual arousal. Measures of penile tumescence monitored continuously throughout the viewing of an erotic film showed a significant negative linear effect of alcohol. There was a noticeable trend in accordance with the induced expectancies about alcohol, especially at the low dose levels, although these results fell short of statistical significance. Thematic Apperception Test (TAT) measures of sexual imagery similar to those used by McClelland et al. (1972) failed to reflect the effects of alcohol or expectancies.

A more detailed investigation of the effects of expectancy and alcohol on sexual arousal was conducted by Wilson and Lawson (1976). Using the 2 x 2 factorial experimental design described by Marlatt, Demming, and Reid (1973), male social drinkers were randomly assigned to one of two expectancy conditions in which they were led to believe that the beverage contained vodka and tonic or tonic only. For half the subjects in each expectation condition the beverage actually contained vodka, the others drank only tonic. The alcohol dose administered was 0.5 g/kg, which resulted in a mean blood alcohol level of 40 mg% (range 30 - 50 mg%). The entire study was conducted under double-blind conditions; deceptions about alcohol allegedly administered were sustained in part through the use of preprogrammed false Breathalyzer readings. Multiple checks of the credibility and efficacy of the expectancy manipulation established that no subject doubted the authenticity of what he was told. Measures of penile tumescence were recorded during a heterosexual and homosexual film respectively.

The results are summarized in Figure 1. Alcohol per se failed
to affect penile tumescence significantly. However, there were
significant effects of the expectancy manipulation on tumescence
during both the heterosexual and homosexual films. Subjects who
believed that they had consumed alcohol manifested significantly
greater sexual arousal than those who believed that they had consumed only tonic water. As in the Briddell and Wilson (1976) study
TAT responses showed no effects of either alcohol or expectancy.
Using these TAT responses as a dependent measure, McClelland et al.
(1972) reported increased sexual fantasies in males during intoxication, a finding that was interpreted to support the notion that
men drink to satisfy a need for personal power. An increase in
sexual fantasies were equated with a sense of heightened personal
mastery. However, the demonstrably inadequate and unvalidated nature of this TAT measure plus other methodological problems in
studies on which this theory is partly based, at least with respect
to sexuality, undermine McClelland et al.'s (1972) position (cf.
Carpenter and Armenti, 1971; Wilson, 1977). The failure of better
controlled research to replicate their findings on the effect of
alcohol on fantasy is particularly damaging.

The conventional understanding of the effect of alcohol on
human activities such as sex and aggression is based on the belief
that alcohol depresses higher brain functions that exercise control or inhibition over such behavior in the sober state. The result is the loss of inhibitions - the disinhibition theory of alcohol's influence on human behavior. As Goodwin (1976) puts it, alcohol "releases" rather than increases sexual desire. Whether
interpreted in quasi-neurological terms of cortical disinhibition
or in the psychoanalytic sense of dissolution of the superego, the
disinhibiting effect of alcohol on sexual and aggressive behavior
has been assumed to be self-evident (cf. Chafetz and Demone, 1962;
Kessel and Wlaton, 1965).

However, the physiological disinhibition hypothesis is embarrassed by the available data. In contrast to the absence of direct
confirmatory evidence, Wilson and Lawson's (1976) results, replicat-
ing those of Lang, Goeckner, Adesso, and Marlatt (1975) with aggression, discredit the notion that alcohol influences sexual and aggressive behavior directly through some physiological mechanism.

In a partial replication and extension of Wilson and Lawson's
(1976) study, Briddell, Rimm, Caddy, Krawitz, Sholis, and Wunderlin
(in press) assessed the effects of alcohol and expectancy on penile
tumescence in response to audio tape-recordings of normal heterosexual and deviant sexual (forcible rape) stimuli in male social
drinkers. Expectancy but not alcohol per se significantly increased
penile tumescence to both normal and deviant sexual stimuli. This
expectancy effect was particularly marked for the forcible rape
tape. Subjects who believed that they had consumed alcohol irre-

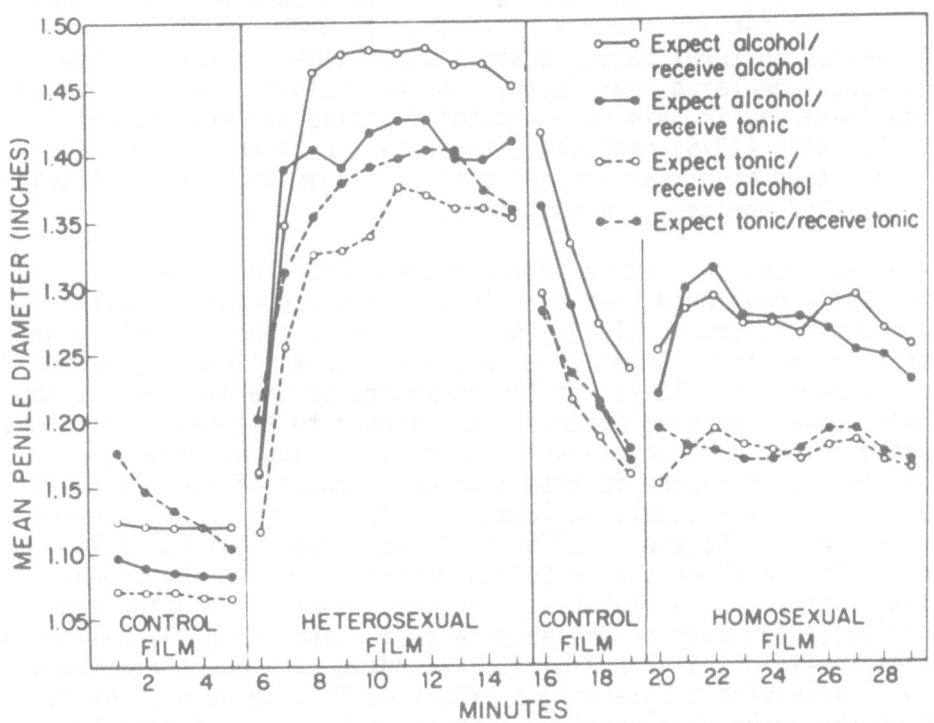

Figure 1: Effects of alcohol and expectancy on male sexual arousal (Wilson & Lawson, 1976).

spective of the actual contents of their drinks showed as much
arousal to the rape scene as to the convention heterosexual recording. When they believed themselves to be sober, arousal to the rape
tape decreased significantly.

Aside from strengthening Wilson and Lawson's (1976) results,
these data bear importantly on the frequently reported association
between sex offenses and alcohol consumption. Gebhard, Gagnon,
Pomeroy, and Christenson (1965), for example, found that alcohol
was more likely to be associated with the more aberrant sex crimes
involving children and the use of force. In order to break the two
most powerful cultural taboos against pedophilia and rape, Gebhard
et al. (1965) declared that "suspension or distortion of rationality
(is required) and in this case alcohol fulfills the requirement."
Similarly, Rada (1975) emphasizes the relation between rape and alcohol and speculated that alcohol acts as a specific (chemical trigger...mediated through...testosterone."

In reviewing the nature of the evidence for these claims,
Wilson (1977) cautioned that correlation is often confused with
cause and that unequivocal data on alcohol as a causal agent in sex
offenses is lacking. Briddell et al.'s (in press) results, among
others, suggest that it is not the chemistry of the alcohol but the
person's social learning history with respect to alcohol that is the
effective determinant of deviant behavior. It appears that the individual's belief system is the major determinant of the effect of
alcohol on sexual arousal, at least at moderate levels of intoxication. Wilson and Lawson's (1976) findings support Valins' (1966)
extension of Schachter's attribution theory of emotion by demonstrating that cognitive labeling can significantly influence emotional responding even in the absence of a pharmacologically-induced
state of physiological plasticity. According to Schachter's (1964)
theory, a significant interaction effect would have been predicted
in which the group that received alcohol would have differed from
the group given alcohol but told that it was tonic water. The two
groups that received tonic water only would not have been assumed
to differ because actual physiological arousal had not been elicited. However, as in the Lang et al. (1975) study, Wilson and
Lawson (1976) found no evidence for any interaction effect. Subjects who expected alcohol but received tonic water showed slightly
more sexual arousal than those who both expected and received alcohol.

At a more molar level of analysis of cross-cultural data,
MacAndrew and Edgerton (1969) have documented that societies exist
whose members' sexual behavior when intoxicated either manifests
nothing that can be interpreted as disinhibited or demonstrates
striking situational specificity, varying widely according to time,
place, and circumstance. "In and of itself," MacAndrew and
Edgerton (1969) concluded, "the presence of alcohol in the body does

not necessarily even conduce to disinhibition, much less inevitably produce such as effect" (p. 88).

If expectancy is a critical factor governing alcohol's influence on sexual responsiveness, how is this effect to be explained? As an alternative to the disinhibition hypothesis, Wilson (1977) has outlined a social learning analysis of alcohol's influence on sexual behavior. Unlike theories of behavior that assume that internalization of behavioral restraints creates a fixed and unitary internal moral agent (e.g., the superego or conscience) that either inhibits behavioral excesses or is temporarily disinhibited by some factor such as alcohol, social learning theory recognizes that self-control is more complex and flexible in nature. Inhibitory functions - what may be called self-evaluative reactions - do not operate automatically, and there are several cognitive and external influences that selectively influence whether or not they are triggered. Since the same behavior is not uniformly self-rewarded or self-punished irrespective of the circumstances under which it occurs, situational specificity results.

Among the different self-generated cognitive processes that selectively trigger or disengage inhibitory self-evaluative functions is the attribution of personal responsibility for actions to someone or something else. Alcohol provides an ideal source for the misattribution of culpability for behavioral transgressions. First, people learn about alcohol what their society "knows" about alcohol. As a result of the not infrequent performance of "disinhibited" actions when intoxicated by influential models, sanctioned - albeit subtly - by society, we 'know' that people will do things under the influence of alcohol that they would never otherwise do. Second, alcohol visibly impairs sensorimotor functions, thereby facilitating the inference by both the person and the public that social behavior is similarly affected. Third, alcohol is usually consumed in relaxed, convivial settings in which members of the opposite sex may be more responsive to sexual advances. In our socio-sexual cultural patterns man frequently meets woman over a drink. As Carpenter and Armenti (1971) pointed out, the pharmacologic effect of alcohol is confounded with sexual opportunity and provocation. The crucial effect of alcohol may not be on specific physiological responses, but defining, by its presence, a set of social role conditions that legitimize actions that otherwise would be looked at askance. In this sense alcohol acts as a cue or discriminative stimulus for sexual behavior.

The Tension Reduction Theory

The thesis that people drink alcohol in order to relieve or escape from aversive states of anxiety, frustration, or tension has a long history as an explanation of alcohol use and abuse (cf.

Cappell, 1974; Conger, 1956; Horton, 1943). This theory derives from the same reasoning as the disinhibition theory. Alcohol is assumed to reduce anxiety by virtue of its pharmacologically depressant effect on cortical control of behavior. Recently, the tension reduction theory has been rejected in some quarters on no better experimental grounds than it was originally uncritically accepted. Thus equivocal and even negative evidence from animal studies has been noted (Cappell, 1974), while anxiety and depression have been observed to increase rather than decrease following excessive drinking by alcoholics in the laboratory setting (McNamee, Mello, and Mendelson, 1968; Nathan and O'Brien, 1971).

Yet the data are far from clear and there are several inconsistencies in the literature to be accounted for. One of the major problems in interpreting the relevant findings is the methodological inadequacy of most of the human studies. In the first place, most studies have been limited to a single (usually subjective) measure of anxiety. However, it is now well-established that the assessment of anxiety requires multiple measurement of overt (avoidance) behavior, psychophysiological arousal, and self-report of distress (cf. Borkovec and O'Brien, 1976; Hodgson and Rachman, 1974; Lang, 1969). These different dimensions of anxiety may be differentially responsive to different forms of influence and change at different speeds. In the second place, the majority of studies have failed to control for various cognitive influences, such as expectation, in assessing alcohol's effects on anxiety.

The tension reduction theory consists of two related assumptions: (1) that alcohol reduces tension; and (2) that this tension-reducing effect motivates drinking. The present paper focuses on the first of these assumptions.

In the first study, Wilson and Abrams (1977) used the 2 x 2 design described above to investigate cognitive versus pharmacological processes in alcohol's effects on social anxiety. Male social drinkers were randomly assigned to one of two expectancy conditions in which they were led to believe that the beverage they consumed contained either vodka and tonic or tonic only. For half of the subjects in each expectancy condition the beverage contained vodka (.5 gm/kg); the others drank only tonic. After their drinks, subjects participated in a brief social interaction with a female confederate before, during, and after which multiple behavioral, physiological, and subjective measures of anxiety were obtained. The interpersonal interaction employed is a clinically relevant laboratory assessment of social anxiety developed by Borkovec, Stone, O'Brien, and Kaloupek (1974).

Figure 2 presents heart rate data from this study. In a conceptual replication of previous studies by Lang et al. (1975), Marlatt et al. (1973), and Wilson and Lawson (1976), a significant

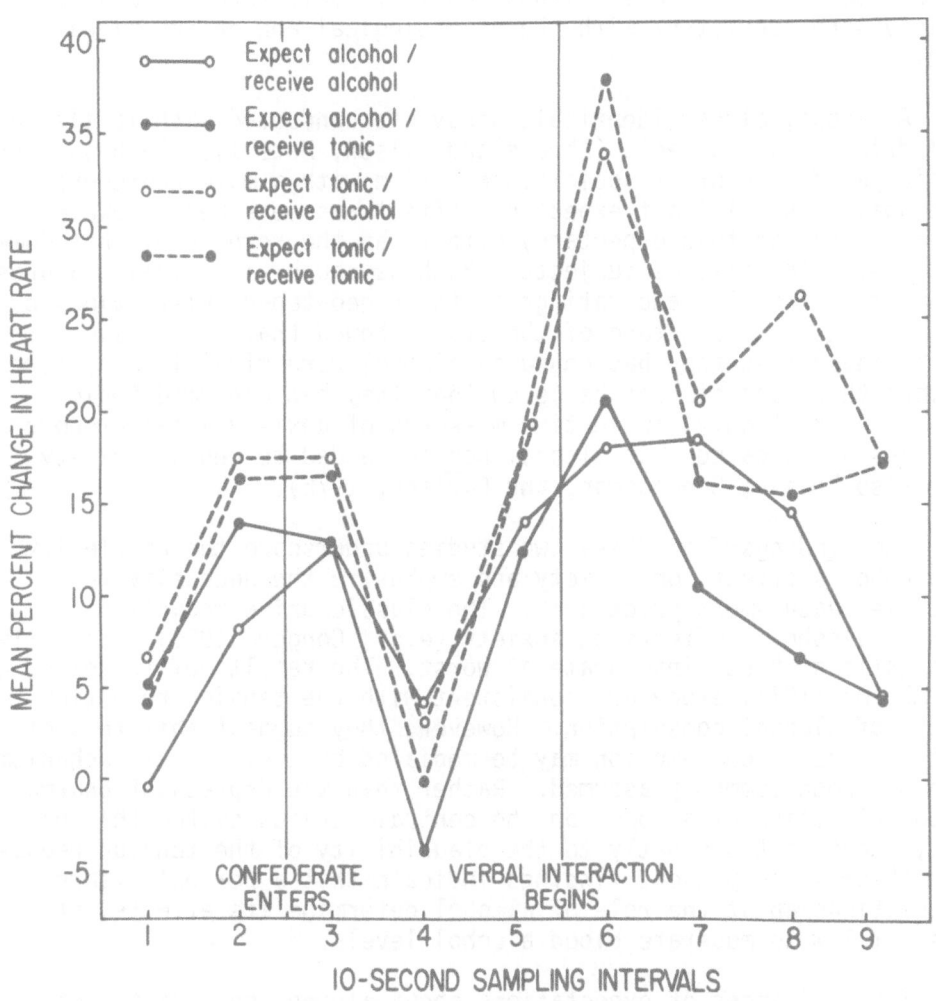

Figure 2: Effects of alcohol and expectancy on
heart rate (Wilson & Abrams, 1977).

expectancy but not alcohol effect was obtained. Subjects who believed that they had consumed alcohol showed significantly less increase in heart rate during this anxiety-eliciting interaction than
those who believed that they had consumed tonic, regardless of the
actual contents of their drinks. Systematic ratings of video-tapes
of these social interactions by raters who were unaware of the nature and purpose of the study similarly showed that subjects' expectation about alcohol significantly influenced their behavioral expressions of anxiety. Subjects who believed that they had consumed
alcohol behaved in a less anxious fashion. Self-report measures of
anxiety were consistent with the physiological and observational
data.

A second, almost identical, study was conducted with female social drinkers as subjects (Abrams and Wilson, Note 1). Each subject
participated in a brief social interaction with a male confederate.
Once again a significant expectancy effect was obtained. However,
the direction of this expectancy effect was the reverse of that demonstrated with males as subjects. Both psychophysiological measures
(heart rate and GSR) and ratings of the video-taped interactions by
judges blind to the nature of the study showed that female subjects
who believed that they had consumed alcohol were significantly more
anxious than subjects who believed that they had consumed tonic
only. Interestingly, subjective measures of anxiety - self-report
measures - indicated that alcohol per se tended to reduce anxiety
(see also Polivy, Schueneman, and Carlson, 1976).

The findings from these two studies underscore the complexity
of alcohol's effects on anxiety and emphasize the necessity for
multiple measurement procedures. The classic drive reduction notion of alcohol's effects on anxiety (e.g., Conger, 1956) is clearly
simplistic at best, inaccurate at worst. The results of the Wilson
and Abrams (1977) study are consistent with the tension reduction
theory of alcohol consumption. However, they suggest that this effect of alcohol consumption may be mediated by a different mechanism
than has been commonly assumed. Rather than the depressant pharmacological action of alcohol on the central nervous system that has
always contributed greatly to the plausibility of the tension reduction theory, the present findings indicate that a person's learned
expectations about the role of alcohol determine its effects, at
least at low to moderate blood alcohol levels.

The influences of expectations about alcohol on anxiety reactions, particularly the fact that there seems to be a sex difference
according to which the same expectancy can have diametrically opposite effects in men and women, suggests a partial explanation for
some of the conflicting evidence on the tension reduction theory wit
humans. Previous research has ignored the situational circumstances
surrounding the consumption of alcohol by focusing more narrowly on
the pharmacological properties of the drug. Washburne (1956) had

earlier cautioned about "a tendency to ignore the important social and psychological variables which help determine behavior in situations where alcohol is being used. Individuals...show wide differences in behavior which can be explained only by incorporating psychological and social situational variables. (These) factors change the actual physiological effects of alcohol, because they can affect the physiological system in just as 'real' a way as chemicals" (p. 122). Yet the cognitive context that influences drinking has been ignored despite Washburne's (1956) caveat. As with other cognitive influences on behavior (Mischel, 1968), cognitions concerning alcohol consumption may differ not only across individuals, depending on their respective social learning experiences with alcohol, but also across settings and at different times even in the same individual. Accordingly, variable outcomes on diverse measures of as broad a concept as "tension" should come as no surprise. Expectations constitute only one component of a social learning analysis of alcohol and its effects on social anxiety (cf. Wilson and Abrams, 1977), but they are great.

EXPECTATIONS AND THE MAINTENANCE OF TREATMENT-PRODUCED CHANGE

The foregoing discussion was restricted to the influence of expectations on the behavior of non-alcoholic subjects in the laboratory setting. However, since behavior is behavior is behavior, and because an adequate theory of alcohol abuse must, as McClelland (1972) points out, explain non-pathological use as well, research on social drinkers is clearly relevant to our understanding of problem drinkers. The specific manner in which expectations may directly affect the assessment and treatment of alcohol abusers is discussed next.

Conceptual Considerations

One of the most striking characteristics common to all addictive disorders is a high relapse rate following periods of treatment-produced improvement. Traditionally, the concepts of craving and loss of control are most often appealed to in explaining the phenomena of relapse in the alcoholic. Psychological craving - the overpowering subjective need for alcohol - is typically advanced in order to account for the first drink after a period of abstinence. Once drinking has been initiated following a period of abstinence, alcohol is said to trigger a physiological addictive process that results in compulsive drinking over which the alcoholic has no voluntary control. This process is usually described as loss of control. In Jellinek's (1960) massively influential view, craving and loss of control were intimately related and constituted the defining characteristics of the "disease" of alcoholism (see also Glatt, 1967; Keller, 1972).

The cogent and now familiar criticisms of the traditional concepts of craving and loss of control need not be rehashed here (cf.
Mello, 1975; Pattison, 1976b; Pomerleau, Pertschuk, and Stinnett,
1976). Rather, the purpose is to discuss briefly some of the components of a cognitive social learning conceptualization of a return
to addictive drinking following a period of abstinence. In terms
of a social learning analysis, the first drink after abstinence is
seen as a function of the faulty labeling of emotional or physical
states and anticipated reinforcing consequences. Numerous sources
of reinforcement attach to the consumption of alcohol, ranging from
feeling good to reducing anxiety, boredom, and frustration. Alcoholics are not deterred by the long-term destructive consequences of
their drinking because behavior is more powerfully controlled by
short-term consequences. The principle of the gradient of reinforce
ment is well-documented across species, situations, and behaviors
(cf. Ainslie, 1975). The alcoholic seems to anticipate the favorabl
short-term effects rather than the delayed negative impact of alcohol (e.g., Tamerin, Weiner, and Mendelson, 1970).

Expectations about the presumed reinforcing effects of alcohol
do not have to be veridical in order to influence behavior. Women
social drinkers, for example, reliably anticipate that alcohol consumption will increase their sexual responsiveness when in fact alco
hol produces a significant linear decreasing effect on sexual arousa
(Wilson and Lawson, Note 2). The abstinent alcoholic anticipates
the positive or negative reinforcement that alcohol produces - or at
least what he or she believes it produces. This anticipation is important since the expectation of reinforcement may be as powerful,
if not more so, than actual reinforcement (cf. Bandura, 1977b).
This desire for reinforcement is assumed to be labeled as craving
for alcohol by the alcoholic. Moreover, recent research has suggested a veridical source of physiological arousal that might be
easily misconstrued by the alcoholic as a craving for alcohol.
Lawson (Note 3) found that alcohol has a dipsogenic effect such that
alcoholics might label genuine thirst sensations as a craving for
alcohol per se.

It is very likely that cognitive factors play an important role
in determining alcohol consumption once the first drink is taken.
Thus, cognitive constructs feature prominently in two recent theoretical analyses of the determinants of relapse in alcoholics. According to the psychobiological theory of Ludwig and his colleagues
(Ludwig and Wikler, 1974; Ludwig, Wikler, and Stark, 1974), both
interoceptive (e.g., the pharmacological effects of alcohol) and
exteroceptive (e.g., environmental cues relevant to previous drinking practices) stimuli are necessary for the elicitation of craving
and alcohol-seeking behavior that results in relapse. However,
these stimuli are not sufficient conditions for relapse. The appropriate "cognitive set" has to be present. In this view, craving is
the "cognitive correlate" of a subclinical, conditioned withdrawal

syndrome that is evoked by internal and external stimuli. This
"cognitive correlate" of a psychophysiological imbalance "permits
the organism to engage in efficient, goal-directed, appetitive behavior," namely alcoholic drinking.

The empirical basis Ludwig et al. (1974) put forward in support
of their theory is severely critiqued by Marlatt (1978). Conceptually, the role of the "cognitive correlate" is vague and inchoate.
As evident from the studies reviewed in the first part of this paper,
expectations that can influence physiological responses and overt
behavior are not necessarily correlates of a physiological conditioned response; they can affect behavior in their own right.
Finally, theories such as those of Keller's (1972) and Ludwig et al.'s
(1974) rely heavily upon an automatic classical conditioning process. Bandura (1977b) has summarized data indicating that conditioned
reactions in humans are largely self-activated on the basis of learned
expectations.

Marlatt (1978) has suggested that there is a common denominator
in the relapse process across different addictive behaviors. He has
termed this the Abstinence Violation Effect (AVE). Briefly, the AVE
is said to consist of two basic cognitive components: A cognitive
dissonance effect and a personal attribution effect (Jones, Kanouse,
Kelly, Nisbett, Valins, and Weiner, 1972). In terms of the former,
the first drink following a period of abstinence arouses conflict or
guilt that is reduced by subsequent drinking and the realignment of
self-image with renewed drinking. According to the latter, the alcoholic attributes a return to drinking to personal failings. This
sense of personal inability to control drinking subsequently lowers
resistance to further temptation.

As Marlatt (1978) notes, the intensity of the AVE will vary
according to factors such as the degree of commitment or effort
expended in maintaining abstinence, the duration of abstinence, and
the importance of the behavior to the individual. To this list can
be added the degree to which the violation is perceived to be gra-
tuitous and excessive. Although the studies are not free from
methodological flaws, tentative support for the AVE can be gleaned
from the obesity literature (Herman and Mack, 1976; Polivy, 1976).
In what has been described as the counterregulatory effect, dieting
subjects subsequently consumed more food after eating a high caloric
snack whereas non-dieting subjects tended to consume less after eating what they perceived to be a high caloric snack. (Unfortunately,
subjects' perceptions of the high calorie snack were not independently assessed). In addition to the AVE (Marlatt, 1978) and the
counterregulatory effect (Polivy, 1976), Mahoney and Mahoney (1976)
have coined the term "cognitive claustrophia" to describe the same
pattern of excessive behavior triggered by a behavioral transgression following a period of self-imposed abstinence.

Self-efficacy Theory

With its early emphasis on the application of principles from the animal conditioning laboratory and its attendant anti-cognitive bias, behavior therapy only grudgingly came to recognize the role of expectancy in clinical behavior change (Mahoney, 1974; Wilson and Evans, in press). Then, although recognized, conceptual analyses applied to other dimensions of behavior change tended to by-pass the role of expectancy. Recently, however, Bandura (1977a) has presented a theoretically-integrated account of the contribution of expectations to the behavior change process. Instead of treating expectancy as a global, static, and unidimensional variable, Bandura (1977a) emphasized the need for specific analyses of different dimensions of expectations over time. Perhaps most importantly, he has drawn a distinction between efficacy and outcome expectations. An outcome expectancy is an individual's estimate that a given behavior will result in a certain outcome. An efficacy expectancy is the belief that one is capable of successfully carrying out the behavior required to produce the outcome.

Consider, then, the nature of the treatment process the alcoholic customarily undergoes, at least in the United States. The majority of therapies inculcate the belief that the alcoholic has an irreversible disease, that (s)he is different from non-alcoholics, that (s)he is uniquely vulnerable to the effects of alcohol, and that (s)he is and will always be unable to exercise voluntary control over consumption once drinking has been initiated. In short, the alcoholic is taught to believe in the traditional disease formulation of alcoholism, if (s)he already did not do so. The expectations that are created by this sort of therapeutic philosophy can have profound implications for thephenomena of relapse.

Recast in terms of social learning theory, the alcoholics' efficacy expectations about coping with alcohol are deliberately minimized. A fundamental tenet of Alcoholics Anonymous, for example, is that the alcoholic must admit his or her total weakness and inadequacy with respect to alcohol and accept wholeheartedly the notion that (s)he must recognize a greater power than him- or herself. As McClelland (1972) has put it, (s)he "must accept the power of God as a substitute for the power of the bottle to enhance his sense of potency" (p. 302). The outcome expectations that are developed stress the certainty of a return to uncontrollable drinking in the event of any subsequent consumption, regardless of the specific socio-psychological circumstances under which the drinking occurs.

The often startling abruptness with which an alcoholic who has been abstinent for a lengthy period of time can revert to addictive drinking has encouraged the theory of the biological inevitability of "first drink then drunk" by default if for no other reason. On the face of it, this represented a pattern of behavior apparently inexplicable in psychological terms. However, this behavioral pattern is consistent with predictions from a cognitive social learning analysis. First, the fact that an alcoholic has been abstinent even for a period of years does not necessarily mean that (s)he has attained a sense of self-efficacy about alcohol. It might be that that person still feels highly vulnerable to alcohol. Or as Pattison (1976a) has put it, there has been no "change in the symbolic meaning of the drinking." Second, an exaggerated belief in the "first drink then drunk" notion can result in a self-fulfilling prophecy. In taking the first drink after abstinence, the alcoholic reaffirms his or her inability to resist temptation along the lines of the attributional process described by Marlatt (1978).

Finally, a major prediction of Bandura's (1977a) theory is that the strength of efficacy expectations will determine whether an individual will initiate coping behavior, what degree of effort will be invested in that behavior, and how long it will be maintained in the face of obstacles and adverse experiences. Inevitably, the alcoholic's post-treatment success is related to these activities. Of course, expectations alone cannot account for the reasons for relapse in all alcoholics. A complete social learning analysis requires consideration, among other factors, of the individual's incentives for sobriety and his or her requisite behavioral skills for coping with the interpersonal difficulties of returning to continued sobriety. Given the necessary behavioral skills and adequate incentives, however, efficacy expectations may be a major determinant of the maintenance of treatment-produced improvement.

Therapeutic Strategies

It follows from the preceding analysis that different therapeutic expectations can be fostered that minimize the likelihood of relapse and maximize the chances of the alcoholic coping successfully with any "slip" by returning promptly to abstinence or, as is the case, a controlled drinking regimen. If this is to be done, the probability of posttreatment relapse should be explicitly addressed during therapy. It is too important a matter to be deferred to the occasion of relapse itself. Forewarned is forearmed and the development of specific strategies designed to facilitate generalization and maintenance of therapeutic change as part of an overall, integrative treatment program is characteristic of behavior therapy.

Bandura (1977a) has outlined how expectations of personal efficacy are derived from four main sources of input: verbal persuasion, physiological change, vicarious experience, and real-life performance. Verbal persuasion (e.g., insight-oriented psychotherapy) is a weak and inefficient method of affecting efficacy expectations and behavior change, to which the evidence on psychotherapy and alcoholism amply attests (Baekeland, Lundwall, and Kissin, 1975).

The techniques of behavior therapy have given promise to greater
efficacy.

Symbolic Methods

Both imaginal and verbal techniques may influence efficacy expectations. Guided imagery techniques include systematic desensitization as a form of self-control (e.g., Goldfried, 1971), symbolic (e.g., Bandura, 1977b), and covert modeling (e.g., Kazdin,
1974). In these techniques the client typically rehearses coping
responses to high risk situations for drinking. In addition, these
techniques should include imaginal rehearsal of adaptive behavior in
the face of a transgression.

This use of coping imagery can be usefully coupled with self-
instructional training as part of a comprehensive cognitive restructuring treatment (e.g., Lazarus, 1974; Mahoney, 1974; Meichenbaum,
1977) that enhances expectations of self-efficacy. It is not so
much the consumption of a drink or two of alcohol that catapults the
alcoholic back into addictive drinking as it is his or her personal
interpretation of the meaning of that drink. This reaction to a
violation of treatment-inspired sobriety can have the effect of
prompting the individual to apply him- or herself more diligently
than ever to the treatment program. The person learns a lesson from
the transgression that is helpful in avoiding similar setbacks in
the future. The slip does not serve as an opportunity to punish
oneself for personal weakness or to provide an excuse for further
drinking. For many clients, however, this behavioral transgression
is construed in a self-defeating manner that directly or indirectly
results in continued drinking.

Specifically in the latter instance, two major classes of irrational, self-defeating thoughts or self-statements have to be identified and corrected. One consists of catastrophizing reactions
(cf. Ellis, 1970). Examples include irrational thinking in which
the client transforms what may be an isolated, relatively minor
lapse into a total treatment failure that has undone the entire
preceding period of abstinence and that dooms him or her to continued abusive drinking in the future. The other involves rationalization processes. For example, the client may conclude that since one
drink has "blown" all his or her progress, (s)he may as well get
completely drunk, then return to a treatment program tomorrow - and
tomorrow, and tomorrow... Another is that a slip is taken as proof
that (s)he "needs" alcohol and that it is futile to fight this need.
Cognitive restructuring should be geared to scotching (sic) these
irrational and self-destructive thoughts and having clients replace
them with constructive self-statements. Examples include "Just because I had a drink does not mean that I've blown the whole program,
"I'd like a drink but I know I don't really need one to cope," "Now
that I've been drinking again means that I must get back on the

program immediately," and "I made a mistake taking that drink but
that does not mean that I'm a hopeless failure; it means I must try
harder next time."

In Vivo Procedures

As self-efficacy theory predicts and the current experimental
evidence demonstrates, performance-based treatment methods are significantly more effective in producing therapeutic changes on subjective and objective measures of psychological functioning than
methods that rely upon verbal, imaginal, or vicarious procedures
(Bandura, 1977a; Wilson, in press a). The fact that the two most
successful behavioral treatment programs for alcoholism yet reported (cf. Azrin, 1976; Sobell and Sobell, 1976) consisted of explicit performance-based treatment techniques is consistent with
this view. The implications for the maintenance of treatment effects as discussed here are more controversial.

The logic of this position dictates that, with appropriate
therapeutic preparation and instruction, the alcoholic client might
be deliberately encouraged to take a drink in the natural environment in the presence of cues that were formerly discriminative stimuli for excessive alcohol consumption. Initially, this might be
best accomplished with the therapist present so as to ensure adherence to therapeutic directions and to monitor carefully the client's
different reactions to the act of drinking. Thereafter, the therapist's presence could be faded, with the client engaging in self-
directed coping behavior.

In many ways, the performance-based treatment described above
is similar to the participant modeling (in vivo exposure) methods
that have recently been applied to phobic and obsessive-compulsive
clients with considerable success (Bandura, 1976; Rachman and Hodgson,
in press). In this procedure the client is systematically exposed
to the full range of actual stimuli that elicit the problem behavior
so as to neutralize maladaptive avoidance responses (i.e., drinking
to intoxication) and to replace them with constructive coping responses.

Although this therapeutic innovation represents a radical departure from the conventional clinical lore governing the treatment
of alcoholics, other cognitive-behavioral psychologists have proposed
essentially similar procedures. Marlatt (1978) has suggested a "programmed relapse" to anticipate the adverse impact of the AVE.
Mahoney and Mahoney (1976) instructed an obese client to eat small
quantities of formerly forbidden foods in order to defuse feelings
of deprivation and restriction. According to the Mahoneys, the
feeling of being "suffocated" by the ban on "forbidden fruits" produced a state of "cognitive claustrophobia". In breaking out of

this sense of psychological constriction, the client typically engaged in extravagant binges. Lastly, Ellis (1970) has long advocated that clients be able to engage in the formerly problematic behavior without catastrophizing and by reacting rationally. The sense of this discussion can be summarized by stating that it is judicious prescription rather than absolute proscription that is called for.

There will be those who will object to the planned rehearsal of potential relapse experiences during therapy on the grounds that it may encourage drinking. A realistic appraisal of the treatment process suggests otherwise. First, anticipating future difficulties and potential posttreatment pitfalls is an integral part of sound clinical practice in general (cf. Wilson and Evans, 1976). Second, there is mounting evidence that even in the majority of abstinence-oriented treatment programs, a number of alcoholic clients do occasionally consume alcohol and yet remain treatment successes (cf. Armor, Polich, and Stambul, 1976). To avoid discussion during therapy of the probability of posttreatment drinking by the alcoholic client is to ignore available evidence and deny reality.

It is important to reiterate that a social learning approach is not synonymous with the treatment goal of controlled drinking. Nor is self-efficacy theory necessarily inconsistent with abstinence as a therapeutic goal. Despite some current practices, simply because therapy is geared towards total abstinence does not necessitate the development of counter-productive expectations that create a self-fulfilling prophecy about the inevitable effects of transgressions from the treatment regimen. Self-efficacy theory is consonent with the goal of controlled drinking but its relevance for controlled drinking at present is how to select those alcohol abusers who are best suited to this treatment objective. The systematic assessment of efficacy expectations promises to be useful in this respect. For example, analyses of clients who initially show controlled drinking behavior should reveal a range of efficacy expectations about continued success in regulating drinking. Controlled drinking as a long-term therapeutic goal might be pursued only with those clients who achieve a certain level of reality-based self-efficacy with regard to drinking. As with other clinical disorders, few if any reliable predictor variables of treatment success exist (e.g., Wilson, in press b). Bandura (1977a), however, has demonstrated that efficacy expectations may be an unusually good predictor of generalized behavior change, superior even to past behavior, for example. Accordingly, the extent to which assessment of efficacy expectations helps predict posttreatment maintenance of controlled drinking or abstinence is an open empirical question that warrants careful research attention.

A CONCLUDING NOTE

The preceding discussion has suggested several ways in which cognitive processes may significantly advance our understanding and treatment of alcohol use and abuse. Given the current climate of enthusiasm for the cognitive connection in psychology in general and clinical practice in particular (Wilson, in press a), it would be well to caution that much of the research cited in this paper is preliminary in nature and that the therapeutic strategies discussed here are frankly speculative. The alternative conceptual analyses and treatment recommendations outlined above are intended not so much to convince as to spur innovative, controlled investigations of hitherto neglected areas.

The real significance of the view expressed here and in Marlatt's (1978) closely-related paper lies in the application of the principles and procedures of experimental-clinical psychology to alcohol use and abuse. The study of alcohol abuse has for too long been dominated by a limited biomedical model and pseudo-psychological concepts of dubious validity. Enough of the conceptual insularity and often protective professionalism of "alcohology" and "alcohologists". Infusing the study of alcohol abuse with the theoretical and methodological rigor of psychological science promises to be one of the most exciting developments in the field to date.

ACKNOWLEDGMENTS

Preparation of this manuscript was completed while the author was a Fellow at the Center for Advanced Study in the Behavioral Sciences, Stanford, California.

The invaluable contributions of David B. Abrams, Dan W. Briddell, and David M. Lawson and the members of the staff of the Rutgers Alcohol Behavior Research Laboratory to the research outlined in this paper are gratefully acknowledged. Some of the research reported here was supported by NIAAA Grant No. 00259.

REFERENCE NOTES

1. Abrams, D. B. & Wilson, G. T. The effects of alcohol on social anxiety in women: Cognitive and pharmacological processes. Unpublished manuscript, Rutgers University, 1978.

2. Wilson, G. T., & Lawson, D. M. Expectancies, alcohol, and sexual arousal in women. Unpublished manuscript, Rutgers University, 1978.

3. Lawson, D. M. The dipsogenic effect of alcohol and the loss of control phenomenon. Unpublished manuscript, Rutgers University, 1978.

REFERENCES

Ainslie, G. Specious reward: A behavioral theory of impulsiveness and impulse control. Psychological Bulletin, 1975, 87, 463-496.

Armor, D. J., Polich, J. M., & Stambul, H. B. Alcohol and treatment Santa Monica, CA: The Rand Corporation, 1976.

Azrin, N. H. Improvements in the community-reinforcement approach to alcoholism. Behaviour Research and Therapy, 1976, 14, 339-348.

Baekeland, F., Lundwall, L., & Kissin, B. Methods for the treatment of chronic alcoholism: A critical appraisal. In R. J. Gibbins, Y. Israel, H. Kalant, R. E. Popham, W. Schmidt, & R. G. Smart (Eds.), Research advances in alcohol and drug problems, Vol. II. Toronto: Addiction Research Foundation, 1975.

Bandura, A. Principles of behavior modification. New York: Holt, Rinehart & Winston, 1969.

Bandura, A. Effecting change through participant modeling. In J. Krumboltz & C. E. Thoresen (Eds.), Counseling methods. New York: Holt, Rinehart & Winston, 1976.

Bandura, A. Self-efficacy: Toward a unifying theory of behavior change. Psychological Review, 1977, 84, 191-215(a).

Bandura, A. Social learning theory. Englewood Cliffs, NJ: Prentice-Hall, 1977(b).

Borkovec, T. D., Stone, N. M., O'Brien, G. T., & Kaloupek, D. G. Evaluation of a clinically relevant target behavior for analogue outcome research. Behavior Therapy, 1974, 5, 504-514.

Borkovec, T. D., & O'Brien, G. T. Methodological and target behavior issues in analogue therapy research. In M. Hersen, R. M. Eisler, & P. M. Miller (Eds.), Progress in behavior modification, Vol. 3. New York: Academic Press, 1976.

Briddell, D. W., Rimm, D. C., Caddy, G. R., Kravitz, G., Sholis, D., & Wunderlin, R. J. The effects of alcohol and cognitive set on sexual arousal to deviant stimuli. Journal of Abnormal Psychology, in press.

Briddell, D. W., & Wilson, G. T. The effects of alcohol and expectancy set on male sexual arousal. Journal of Abnormal Psychology, 1976, 85, 225-234.

Cappell, H. An evaluation of tension models of alcohol consumption. In R. J. Gibbins, Y. Israel, H. Kalant, R. E. Popham, W. Schmidt, & R. G. Smart (Eds.), Research advances in alcohol and drug problems, Vol. II. New York: Wiley, 1974.

Carpenter, A., & Armenti, N. P. Some effects of ethanol on human sexual and aggressive behavior. In B. Kissin & H. Begleiter (Eds.), The biology of alcoholism. New York: Plenum Press, 1971.

Chafetz, M. E., & Demone, H. W. Alcoholism and society. New York: Oxford University Press, 1962.

Conger, J. J. Alcoholism: Theory, problem and challenge. II. Reinforcement theory and the dynamics of alcoholism. Quarterly Journal of Studies on Alcohol, 1956, 17, 296-305.

Ellis, A. The essence of rational psychotherapy: A comprehensive approach to treatment. New York: Institute for Rational Living, 1970.

Franks, C. M. Alcoholism. IN C. G. Costello (Ed.), Symptoms of psychopathology. New York: Wiley, 1970.

Franks, C. M., & Wilson, G. T. Annual review of behavior therapy: Theory and practice, Vol. I. New York: Brunner/Mazel, 1973.

Franks, C. M., & Wilson, G. T. Annual review of behavior therapy: Theory and practice, Vol. V. New York: Brunner/Mazel, 1977.

Gebhard, P. H., Gagnon, J. H., Pomeroy, W. B., & Christenson, C. B. Sex offenders. New York: Harper & Row, 1965.

Glatt, M. M. The question of moderate drinking despite "loss of control." British Journal of the Addictions, 1967, 62, 267-274.

Goldfried, M. R. Systematic desensitization as training in self-control. Journal of Consulting and Clinical Psychology, 1971, 37, 228-234.

Goodwin, D. Is alcoholism hereditary? New York: Oxford University Press, 1976.

Herman, C. P., & Mack, D. Restrained and unrestrained eating. Journal of Personality, 1975, 43, 647-660.

Hodgson, R., & Rachman, S. II. Desynchrony in measures of fear. Behaviour Research and Therapy, 1974, 12, 319-326.

Horton, D. The functions of alcohol in primitive societies: A cross-cultural study. Querterly Journal of Studies on Alcohol, 1943, 4, 201-320.

Jellinek, E. M. The disease concept of alcoholism. New Brunswick, NJ: Hillhouse Press, 1960.

Jones, E. E., Kanouse, D. E., Kelley, H. H., Nisbett, R. E., Valins, S., & Weiner, B. Attribution: Perceiving the causes of behavior. Morristown, NJ: General Learning Press, 1972.

Kazdin, A. E. Effects of covert modeling and modeling reinforcement on assertive behavior. Journal of Abnormal Psychology, 1974, 83, 240-252.

Kazdin, A. E., & Wilson, G. T. Evaluation of behavior therapy: Issues, evidence and research strategies. Cambridge, Mass: Ballinger, 1977.

Keller, M. On the loss-of-control phenomenon in alcoholism. British Journal of Addiction, 1972, 67, 153-166.

Kessel, N., & Walton, H. Alcoholism. Hammondsworth, England: Penguin Books, 1965.

Lang, A. R., Goeckner, D. J., Adesso, V. G., & Marlatt, G. A. Effects of alcohol on aggression in male social drinkers. Journal of Abnormal Psychology, 1975, 84, 508-518.

Lang, P. J. The mechanics of desensitization and the laboratory study of fear. In C. M. Franks (Ed.), Behavior therapy: Appraisal and status. New York: McGraw-Hill, 1969.

Lazarus, A. A. Desensitization and cognitive restructuring. Psychotherapy: Theory, Research and Practice, 1974, 11, 98-102.

Ludwig, A. M., & Wikler, A. "Craving" and relapse to drink. Quarterly Journal of Studies on Alcohol, 1974, 35, 108-130.

Ludwig, A. M., Wikler, A., & Stark, L. H. The first drink: Psychobiological aspects of craving. Archives of General Psychiatry, 1974, 30, 539-547.

MacAndrew, C., & Edgerton, R. B. Drunken comportment. Chicago: Aldine, 1969.

Mahoney, M. J. Cognition and behavior modification. Cambridge, Mass: Ballinger, 1974.

Mahoney, K., & Mahoney, M. J. Cognitive factors in weight reduction. In J. D. Krumboltz & C. E. Thoresen (Eds.), Counseling methods. New York: Holt, Rinehart and Winston, 1976.

Marlatt, G. A. Craving for alcohol, loss of control, and relapse: A cognitive-behavioral analysis. In P. E. Nathan & G. A. Marlatt (Eds.), Alcoholism: New directions in behavioral research and treatment. New York: Plenum Press, 1978.

Marlatt, G. A., Demming, B., & Reid, J. Loss of control drinking in alcoholics. Journal of Abnormal Psychology, 1973, 81, 233-241.

Marlatt, G. A., & Nathan, P. E. Behavioral assessment and treatment of alcoholism. New Brunswick, NJ: Rutgers Center for Alcohol Studies, 1978.

McClelland, D. C. Examining the research basis for alternative explanations of alcoholism. In D. C. McClelland, W. N. Davis, R. Kalin, & E. Wanner (Eds.), The drinking man. New York: The Free Press, 1972.

McClelland, D. C., Davis, W. N., Kalin, R., & Wanner, E. The drinking man. New York: The Free Press, 1972.

McNamee, H. B., Mello, N. K., & Mendelson, J. H. Experimental analysis of drinking patterns of alcoholics: Concurrent psychiatric observations. American Journal of Psychiatry, 1968, 124, 1063-1069.

Meichenbaum, D. Cognitive behavior modification. New York: Plenum Press, 1977.

Mello, N. A semantic aspect of alcoholism. In H. D. Cappell, & A. E. LeBlanc (Eds.), Biological and behavioral approaches to drug dependence. Toronto: Addiction Research Foundation, 1975.

Mischel, W. Personality and assessment. New York: Wiley, 1968.

Nathan, P. E., & Briddell, D. W. Behavioral assessment and treatment of alcoholism. In B. Kissin & H. Begleiter (Eds.), The biology of alcoholism, Vol. 5. New York: Plenum Press, 1977.

Nathan, P. E., & O'Brien, J. S. An experimental analysis of the behavior of alcoholics and non-alcoholics during prolonged experimental drinking: A necessary precursor of behavior therapy? Behavior Therapy, 1971, 2, 455-476.

O'Leary, K. D., & Wilson, G. T. Behavior Therapy: Application and outcome. Englewood Cliffs, NJ: Prentice-Hall, 1975.

Pattison, E. M. A conceptual approach to alcoholism treatment goals. Addictive Behaviors, 1976, 1, 177-192 (a).

Pattison, E. M. Nonabstinent drinking goals in the treatment of alcoholics. In R. J. Gibbins, Y. Israel, H. Kalant, R. E. Popham, W. Schmidt, & R. G. Smart (Eds.), Research advances in alcohol and drug problems, Vol. III. New York: Wiley, 1976 (b).

Polivy, J. Perception of calories and regulation of intake in restrained and unrestrained subjects. Addictive Behaviors, 1976, 1, 237-244.

Polivy, J., Schueneman, A. L., & Carlson, K. Alcohol and tension reduction: Cognitive and physiological effects. Journal of Abnormal Psychology, 1976, 85, 595-600.

Pomerleau, O., Pertschuk, M., & Stinnett, J. A critical examinatior of some current assumptions in the treatment of alcoholism. Journal of Studies on Alcohol, 1976, 37, 849-867.

Rachman, S., & Hodgson, R. Obsessions and compulsions. Englewood Cliffs, NJ: Prentice-Hall, in press.

Rada, R. T. Alcohol and rape. Medical Aspects of Human Sexuality, 1975, 9, 48-65.

Schachter, S. The interaction of cognitive and physiological determinants of emotional state. In L. Berkowitz (Ed.), Advances in experimental social psychology. New York: Academic Press, 1964.

Skinner, B. F. The steep and thorny way to a science of behavior. American Psychologist, 1975, 30, 42-49.

Sobell, M. B., & Sobell, L. C. Second year treatment outcome of alcoholics treated by individualized behavior therapy: Results Behaviour Research and Therapy, 1976, 14, 195-216.

Tamerin, J. S., Weiner, S., & Mendelson, J. H. Alcoholics' expectancies and recall of experiences during intoxication. American Journal of Psychiatry, 1970, 126, 1697-1704.

Valins, S. Cognitive effects of false heart-rate feedback. Journal of Personality and Social Psychology, 1966, 4, 400-408.

Washburne, L. Alcohol, self and the group. Quarterly Journal of Studies on Alcohol, 1956, 17, 108-123.

Wilson, G. T. Alcohol and human sexual behavior. Behaviour Research and Therapy, 1977, 15, 239-252.

Wilson, G. T. Cognitive behavior therapy: Paradigm shift or passing phase? In J. P. Foreyt & D. Rathjen (Eds.), Cognitive behavior therapy: Research and application. New York: Plenum Press, (a).

Wilson, G. T. Methodological considerations in treatment outcome research on obesity. Journal of Consulting and Clinical Psychology, in press (b).

Wilson, G. T., & Abrams, D. B. Effects of alcohol on social anxiety and physiological arousal: Cognitive versus pharmacological processes. Cognitive Therapy and Research, 1977, in press.

Wilson, G. T., & Evans, I. M. Adult behavior therapy and the therapist-client relationship. In C. M. Franks & G. T. Wilson (Eds.), Annual review of behavior therapy: Theory and Practice, Vol. IV. New York: Brunner/Mazel, 1976.

Wilson, G. T., & Evans, I. M. The therapist-client relationship in behavior therapy. In R. S. Gurman & A. M. Razin (Eds.), The therapist's contribution to effective psychotherapy: An empirical approach. New York: Pergamon, in press.

Wilson, G. T., & Lawson, D. M. Expectancies, alcohol and sexual arousal in male social drinkers. Journal of Abnormal Psychology, 1976, 85, 587-594.

Wolpe, J. Psychotherapy by reciprocal inhibition. Stanford: Stanford University Press, 1958.

CRAVING AND LOSS OF CONTROL

Ray Hodgson, Howard Rankin, and Tim Stockwell

Addiction Research Unit, Institute of Psychiatry

London

CRAVING AND LOSS OF CONTROL

At the heart of the disease model of alcoholism is a poorly understood and poorly defined concept usually called "loss of control" which, for most psychologists, is on its last legs if not actually down and out. Our aim in this paper is to try to bring it in from the cold, dust it down and generally make it more presentable to the behavioural scientist. Let us begin with an alcoholic's view of craving and loss of control:

> After six years of not taking one drop of alcohol, I got to arguing that one small sherry could not possibly do any harm and so one night I took that fatal first drink. To my utter disillusionment within a fortnight I was drinking two bottles of gin a day...It had never for one moment entered my head that I could ever find myself in that same desperate position again. The old craving had fully reestablished itself and I had lost all means of control...Once an alcoholic always an alcoholic is one hundred percent true and if I ever take a drink, say, in ten, fifteen or twenty years time I know with my heart and with my head that I should always react in exactly the same way...Is alcoholism a disease? I know to my cost that it is (Williams, 1967)

Here we have the whole concept described very clearly and we see that, for this alcoholic, one drink appears to set off a chain reaction resulting in the loss of all means of control. It is also evident that this chain reaction is considered to be an irreversible disease. This view is often known by the catch phrase "one drink, one drunk" and it is this version of loss of control that we will consider first.

341

ONE DRINK, ONE DRUNK

In 1952 Jellinek published his widely quoted description of loss of control in the Quarterly Journal of Studies on Alcohol; his view of the phenomenon is roughly the same as the alcoholic's view presented above. He states that:

> Loss of control means that any drinking of alcohol starts a chain reaction which is felt by the drinker as a physical demand for alcohol. This state, possibly a conversion phenomenon, may take hours or weeks for its full development; it lasts until the drinker is too intoxicated or too sick to ingest more alcohol.

This view has been so well publicised that it is now part of the folk lore of our society. There is a tendency to believe that as soon as one shot of whisky hits the central nervous system, a switch is thrown and self control processes are shut down for the duration of the ensuing drinking binge. Consider, for example, the case of Mr. Powell (quoted by Keller, 1972), a laborer who was arrested for public drunkeness. On his way to the trial in Travis County, Texas, he had stopped at a bar, consumed one drink and only one drink. On discovering this fact, the counsel for the prosecution successfully made use of it by arguing that Mr. Powell had shown some control over one drink and therefore could not claim to be the helpless victim of a disease, the assumption being that a true alcoholic has no control at all. Most people involved with alcoholics know that this is a wrong assumption; whether or not a single drink leads to a drinking binge depends upon the alcoholic's mood, his situation and his reason for taking the drink. Within a hospital or laboratory environment, for instance, when drinking is part of an experiment it has been shown that consuming one single drink does not lead to craving and loss of control. For example, two well-designed experiments assessed the effects of single drink and placebo priming doses on drinking behaviour (Engle and Williams 1972; Marlatt, Demming and Reid, 1973). Engle and Williams demonstrated that neither priming dose increased the tendency to consume an available drink while Marlatt, Demming and Reid found no difference between the alcohol and placebo priming doses on later consumption during an alcohol taste test. We don't want to spend too long on the "one drink, one drunk" view of loss of control lest we will be accused of burning straw men. Today most psychologists and psychiatrists with experience with alcoholic drinking behavior no longer believe it to be true. One possible explanation of the phenomenon is that a small drink containing one ounce of vodka does not trigger loss of control but that a few drinks, or a few hours of drinking, can switch off some central controlling mechanism. We will now consider this possibility in the light of experimental results published during the last ten years.

SOME DRINKS, ONE DRUNK

Consider an alcoholic who is allowed to drink up to a bottle of spirits every day, for one month or more, in a hospital environment. One possible outcome, predicted by the loss of control notion, is that once he starts he will have no control and will be unable to reduce or stop his drinking when drink is available. In fact, there are many reports of alcoholics reducing and stopping their drinking during this type of prolonged drinking program. For example, Mello and Mendelson (1971) report that when alcoholic subjects are given the opportunity to work for alcohol in an operant situation they frequently stay abstinent for one whole day during which time they work in order to obtain enough alcohol to keep them going for a few more days. On these abstinent days blood alcohol levels are minimal or zero, withdrawal symptoms are clearly observed, and yet the alcoholic does not drink. He prefers to spend a whole day working to accumulate a new stock of alcohol. It would appear that these planned days of abstinence are better described as periods of self-control rather than loss of control.

Further evidence of control is provided by Gottheil and his colleagues (1972), who studied the drinking decisions of 25 alcoholics. Subjects were allowed up to 2 ounces of 40% ethanol per hour, throughout the day, for 4 weeks. Of relevance to the present discussion is the fact that nine subjects started to drink and yet were totally abstinent during the last week of the study, even though drink was still available. Four of these subjects drank heavily for at least one week and yet they still decided to be abstinent. It has also been reported that alcoholics taper their drinking towards the end of a period of prolonged drinking, presumably in order to avoid withdrawal symptoms when drinking must end.

We conclude from this research that loss of control is certainly not an inevitable consequence of consuming drink even for the severely dependent alcoholics. If the concept is to be retained then we must speak of a relative loss of control or an increased probability of consuming drink.

Of course, we have no idea whether subjects in the prolonged drinking experiment did have any desire to continue drinking on those days when they decided not to drink. It is possible for an alcoholic to crave a drink without drinking just as it is possible for a dieter to crave a chocolate biscuit without indulging. We must, therefore, modify the hypothesis to be tested. We know that the consumption of alcohol, even in heavy doses, does not inevitably lead to further drinking, even in the severely dependent alcoholic. We must now ask whether the consumption of alcohol increases the desire to consume more alcohol.

Instead of speaking of the urge to drink, the desire to drink, the compulsion to drink, or the disposition to drink, we will use the term "craving" even though this word has often been rejected by behavioral researchers. We will use craving in the same way that psychologists often use the label "fear," i.e., to refer to a multi-dimensional construct involving subjective, behavioral, and physiological response systems which are partially coupled (e.g., Hodgson and Rachman, 1974; Lang, 1971; Rachman and Hodgson, 1974). Viewed in this way, craving is simply a label referring to a cluster of responses. One of our first tasks must be to anchor the concept to objectively measurable responses.

"One drink, one drunk" turns out to be an invalid notion and even a period of heavy drinking is not inevitably followed by loss of control. We will now ask whether the cognitive and physiological effects of drinking alcohol can become cues which elicit or at least influence further craving.

SOME DRINKS, SOME CRAVING

As part of treatment we recently persuaded an alcoholic who expressed a strong desire for alcohol to sit with an open bottle of whisky and to sniff it without drinking it. He did this for half an hour and did not drink. However, he did express anger, he did stare at the bottle and then stare out of the window, he did beg for just one drink and furthermore his hands were shaking and his pulse increased from 90 to 125. We call this state high craving whether it is triggered by drink or by another set of internal or external cues. A very slight craving may involve no measurable physiological component and only minimal behavioral changes. We must stress that we use the word craving simply to label a system of partially interrelated responses. We must determine which responses are involved, to what extent they are coupled and under what circumstances they tend to vary independently.

Ludwig, Wikler and Stark (1974) tested the possibility that the consumption of alcohol can increase the subjective, behavioral and physiological components of craving. Their subjects were 24 hospitalized alcoholics and each subject was tested for craving after consuming a placebo dose, a low dose (0.6ml/kg.wt), or a high dose (1.2ml/kg.wt) of alcohol. Half of the subjects were also exposed to strong alcohol cues (i.e., taste, sight and smell of alcohol). Subjects were informed that the study was designed "to learn more about alcoholism through the administration of sedative, stimulant, and alcohol-like drugs" and, so, only those subjects who were exposed to alcohol cues had reason to believe that they were actually drinking alcohol. The main finding of this experiment was a significant increase in craving after both low and high doses but only for the group led to believe it had consumed alcohol. Ludwig and his colleagues used an operant measure of craving - how many button presses the alcoholic was prepared to make in order to obtain a drink at the rate of approximately one double per 600 presses. Changes in hand tremor and subjective desire for the drink also suggested that these components of craving were influenced in the same way as the behavioral component.

Funderburk and Allen (1977) also made use of an operant measure in their investigation of craving the morning after consumption of high doses of alcohol. In this study the assumption was made that subjects who desire alcohol will prefer an immediate rather than a delayed drink and so reduction in delay was made contingent upon number of button presses. Pressing a button 300 times would allow the alcoholic to consume 2 ounces of alcohol (just over one double) at 11:30 a.m. whereas pressing 5,200 times (for example) would allow him to consume the same drink at 8 a.m. The study lasted 15 days, during which time the daily dose of alcohol was increased from 16 ounces to approximately 30 ounces of 47.5% alcohol. Although only four subjects were tested, there was a significant dose effect: Alcoholics were more likely to work in order to obtain an immediate rather than a delayed drink if they had consumed appxoximately one bottle of spirits the day before.

One part of our own research program has been directed to this issue. We have attempted to test the hypotheses that drink can prime craving and also that, if present, this phenomenon will be a function of severity of dependence. Before testing these hypotheses, we decided to look for a simple measure of craving which would be suitable for both observational and experimental investigations. Our first attempt to validate a measure, based upon amount of drink consumed, was totally unsuccessful because our severely dependent alcoholics usually consumed all of the drinks that we made available. We then hypothesized that speed of drinking, especially speed of consuming the first drink, might be a reasonable measure. In order to validate this behavioral measure we visited drinking alcoholics in their own homes and persuaded them to resist alcohol for either a three-hour or a half-hour period (i.e., a high craving or a low craving condition). There was a very significant difference in the time taken to consume one glass of vodka (30 mls) in the high and low craving conditions.

Having validated a behavioral measure of craving we were now in a position to test the hypothesis that a dose of alcohol can have a priming effect on craving. The priming dose was administered in the morning (at approximately 10 a.m.) and craving was tested 3 hours later, by which time blood alcohol levels had returned to near zero levels. Twenty volunteer in-patient alcoholics were tested under high, low or no priming dose conditions, on 3 separate days, after at least ten days of abstinence. In the "high primer" condition 150 mls of vodka were consumed with tonic while 15 mls of

vodka were consumed in the "low primer" condition. No drink at
all was administered in the "no primer" condition. Three hours
after the priming dose subjects rated their anxiety and desire for
a drink and pulse was also measured. Subjects were then told that
they would be given the option of consuming more alcohol. Five
drinks were lined up, each containing 30 mls of vodka with 30 mls
of tonic and the following instructions were given: "I am leaving
you now for half an hour. Drink as little or as much as you like
in that time. In half an hour I will return to ask you how you
feel and also to take a few measurements. Occasionally I will look
through this one-way screen to make sure that you are O.K." Unobtrusive measures of time between sips, time between glasses and
time to consume each glass were noted throughout the half hour period.

     Subjects were all hospitalized alcoholics who volunteered to
take part in the experiment. They were classified by an independen†
psychiatrist as severely or moderately dependent. Severity of dependence was defined very simply as "frequent drinking to escape an‹
avoid withdrawal symptoms during the six months prior to admission.'
The psychiatrist (Dr. Griffith Edwards) is very experienced in
clinical judgments of this kind. Moreover, a behavioral measure of
severity of dependence derived from speed of drinking and amount
consumed correlated highly with the psychiatric rating of dependence. Eleven of the 20 volunteers were classified as severely dependent.

RESULTS

     In order to cope with the repeated measures design, a multi-
variate analysis of variance was performed.

Differences between Severely and Moderately Dependent Groups

Ignoring Priming Dose Conditions

     The severely dependent group reported greater desire for a
drink ($p < .03$) and they consumed significantly more alcohol during
the behavior tests ($p < .007$). On the whole they were quicker to
take the first sip ($p < .01$) and also the first glass ($p < .008$).

Differences between Priming Dose Conditions

Ignoring Severity of Dependence

     The only difference between conditions, ignoring severity of
dependence, was a higher pulse three hours after consuming the high

dose than after both the "low dose" ($p < .08$) and the "no dose" ($p < .01$).

Interactions between Priming Dose Conditions

and Severity of Dependence

There were significant interaction effects on both number of sips and also time taken to consume the first glass. The severely dependent subjects consumed the first afternoon drink in fewer sips after the high priming dose than after no primer. This was not the case for the moderately dependent subjects and this interaction was statistically significant ($p < .01$).

The major finding of the present investigation is a significant interaction on our behavioral measure of craving, speed of consuming one drink. The severely dependent subjects consumed the first afternoon drink quicker in the high priming dose condition than in the other two conditions. This trend was reversed in the moderately dependent group and the interaction was significant for both the high versus no primer ($p < .06$) and the high versus low primer condition ($p < .04$). Another way of expressing this interaction is in terms of the number of subjects in each group who show a 10 percent change in the predicted direction when the high priming condition is compared with either of the other two conditions. Only 1 out of 9 moderately dependent subjects displayed this effect compared with 9 out of 11 severely dependent subjects.

Because we tested only one severely dependent but five moderately dependent women the groups are somewhat confounded with sex. To test the possibility that men show the effect and women do not, we looked at differences between men and women within the moderately dependent group. There were no significant differences; neither male nor female subjects tended to speed up after a high priming dose.

DISCUSSION

Although the severely and moderately dependent groups were classified on the basis of a psychiatric assessment, it is clear that this assessment has criterion validity. As we predicted there were large differences between the groups, across all conditions, in amount of alcohol consumed and in speed of drinking.

The very clear priming effect of 3 double vodkas on the behavioral measure of craving was not matched by a significant change in subjective desire. There are a number of reasons why this result did not surprise us. First, a number of subjects stated that it

was very difficult to rate their desire. Secondly, subjects appeared to respond in a stereotyped way so that there was very little variability across occasions. Finally, one subject reported
that he was surprised how much he wanted to drink once he started
and that his prior subjective rating was not a good index of craving. Note that the craving which results from the consumption of
three double vodkas is very slight and that we are using the term
to represent a dimension ranging from very slight to extreme.

The statistically significant increase in craving three hours
after the priming dose, as indexed by our behavioral measure, would
be predicted by a disease model of alcoholism but is equally consistent with a learning theory or behavioral model. In order to develop such a model we must first ask what pattern of reinforcement
history differentiates the severely dependent from the less dependent alcoholic. One crucial difference was embodied in our simple
criterion for selecting the severely dependent subjects, repeated
and frequent consumption of alcohol in order to escape and avoid
withdrawal symptoms. Such a learning experience involves a set of
discriminative stimuli and also a powerful reinforcer so that,
after hundreds of repetitions, the severely dependent alcoholic
will tend to experience a compulsion to drink when exposed to the
cognitive and physiological cues associated with stopping drinking.
Both severely and moderately dependent subjects had a raised pulse
3 hours after consuming the high dose, indicating that there was a
physiological change; we would hypothesize that this altered physiological state is a component of craving but only for the severely
dependent alcoholic for whom it has become a discriminative stimulus. The results obtained by Ludwig, Wikler and Stark (1974) suggest that this physiological state is only a component of craving
when the alcoholic believes that alcohol has been consumed. A
further prediction from such a model would be that a whole day of
programmed heavy drinking within a hospital environment would result in increased craving when BALS had returned to zero the following morning.

Jellinek always stressed that a disease model of alcoholism
should only be a working hypothesis to be rejected if unsupported
by the objective evidence. It is now clear that a behavioral explanation is equally tenable as a working hypothesis and that such
an explanation generates a number of testable predictions about
the nature of dependence and methods of treatment. The optimistic
prediction generated by a learning theory approach is that craving
may be a learned compulsion which can be extinguished and not the
symptom of an irreversible disease.

REFERENCES

Engle, K. B. & Williams, T. K. Effects of an ounce of vodka on alcoholics' desire for alcohol. Quarterly Journal of Studies on Alcohol, 1972, 33, 1099-1105.

Funderburk, F. R. & Allen, R. P. Alcoholics' disposition to drink. Journal of Studies on Alcohol, 1977, 38, 410-424.

Gottheil, E., Murphy, B. F., Skoloda, T. E. & Corbett, L. O. Fixed interval drinking decisions. Quarterly Journal of Studies on Alcohol, 1972, 33, 325-340.

Hodgson, R. J. & Rachman, S. Desynchrony in measures of fear. Behavior Research and Therapy, 1974, 12, 319-326.

Jellinek, E. M. Phases of alcohol addiction. Quarterly Journal of Studies on Alcohol, 1952, 13, 673-684.

Keller, M. On the loss-of-control phenomenon in alcoholism. British Journal of Addictions, 1972, 67, 153-166.

Lang, P. J. The application of psychophysiological method to the study of psychotherapy and behavior modification. In A. E. Bergin & S. L. Garfield (Eds.), Handbook of psychotherapy and behaviour change. New York: Wiley, 1971.

Ludwig, A. M., Wikler, A. & Stark, L. H. The first drink. Archives of General Psychiatry, 1974, 30, 539-547.

Marlatt, G. A., Demming, B. & Reid, J. B. Loss of control drinking in alcoholics: An experimental analogue. Journal of Abnormal Psychology, 1978, 81, 223-241.

Mello, N. K. & Mendelson, J. H. Drinking patterns during work contingent and non-contingent alcohol acquisition. In N. K. Mello & J. H. Mendelson (Eds.), Recent advances in studies of alcoholism. Washington, D. C.: National Institute of Mental Health, 1970.

Rachman, S. & Hodgson, R. J. Synchrony and desynchrony in fear and avoidance. Behavior Research and Therapy, 1974, 12, 311-318.

Williams L. Alcoholism explained. London: Evans Brothers, Ltd., 1967.

RELATIONSHIP OF SOCIAL FACTORS TO ETHANOL SELF-ADMINISTRATION IN ALCOHOLICS

Roland R. Griffiths, George E. Bigelow and Ira Liebson

Department of Psychiatry, Baltimore City Hospitals and

The Johns Hopkins University School of Medicine

This paper will review a series of experiments from our residential drug research laboratory at Baltimore City Hospitals which have investigated the relationship of social factors to ethanol self-administration in alcoholics. The primary focus of the first section of the paper is on the rationale, methods and results of this series of experiments conducted over the last five years. The final section of the paper discusses the results of this research in relation to other experimental studies of alcoholism.

GENERAL METHOD

Subjects

All of the subjects in these experiments were chronic alcoholics. Subjects generally reported long histories of problem drinking, repeated hospitalization for alcoholism and having experienced symptoms of physical dependence on ethanol. Volunteers were detoxified and their informed consent obtained in writing before participation.

Setting

All of the experiments were conducted on an 8-bed behavioral pharmacology research ward at Baltimore City Hospitals. Figure 1 shows the floor plan of the research ward. Subjects in each experiment participated successively, not simultaneously. Such successive participation increased the independence of each subject's data. Other residents participated in different behavioral

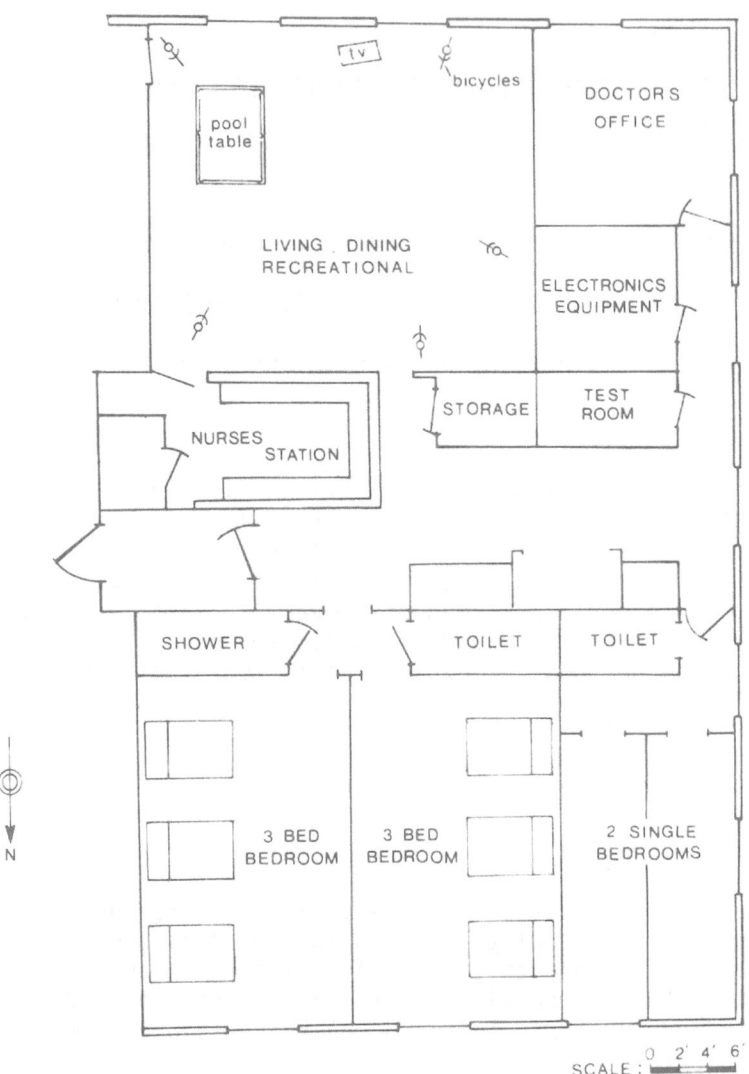

Figure 1: Floor plan of eight-bed behavioral
  pharmacology research ward at Baltimore City
  Hospitals.

studies involving drugs. Various recreational, reading and craft
materials were continuously available to subjects. General ward
behavior was maintained via a point economy in which points were
earned for various personal and ward maintenance activities, spent
for minor ward privileges, and sacrificed for ward rule violations.

Procedures

Subjects received explicit instructions concerning ward rules.
Subjects were also informed about the general experimental conditions under which they would participate, such as the number of
drinks available each day and the times of availability. However,
subjects were given only very general information about the purpose
of the experiments. They were told that the studies on the research unit involved how drugs and alcohol affect people's moods
and behavior. Other than this vague explanation of purpose, subjects were given no instructions or explanations of what they were
"supposed" to do or of what outcomes might be expected. To reduce
the possibility that subjects would receive instructions or explanations which might confound results, ward staff were explicitly instructed to refrain from discussing experiments with subjects,
except to provide an objective description of the routines and
procedures which the subject was required to follow.

A more detailed description of the major methodological principles underlying this human drug self-administration research has
been presented previously (Bigelow, Griffiths and Liebson, 1975a,
1975b).

EXPERIMENT I: EFFECTS OF ETHANOL ON RATE OF SOCIALIZING IN

ALCOHOLICS

Informal observation of alcoholic subjects on our residential
research ward suggested that subjects became more sociable after
ingesting ethanol. This observation was interesting since there
appeared to be confusion among clinicians about ethanol's effects
on social behavior, and a review of the research literature indicated that there had been no quantitative evaluations of this effect. Therefore, the first experiment (Griffiths, Bigelow and
Liebson, 1974a) was undertaken simply to evaluate the effects of
ethanol self-administration on the social behavior of alcoholics.

Method

During daily sessions of five to eight hours duration, staff
members rated subject social behavior at variable intervals (with a

mean interval of 15 minutes). Intervals were timed with electronic
equipment which sounded a tone through a speaker at the nurses'
station. Staff terminated the tone by manual operation of a
switch, recorded the time on the data sheet, and then rated subject
behavior as either: 1) interaction with others, or 2) no interaction. An interaction was defined as a behavior which required the
presence of or involved another person. For example, playing pool
or cards with other people were rated as interaction while playing
pool or cards alone were rated as no interaction. Mere physical
proximity to others (e.g., eating a meal at the same table) was not
sufficient to qualify as a social interaction.

In a mixed order over consecutive days, subjects were assigned
to either an ethanol-available or no ethanol-available condition.
On ethanol days, 12 drinks were made available. Each drink consisted of one ounce of 95-proof ethanol (11.14 g ethanol) in two
ounces of orange juice. Individual drinks were dispensed upon request by the research ward staff and subjects could consume the
drinks at whatever rate they chose. In a second phase of the
study, several of the subjects were exposed to conditions in which
ethanol days or non-ethanol days were scheduled for a period of
successive days.

Results

Figure 2 shows that, for all subjects under both the random
and successive scheduling conditions, the frequency of social interactions was consistently greater on ethanol days than on non-
ethanol days.

EXPERIMENT II: EFFECTS OF ETHANOL ON CHOICE BETWEEN

MONEY AND SOCIALIZING

Although Experiment I demonstrated that ethanol produced increases in socializing, the study provided no information about the
specificity of that effect. Other research with alcoholics on our
ward has shown that ethanol also increases non-social behaviors
such as cigarette smoking (Griffiths, Bigelow and Liebson, 1976).
Furthermore, the effect of ethanol on socializing is not specific
to ethanol since we have also shown that d-amphetamine also increases socializing on our residential ward (Griffiths, Stitzer,
Corker, Bigelow and Liebson, in press).

Experiment II was undertaken to provide more information about
the specificity of the effects of ethanol on socializing. The experiment (Griffiths, Bigelow and Liebson, 1975) represented a
systematic replication and extension of Experiment I to investigate

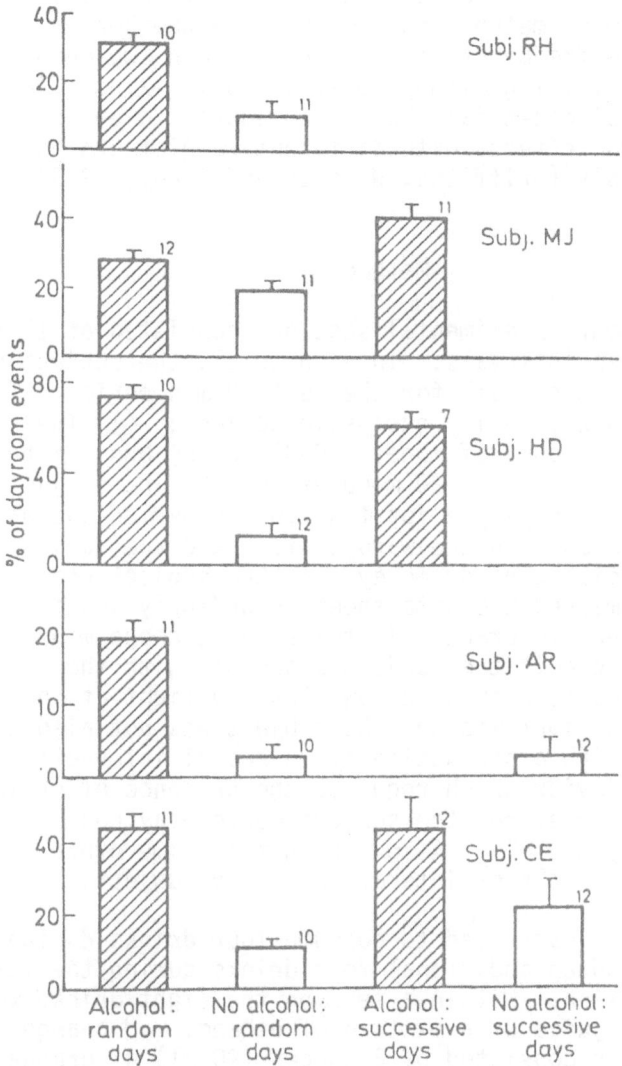

Figure 2: Social interactions expressed as the percentage of events sampled in the dayroom. Means(bars), standard errors of means(brackets), and number of days in each condition are presented for all subjects. Absense of bar and number means subject did not participate in that condition. (From Griffiths, Bigelow, & Liebson, 1974a)

further the effect of ethanol on the social behavior of alcoholics and especially to assess whether ethanol altered the reinforcing potency of social access. Specifically, Experiment II utilized a discrete-trial choice methodology to evaluate whether ethanol would alter the relative frequency of choosing between two mutually exclusive options involving either money or socializing. The advantages of using discrete-trial choice procedures to evaluate the specificity of drug effects with different reinforcers have been discussed previously (Griffiths, Wurster and Brady, 1975).

Method

Daily four-hour experimental sessions consisted of 12 trials, occurring at 20 min intervals. In each trial, the subject chose what condition would prevail for the next 20 min period. The subject chose between 2 mutually exclusive options: (1) The subject could have a small amount of money (10-35 cents) and neither talk nor interact with anyone for 20 minutes, or (2) The subject could talk and interact with people for that 20 min period but have no money. A trial began when the ward staff asked the subject whether he wanted socializing or money. If the subject chose socializing, the staff marked the data sheet accordingly and the subject was free to talk and interact. If the subject chose money, staff would immediately credit the subject's account with the appropriate amount of money and turn on a yellow flashing light at the nurses' station. When the light was on, the subject was not allowed to talk or interact with other patients. A social interaction was defined as any behavior which required the presence of or involved another person. Therefore, the subject could play pool, cards or games, but only by himself. Also during this time other patients were not allowed to talk or interact with the subject.

Subjects were instructed to consume four drinks during the first two hours and an additional four drinks during the last two hours of each session. Within a session the eight drinks were either all orange juice or a mixture of ethanol and orange juice. Orange juice drinks consisted of 3 ounces (90 ml) of orange juice. Ethanol drinks consisted of one ounce of 95-proof ethanol (11.14 g ethanol) in 2 ounces of orange juice. Individual drinks were dispensed upon request by the research ward staff; subjects could consume the drinks at whatever rate they chose while participating in other ward activities. The availability of drinks was not affected by whether the subject chose socializing or money on a given trial.

Results

For all subjects, the average percent choice of socialization over money was significantly greater during sessions involving ethanol self-administration than during sessions involving orange juice self-administration. Figure 3 presents daily session and averaged data for both conditions for all 4 subjects. Inspection of the daily session data for all 4 subjects reveals a clear difference between ethanol and orange juice sessions, despite what appears to be a tendency for the choice baseline to drift over successive sessions in some subjects.

The results of Experiment II suggest that ethanol increases the relative reinforcing potency of social access to an extent sufficient to alter alcoholics' choice behavior. Overall, Experiment II systematically replicated the results of Experiment I using a different measure of social behavior by demonstrating that ethanol increases social behavior in alcoholics. Taken together, the two studies demonstrate the robustness of ethanol's effect on socializing since they involved widely different methodologies.

EXPERIMENT III: SUPPRESSION OF ETHANOL SELF-ADMINISTRATION

WITH A BRIEF SOCIAL-ACTIVITY TIME-OUT PROCEDURE

Experiments I and II demonstrated a clear functional relationship between ethanol and socializing: Moderate to high doses of ethanol produced an immediate increase in the amount of socializing by alcoholics. The results of these experiments suggest an interesting interpretation of drinking in the alcoholic. From a behavioral viewpoint, it is commonly assumed that ethanol is the principal reinforcer which maintains drinking in the alcoholic. Since it is well established that, in many situations, social interactions are powerful reinforcers for human behavior, the results of Experiments I and II suggest that an additional factor that might maintain ethanol self-administration is that drinking results in increased social interactions. Indeed, in an interview situation, both normals and alcoholics agreed that they drank at least in part to gain access to the conviviality of the neighborhood bar (Nathan and O'Brien, 1971).

These data (and those of other researchers) make it all the more important to determine the extent to which social factors can be manipulated to control drinking. To this end, behavioral time-out procedures have been effectively used to suppress undesirable behaviors in various clinical populations. Such time-out procedures involve scheduling, as an immediate consequence of the undesirable behavior, an interval during which other reinforcers are unavailable. Therefore, we undertook a series of studies to examine the effectiveness of social and activity time-out procedures to suppress ethanol self-administration. Our first attempt to explore the usefulness of brief time-out procedures was a

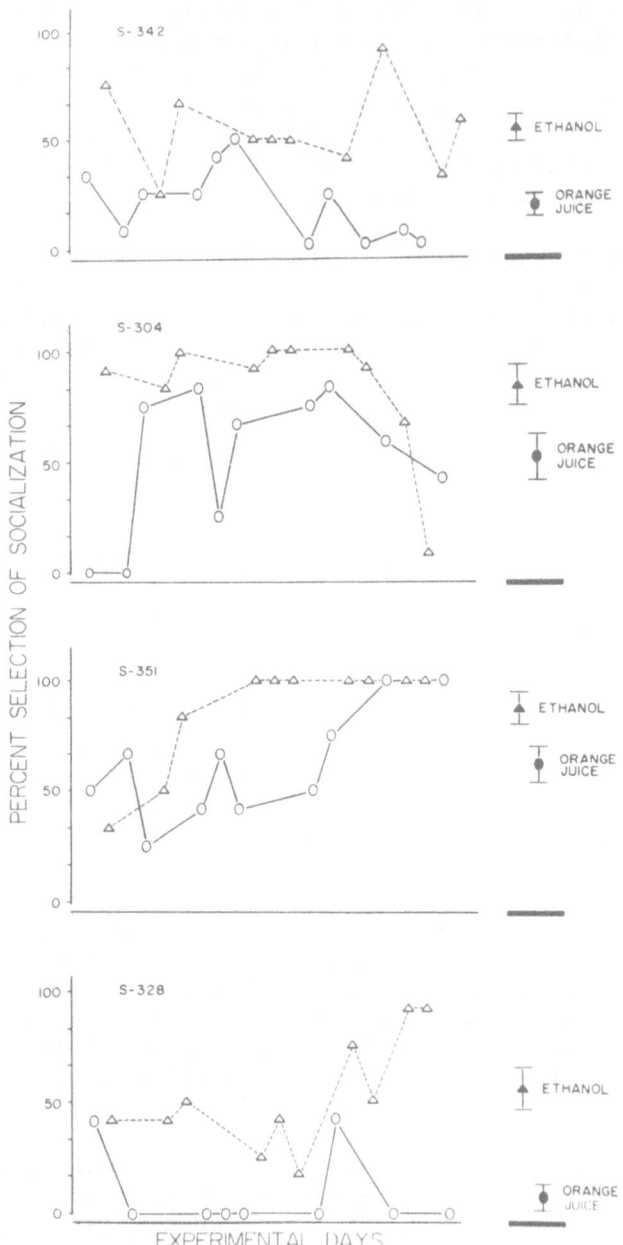

Figure 3: Choice between socialization and money:
percent selection of socialization. Date are presented for each daily session for all 4 subjects.
Means are shown at the far right (filled symbols),
with standard error of the means indicated by
brackets (From Griffiths, Bigelow, & Liebson, 1975).

single case study at the University of Minnesota Hospitals (Pickens, Bigelow and Griffiths, 1973). An alcoholic was admitted to the Psychiatry Ward and given free access to drinks containing one ounce of Bourbon and one ounce of water. Figure 4 shows that, for an initial six-week baseline period, the number of drinks per day remained relatively constant. After six weeks, a set of contingencies was introduced to determine whether drinking could be reduced. To obtain each drink the patient was required to carry on a conversation with a staff person for 1 minute but, once the drink was obtained, the subject was required to stay in his room for a 10-minute period of social and physical isolation. As shown in Figure 4, over the next six weeks, daily drinking initially decreased gradually, and then more abruptly, until drinking ceased altogether.

Since these pilot results were encouraging, we subsequently undertook a larger, more rigorously controlled experimental study to determine the utility of social-activity time-out in suppressing drinking (Bigelow, Liebson and Griffiths, 1974).

Method

A total of 12 to 24 drinks (number varied between individual subjects) were made available daily. Each drink consisted of one ounce of 95-proof ethanol (11.14 g ethanol) in 2 ounces of orange juice. Drinks were dispensed upon request by the research ward staff. The effect of contingent time-out was evaluated using a within-subject experimental design in which subjects were first exposed to baseline control conditions for a number of days, then to the experimental time-out condition for a number of days and, finally, some subjects were re-exposed to the control condition. During the initial baseline control condition subjects consumed their drinks in the main ward social area and were free to participate in other activities while they drank. In the following social-activity time-out condition, subjects were required to sit for either 10 or 15 minutes in an isolation booth immediately upon receiving each drink. The isolation booth was a three-sided booth with a curtain across the fourth side. The effect of the time-out procedure was to eliminate virtually all concurrent behavioral alternatives as an immediate consequence to receiving an alcoholic drink.

Results

Contingent time-out suppressed drinking to about one-half of control levels in the ten subjects who participated in this experiment. Although the magnitude of contingent isolation's suppressive effect upon drinking varied somewhat across subjects, individ-

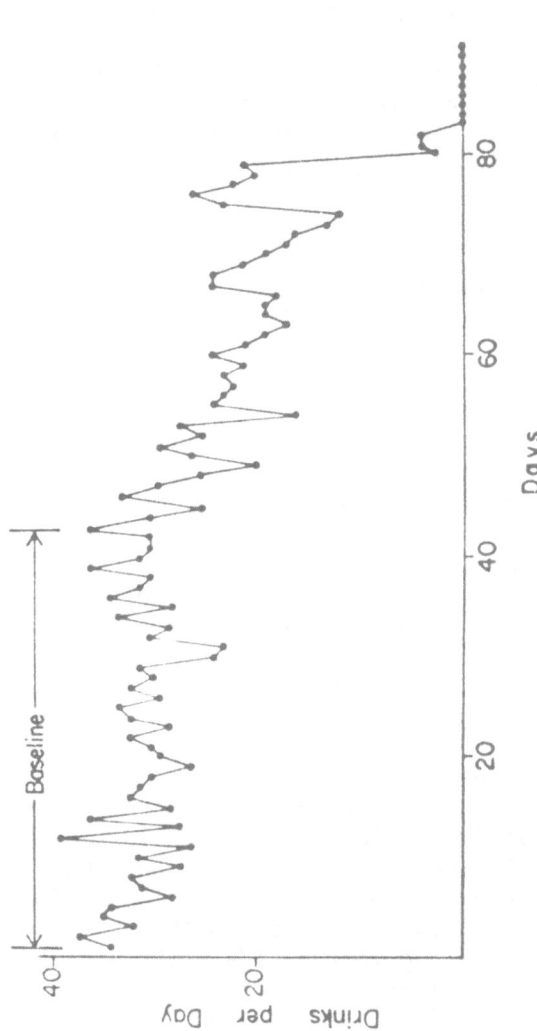

Figure 4: Effects of socialization-isolation contingency on number of drinks per day in one pilot subject. Contingency was introduced after a six-week baseline period (From Pickens, Bigelow, & Griffiths, 1973).

ual data indicate that the average effect was also the modal effect. As shown in Figure 5, the ten subjects consumed 94.6 percent of the available drinks during the initial baseline period. During the subsequent contingent isolation period, however, they consumed only 52.1 percent of the available drinks. The six subjects who were subsequently returned to baseline conditions again consumed 91.9 percent of the available drinks.

EXPERIMENT IV: COMPARISON OF SOCIAL TIME-OUT AND ACTIVITY

TIME-OUT PROCEDURES IN SUPPRESSING ETHANOL SELF-ADMINISTRATION

Experiment III demonstrated that a brief period of social and activity time-out was effective in suppressing drinking to one-half of baseline control levels. It was not known, however, to what extent the suppression produced by the social-activity time-out procedure was due to the physical restriction component vs. the loss of socialization component. As a result, two interrelated studies were undertaken (Griffiths, Bigelow and Liebson, 1974b, in press) to determine whether drinking could be suppressed by time-out from social interactions alone and to compare directly the efficacy of social time-out procedures with activity time-out procedures.

Method

The schedule of ethanol availability was constant throughout the experiment. Seventeen drinks were available during each daily 11-hour session; each drink consisted of one ounce of 95-proof ethanol (11.14 g ethanol) in 2 ounces of orange juice. Drinks were dispensed upon request by ward staff provided that a minimum of 40 minutes had elapsed since dispensing the previous drink.

The experiment utilized a within-subject experimental design to examine drinking under four different conditions: (1) Baseline with no time-out; (2) Social time-out; (3) Activity time-out; (4) Social and Activity time-out.

Baseline with no Time-out

During the baseline condition subjects could consume the drinks in the main ward social area and were free to participate in all ward activities, except when explicitly restricted as noted below.

Social Time-out

During the time-out from Social Interaction condition, subjects

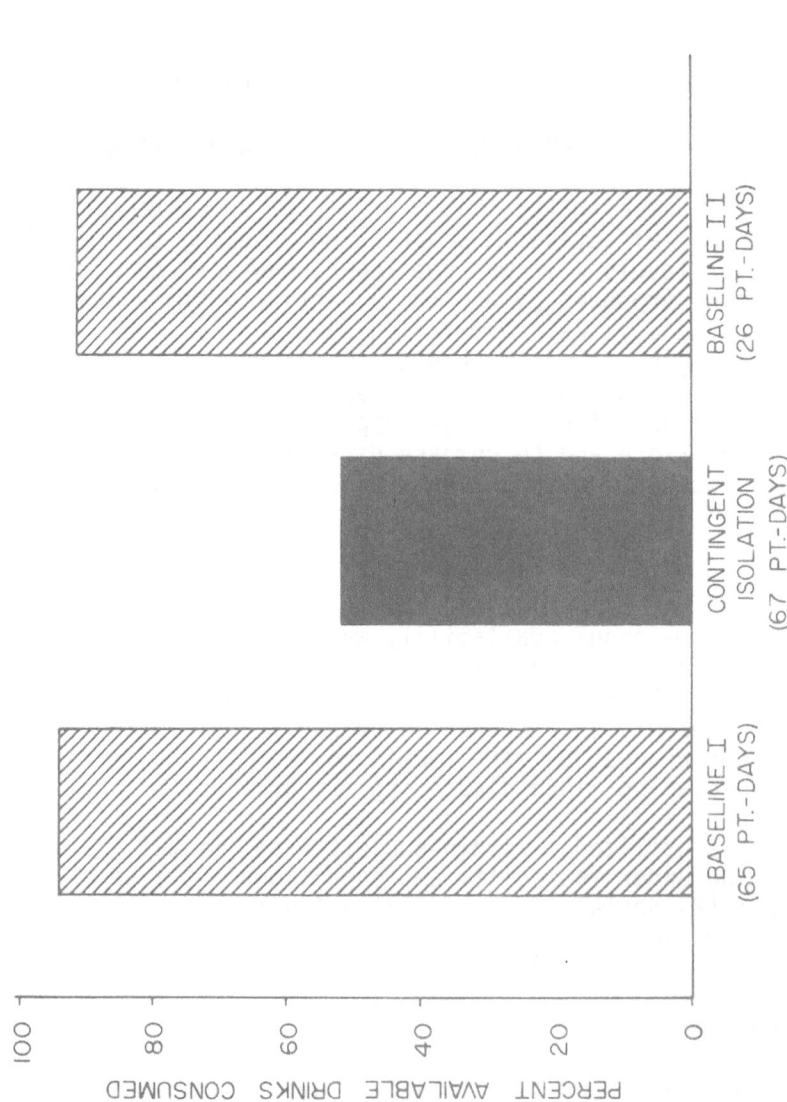

Figure 5: Amount of ethanol self-administration during periods of baseline control and social-activity time-out in 10 subjects. Ethanol self-administration was suppressed when 10 or 15 minutes social-activity isolation was contingent on receiving each alcoholic drink (From Bigelow, Griffiths, and Liebson, 1975a).

could consume their drinks as before; however, each drink resulted
in a 40-minute time-out from social interactions. Specifically,
each time the ward staff dispensed a drink to a subject they would
turn on a yellow flashing light at the nurses' station. While the
light was on the subject was not allowed to talk, gesture or interact with patients or staff.

Activity Time-out

During the contingent Activity time-out condition, subjects
could consume drinks as described above; however, the dispensing
of each drink resulted in a 40-minute time-out from engaging in
most activities except socializing. For the 40-minute period following receiving each drink, the subject was required to sit in a
designated chair. Also during this time, the subject was not allowed to have any materials or equipment (e.g., no food, cards,
games, reading materials, pencil, paper, radio, etc.), although
he was permitted to smoke cigarettes. However, the subject was
free to socialize (e.g., talk, interact, gesture) with patients
and staff during this period.

Social and Activity Time-out

During the Social and Activity time-out condition subjects
could consume drinks as described above; however, the dispensing
of each drink resulted in a 40-minute time-out from both socializing and engaging in other activities. This condition involved a
combination of the social time-out and activity time-out procedures described above.

Sequencing of Experimental Conditions

Two different procedures, a continuous exposure procedure and
an intermittent exposure procedure, were used for examining the
several experimental conditions. The continuous procedure involved exposing subjects to different conditions for a number of
consecutive days. With the intermittent procedure, the experimental conditions changed on a daily basis in a mixed order.

Varying the Background Levels of Available Privileges

To obtain further information about the effectiveness of
social time-out in suppressing drinking, the social time-out manipulation in four subjects was systematically replicated across
different conditions which varied with respect to the number of
ward privileges available.

Results

Figure 6 graphically summarizes the effects on drinking of the three contingent time-out manipulations using the two experimental methodologies. The 14 subjects tested with the Social time-out procedure showed an average suppression to 71 percent of baseline drinking. The Activity time-out condition produced a somewhat greater suppression, decreasing ethanol intake to an average of 36 percent of baseline in the eight subjects tested. Finally, the combined Social and Activity time-out produced the greatest suppression, decreasing ethanol intake to an average of 24 percent of baseline intake in the seven subjects tested. As shown in Figure 6, these average data are similar to those observed in individual subjects. Although there was consistency across subjects with respect to the relative suppressive effect of the various time-out conditions, there was nonetheless substantial variability among subjects with respect to the absolute amount of suppression observed. For instance, Social time-out was actually associated with slight increases in the drinking of three subjects, while it produced 80-100 percent suppression in three other subjects. Finally, Figure 6 shows that the continuous and the intermittent procedures for sequencing conditions produced similar results.

Manipulating the number of ward privileges available directly affected the effectiveness of social time-out in suppressing drinking. Within individual subjects, social time-out was increasingly effective in suppressing drinking as the available privileges were increasingly restricted. Figure 7 shows this effect in two subjects.

EXPERIMENT V: CONTROL OF DRINKING BY SOCIAL FACTORS IN THE

NATURAL ENVIRONMENT

Experiments III and IV demonstrated control over drinking by manipulating social and activity consequences to drinking. It is important to recognize that these experimental observations took place in a residential hospital ward situation which represents a closed social system. Accordingly, it seemed of significant clinical interest to determine whether social control over drinking could be extended beyond the ward situation. Experiment V represents an initial effort to control drinking on the residential ward by two subjects by arranging contingent access to social factors outside of the ward environment.

Method

Two alcoholic subjects were selected for participation in

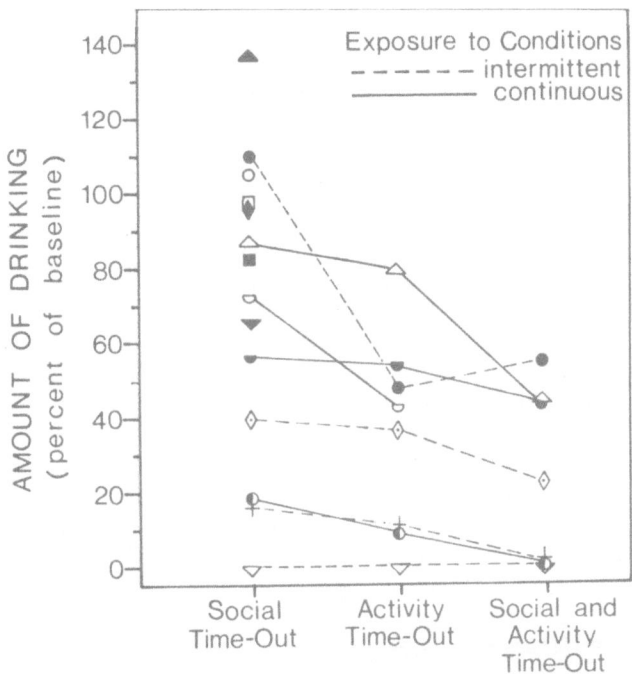

Figure 6: Effect of three different time-out procedures and two different experimental methodologies on amount of drinking by 14 subjects.
Different symbols represent different individual
subjects. Lines connect data points from individual subjects. For each subject the mean amount
of drinking during different time-out conditions
is expressed as percent of the mean amount of
drinking during all the baseline conditions for
that subject. The four symbols connected by
dashed lines represent subjects who were exposed
to different conditions using the intermittent
exposure procedure; the remaining ten symbols
(either single symbols or symbols connected
with solid lines) represent subjects exposed
to different conditions using the continuous
exposure procedure (From Griffiths, Bigelow,
& Liebson, in press).

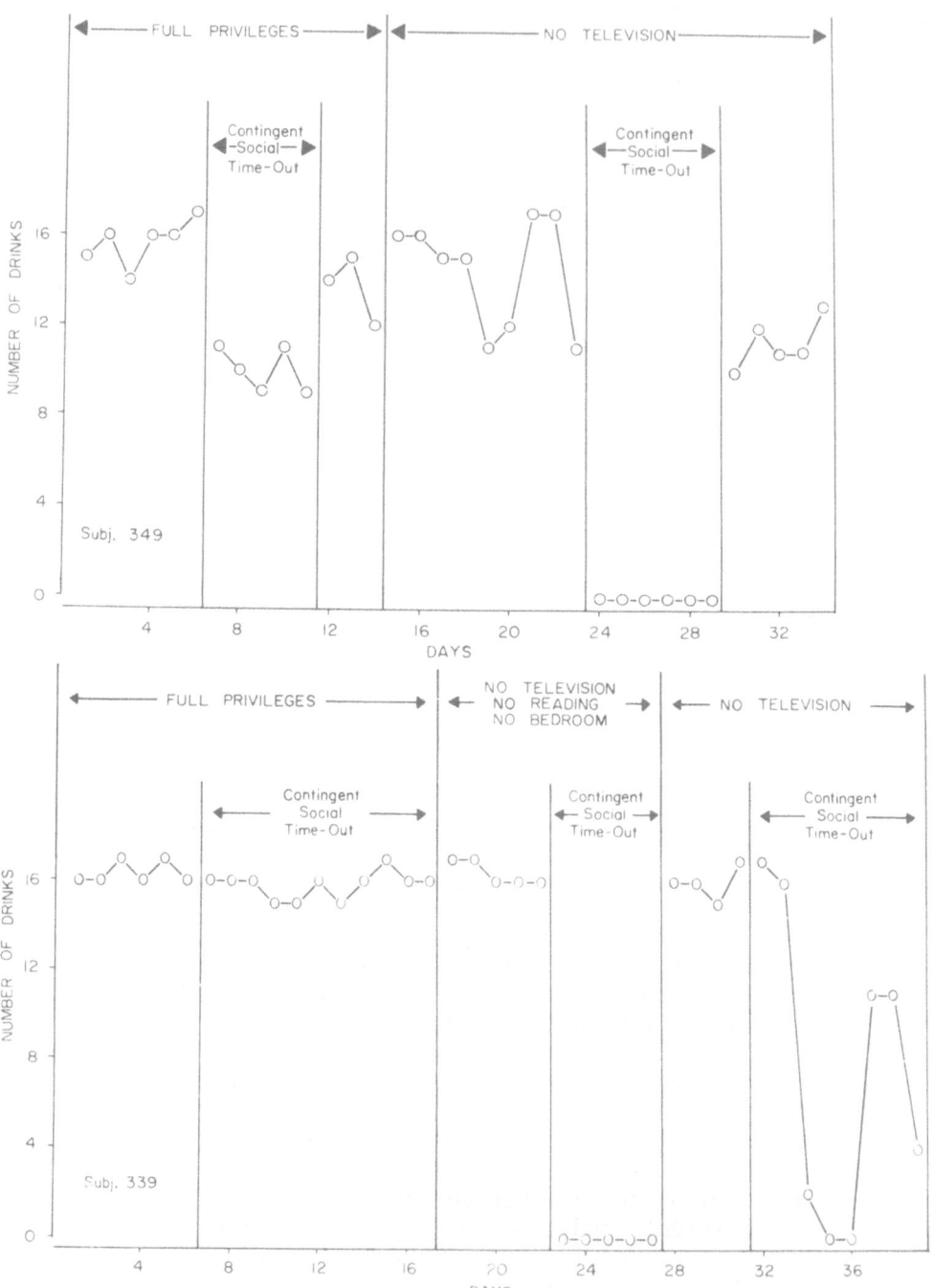

<u>Figure 7</u>: Daily number of drinks consumed by two representative subjects. Note that the effectiveness of social time-out in suppressing drinking was directly affected by the level of privileges available (From Griffiths, Bigelow, & Liebson, 1974b).

this study. Both had available what appeared to be a potent social
reinforcer within their natural environments. One individual
wanted to earn a weekend pass from the hospital to attend a family
gathering. The other subject wanted to receive special visiting
privileges for his girlfriend and to earn a weekend pass to visit
her. For both subjects the availability of these natural social
reinforcers was made contingent upon drinking moderately for a
specified period. During daily drinking sessions, either 12 or 15
drinks were available to subjects 143LP and 132ZP, respectively.
Each drink consisted of one ounce of 95-proof ethanol (11.14 g
ethanol) in two ounces of orange juice. Individual drinks were
dispensed upon request by the research ward staff; subjects could
consume the drinks at whatever rate they chose. Subjects were
told that, by taking five or less drinks per day for a specified
number of days, they would earn prearranged social privileges.

Results

Figure 8 shows the number of drinks consumed across consecutive days for one subject. During the first seven days, when no
contingency was attached, the subject consistently drank all 12
drinks available. During the first social contingency period, the
subject earned a one-day visiting privilege for his girlfriend for
each day he drank five or fewer drinks. Excessive drinking occurred only once during this period; it occurred immediately following an argument with his girlfriend. During the second social
contingency period, the subject was able to earn an overnight pass
to visit his girlfriend by drinking moderately for ten consecutive
days. As shown in the figure, excessive drinking did not occur
during this period.

Social contingencies were also effective in controlling the
drinking of the second subject. When no contingencies were attached to his drinking he consumed 92 percent of available drinks.
When he was given the opportunity to earn a weekend pass home if he
stayed within moderate drinking limits (five or fewer drinks per
day), his drinking promptly stayed within that limit and remained
there for the required nine consecutive days, even though up to 15
ounces of ethanol were available upon request.

GENERAL DISCUSSION

In considering the relationship between social factors and
ethanol consumption, it is important to recognize that this general issue embodies at least three distinct and relatively independent questions: (1) Does ethanol influence alcoholics' social
behavior? (2) Do social factors influence alcoholics' ethanol
consumption? (3) Do alcoholics and non-alcoholics differ with

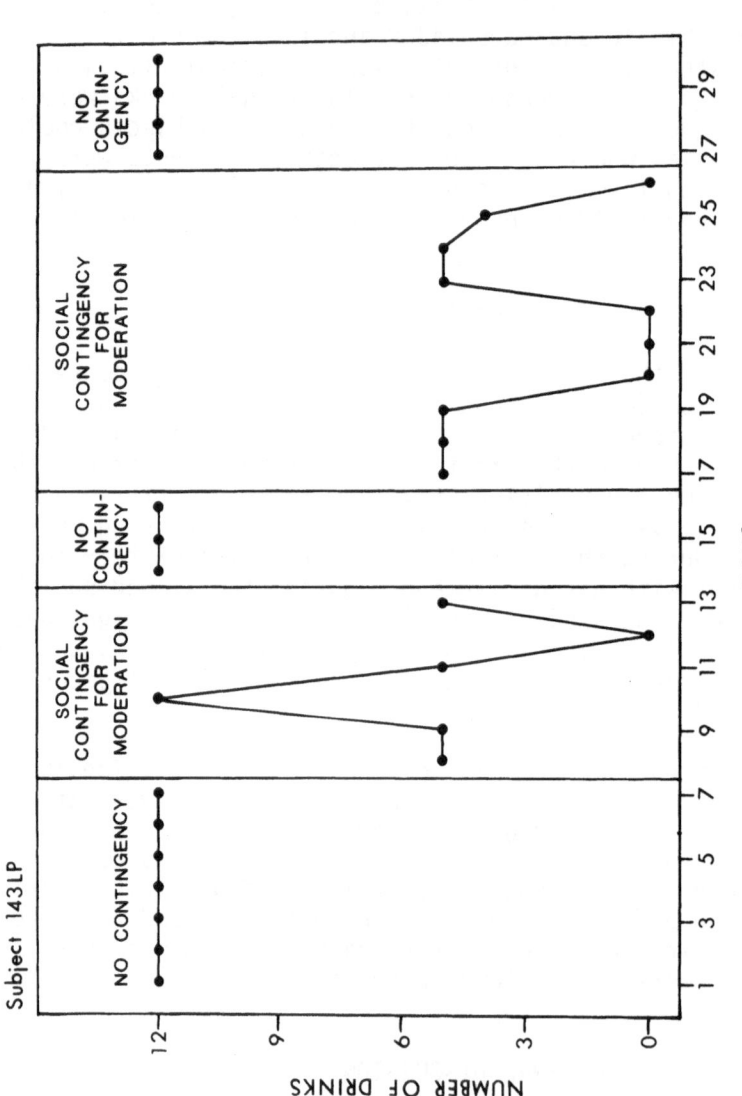

Figure 8: Daily number of drinks consumed by one subject. Drinking was reduced during periods in which a social contingency for moderate drinking was in effect. The social contingency provided for access to social reinforcers outside of the ward environment contingent upon moderate drinking on the ward

respect to their social behavior or with respect to the interaction of social factors and ethanol consumption? Our studies, reviewed above, have addressed the first two of these questions. In this discussion, we will review the relevant experimental literature and will attempt to reach conclusions concerning the current status of knowledge about each of the three major questions listed above.

Ethanol Effects on Alcoholics' Social Behavior

A number of early studies of experimental intoxication in alcoholics have, on the basis of non-quantitative observational data, suggested that ethanol typically sustains or increases alcoholics' rates of social interaction except under conditions of unusually high levels of intoxication. Diethelm and Barr (1962) noted that alcoholics "talked much more freely" during psychotherapeutic interviews while acutely intoxicated (intravenous doses of 60 to 70 cc of 95% ethanol) than while sober. Docter and Bernal (1964) noted that during a 14-day period of chronic intoxication (daily oral doses of 22 or 27 ounces of 80-proof ethanol) alcoholics became assertive and talked extensively and spontaneously while, without ethanol, they were silent and uncommunicative. Mendelson (1964) observed a high degree of social interaction prior to and during periods of chronic ethanol administration (daily doses up to 30 ounces of 86-proof ethanol). Decreases in social interactions were noted in those subjects given 40 ounces of 86-proof ethanol daily. At this level of intoxication, subjects also stopped a number of other activities (reading, watching television and listening to the radio) which had previously been of great interest. During a subsequent phase of alcohol withdrawal, social interactions were also decreased. Finally, a study of chronic experimental ethanol self-administration by alcoholics (McNamee et al., 1968) has observed that levels of social interaction are high throughout the drinking period, with little tendency toward social isolation except during times when alcohol consumption reached peak values. As with previous studies, the post-drinking period of five days was associated with a marked reduction in social interaction.

Our own studies, which have relied upon quantitative measurement of social behavior rather than upon observational impression, have shown that ethanol increases both alcoholics' rate of social interaction and their relative preference for socializing as opposed to monetary reinforcement (Griffiths et al., 1974a; 1975). A series of studies by Nathan and colleagues (Nathan et al., 1971; 1972; Nathan and O'Brien, 1971; Goldman et al., 1973) represent the second major source of objective quantitative measures of alcoholics' social behavior during drinking and non-drinking conditions. In these studies alcoholics have been permitted chronically to self-administer ethanol within a residential laboratory. Although

none of these studies has shown a reliable, replicable effect of
ethanol upon socializing, several experimental observations support
the view that drinking increases alcoholics' socializing. A socio-
gram measure of social behavior recorded by observers has occasionally shown increases during drinking periods (Nathan et al., 1971,
p. 356). Ratings of subjects' physical location in the ward area
have indicated that subjects spend more time out of room isolation
during ad libitum drinking periods than during pre- or post-drinking periods (Goldman et al., 1973, p. 816). Finally, the number of
points that the alcoholics choose to spend for socializing has
sometimes been higher during the first isolation period of the
drinking phase than during either the pre-drinking or post-drinking phases (Nathan et al., 1971, p. 365). It should be noted that
there are several procedural aspects of Nathan et al.'s studies
which may account for their relative insensitivity to ethanol's
effects upon social behavior: (1) Although there have been significant procedural differences across different experiments, in most
of the studies the location of the operant task which was required
to earn ethanol was in the subject's private room; (2) in many
studies ethanol has not been dispensed in the social area of the
laboratory, but only in the subject's private room; (3) in most
experiments, both ethanol and access to social areas could be purchased only with points earned by operant responding and, therefore, spending points on ethanol conflicted with spending points on
socializing. It is possible that this conflict between ethanol and
socializing reduced the effects of ethanol on socializing; (4) finally, in some of the experiments subjects have been permitted to
share drinks (Nathan and O'Brien, 1971, p. 466). During the latter
portion of the drinking period of these studies, subjects do not
have a surplus of points to buy ethanol and must work for every
drink (Nathan et al., 1971, p. 365). It is possible that subjects
avoided the social areas of the ward and stayed in their rooms
simply to avoid the unpleasant social pressure of other alcoholics
trying to "bum" a drink from them.

The vast majority of reports, based upon both objective measures and observational impressions, indicate that ethanol increase
the social behavior of alcoholics. The facilitation of social behavior has been observed in both acute and chronic intoxication
studies; it has been observed after both experimenter-administered
and self-administered ethanol; it has been observed with both
intravenous and oral ethanol administration; and it has been observed at a variety of levels of intoxication. Decreases in socializing have occurred only at very high doses of experimenter-
administered ethanol or at occasional peak levels of self-administered ethanol. Additional controlled parametric studies will be
necessary to characterize completely the effect of dose and
chronicity of ethanol intake on social behavior. Finally, it
should be noted that the demonstrations that ethanol facilitates
alcoholics' social behavior are based upon measures of the quantity

of undifferentiated social behavior occurring. Further understanding of the functional significance of these changes requires analysis of the qualitative nature of this social behavior.

Social Influences Upon Alcoholics' Drinking

The question of whether alcoholics' drinking is influenced by social factors has broad implications concerning the etiology, maintenance and treatment of chronic alcoholism. To the extent that alcoholics' drinking is influenced by social factors (or by other environmental factors), this would support the notion that alcoholics' drinking is an environmentally regulated, graded phenomenon which might under some conditions occur at a moderate level. To the extent that alcoholics' drinking is influenced by social factors, this would point to these factors as potentially being involved in the development and maintenance of inappropriate alcoholic drinking habits and as potentially valuable in treatment efforts to modify inappropriate drinking. Such a relationship would provide a scientific foundation for both prevention and treatment efforts at the social influence level.

Clinically, alcoholics have been characterized as social isolates with little motivation, capacity or ability to sustain social relationships (Mowrer and Mowrer, 1945; Zwerling and Rosenbaum, 1959). However, observational impressions from early studies of experimental alcohol self-administration by alcoholics suggested that social factors exert a significant controlling influence over alcoholics' drinking. Mendelson et al. (1966) observed the social interactions within a small group of alcoholics who worked for and drank from a single commonly-shared alcohol reservoir. These investigators concluded that ethanol itself might be a less potent determinant of alcoholics' drinking than are the social factors inherent in the drinking situation. Several subsequent observational studies (Steinglass et al., 1971; Steinglass, 1975; Weiner et al., 1971) involving dyadic and group drinking situations have also emphasized the significant involvement of social relationships in ethanol consumption.

Nathan and colleagues were the first to introduce into experimental studies of alcoholics' drinking controlled manipulation of social factors in an effort to assess their influence upon drinking. In a series of studies, Nathan et al. (1971; 1972) have permitted alcoholics to self-administer virtually unlimited quantities of ethanol within a residential laboratory setting while conditions varied every two or three days between socialization versus isolation. During socialization periods, subjects had free access to the ward social areas. During isolation periods, subjects were restricted to their private rooms; however, in some studies subjects could work for and purchase access to social areas during the

scheduled isolation periods. Overall, this series of studies found
no reliable effect of the socialization versus isolation conditions
upon alcoholics' drinking.

Cohen, Liebson and colleagues (1971) were the first to demonstrate that manipulating social and environmental factors could influence the amount of ethanol consumed by alcoholics. These investigators examined the effect of "enriched" versus "impoverished"
conditions upon alcoholics' drinking in a residential laboratory.
In the "enriched" condition subjects had access to the usual hospital privileges, while in the "impoverished" condition they were
banned from the ward social area and were denied telephone and
visitor privileges. (The two conditions differed markedly in both
their social and non-social aspects.) The effects of these conditions upon drinking were studied under both contingent and non-
contingent procedures. With the non-contingent procedure, when
subjects were chronically exposed to either the "enriched" or "impoverished" condition independently of the extent of their drinking,
drinking occurred at a high level and was similar under both conditions. This result is similar to the lack of effect noted by Nathar
et al. (1971; 1972) when socialization versus isolation conditions
were varied non-contingently. In contrast, when a contingent procedure was used in which exposure to the "impoverished" condition
occurred only if the subject drank more than a specified moderate
amount, drinking generally occurred only at this moderate level.

Our own studies, reviewed in detail earlier in this paper,
pursued the analysis of contingent variations in social conditions
as determinants of alcoholics' drinking (Bigelow et al., 1974;
1974b, in press). Those studies demonstrated clearly that contingent social consequences can exert a very powerful controlling influence over alcoholics' drinking. These demonstrations help
establish the scientific foundation for efforts to treat alcoholism
via rearrangement of social relationships (e.g., Miller, 1972;
Hunt and Azrin, 1973).

It is clear that social access is a reinforcer for which alcoholics will modify their drinking. However, it is not clear whether ethanol-induced socializing acts as a reinforcer to maintain
alcoholics' drinking. The fact that alcoholics will continue to
drink heavily under conditions of social impoverishment and isolation certainly demonstrates that facilitation of social behavior is
not a necessary element for the maintenance of drinking. However,
data reviewed in the preceding section show that, under many conditions, ethanol does facilitate the social behavior of alcoholics.
Therefore, it remains possible that social facilitation can be a
sufficient element for the maintenance of drinking and that such
ethanol-induced facilitation may be of significance in the etiology
and maintenance of alcoholism.

At this time the sensitivity of alcoholics' drinking to social influences has been demonstrated only with rather gross quantitative variations in overall social access. Additional research is necessary to assess the potencies of qualitative variations in social influence techniques in order to identify those techniques which are most potent and which therefore might offer the greatest therapeutic potential. Factors requiring attention include social reinforcement, social punishment, persuasion, modeling, contingent vs. non-contingent procedures, and effects of scheduling parameters (e.g., immediacy vs. delay of social consequences).

Social Factors and Drinking: Alcoholics vs. Non-Alcoholics

Are the relationships between social factors and drinking different for alcoholics and non-alcoholics? Are alcoholics, perhaps, uniquely sensitive or uniquely insensitive to social influences or to ethanol's social-facilitator effect, and might these differences contribute to the etiology and/or maintenance of alcoholism? Unfortunately, these questions are impossible to answer with certainty at this time; adequate data are simply not available. O'Leary et al. (1976), in their review of social skill acquisition and psychosocial development of alcoholics, conclude that the nature of the relationships between social factors and alcohol consumption is the same for both alcoholics and non-alcoholics, but that alcoholics are deficient in social skill behavior. In this section we will review behavioral research on social factors and drinking which suggests that alcoholics and non-alcoholics probably do not differ with respect to the causal relationships which hold between social factors and their drinking.

The literature indicates that ethanol has similar effects on the social behavior of both alcoholics and non-alcoholics. We have previously discussed the literature indicating that ethanol typically increases social behavior in alcoholics. That ethanol also increases the social behavior of non-alcoholics is common knowledge derived from widespread popular experience. The correctness of this common knowledge is corroborated by laboratory studies of acute ethanol administration to normal subjects. For example, Smith et al. (1975) studied the effect of ethanol upon the verbal interaction of pairs of normal volunteers during a free discussion period and found that ethanol increased amount of verbal interaction. Similar results have been obtained in our own laboratory.

The literature also supports the general conclusion that social influences can affect drinking in both alcoholics and non-alcoholics. In a previous section we reviewed this literature for alcoholics. Not surprisingly, studies have demonstrated that drinking in non-alcoholics is also influenced by a variety of social factors. Marlatt and colleagues have shown the influence of social variables

on amount of drinking by non-alcoholics in a simulated taste-rating
task involving alcoholic beverages. Thus, Caudill and Marlatt
(1975) demonstrated that amount of drinking by non-alcoholics is
subject to modeling of drinking by a confederate. While Garlington
and Dericco (1977) reported the same social modeling phenomenon in
a more naturalistic drinking situation, in a related study, Higgins
and Marlatt (1975) have reported that anticipated interpersonal
evaluation (expecting to be observed and rated on attractiveness by
opposite-sex peers with whom one will subsequently interact) increased the drinking of non-alcoholics. And Marlatt et al. (1975)
found that provocation to anger via social insult increased the
drinking of non-alcoholics, but that this effect was eliminated if
subjects were permitted to retaliate against the provocateur. Perhaps the greatest significance of these studies in terms of comparing alcoholics with non-alcoholics is that they demonstrate that
non-alcoholics will drink in response to social and emotional
stresses (i.e., anxiety, anger). It is often suggested that one of
the abnormal characteristics of alcoholics is their disposition to
drink in response to emotional stresses such as anger or anxiety.
These studies show that this same phenomenon occurs in non-alcoholics.

Only a few studies have measured and compared the interrelationship between social behavior and drinking in alcoholics and non-
alcoholics. One such study (Miller et al., 1974) reported a difference between alcoholics and non-alcoholics with respect to relationships between social factors and drinking. That study consisted of brief sessions during which subjects could perform a
simple operant task to earn alcoholic beverages. Some sessions
were preceded by a social stress manipulation - performing in a
series of simulated situations requiring assertive behavior and
then being informed by the experimenters that reactions in these
situations had been poor. Results revealed a statistical interaction between this social stress manipulation and alcoholism.
Alcoholics increased their drinking under stress whereas non-
alcoholics did not. It should be noted, however, that only the
interaction of diagnosis and stress had a statistically significant
effect upon drinking. The actual amounts drunk in the various
conditions did not appear significantly different, and the procedure did not detect a difference between alcoholics and non-alcoholics in amount of drinking. These qualifying comments concerning
the limited sensitivity of the drinking measure in Miller et al.'s
otherwise excellent study are added because subsequent studies have
obtained results somewhat in conflict with its general conclusions.
Whereas Miller et al. concluded that social stress increased the
drinking only of alcoholics, two studies by Marlatt and colleagues
have shown very pronounced effects of social stress on drinking by
non-alcoholics. Higgins and Marlatt (1975) showed that anticipated
social evaluation increased non-alcoholics' drinking while
Marlatt et al. (1975) showed that social provocation to anger

increased non-alcoholics' drinking. Thus, it appears that the overall data in the area of social-stress effects upon drinking fail to support the notion that alcoholics and non-alcoholics are differentially affected by stress.

Another study (Nathan and O'Brien, 1971) directly compared social factors in alcoholics and demographically-similar non-alcoholics during a period of experimental ethanol self-administration in a residential laboratory. The groups, drawn from the skid-row population of South Boston, differed markedly in their drinking patterns and drinking histories. Four alcoholics participated simultaneously as one group and four non-alcoholics participated simultaneously as the second group. The design consisted of a 9-day pre-drinking period, an 18-day drinking period, and a 6-day post-drinking period, with alternating 3-day periods of socialization versus isolation throughout. During socialization periods subjects had free access to the ward social areas; during isolation periods they were restricted to their private bedrooms unless they earned relief from isolation. Relief from isolation for 15 minutes could be earned via 15 minutes of operant work. This same operant task was required to earn ethanol during the drinking period, with 15 minutes of work earning 30 cc (1 ounce) of 86-proof ethanol. Operant work for either ethanol or relief for socialization could be performed only in subjects' private bedrooms; ethanol was dispensed only in the private bedrooms although, once obtained, it could be consumed elsewhere. The results of this study showed that alcoholics worked for and consumed almost twice as much ethanol as the non-alcoholics. A major finding of this study with respect to social factors and drinking was that alcoholics spent most of their time in their private bedrooms even during socialization periods and seldom purchased relief from isolation, whereas non-alcoholics spent much more time in the ward social area and purchased relief from isolation much more often. It is important to recognize that this result does not necessarily imply that alcoholics and non-alcoholics differ with respect to social factors. Several procedural aspects of the study which were discussed in a previous section may have increased the likelihood that the alcoholics would remain in their private rooms and not socialize. Specifically, both the operant task for earning ethanol and the ethanol dispenser were located in the subject's private room. Since the alcoholics drank twice as much ethanol as the non-alcoholics, they necessarily spent more time in their private rooms working for and obtaining ethanol. Another procedural aspect that may have biased the alcoholics against socializing is that the subjects were permitted to share their drinks; alcoholics may have stayed in their rooms to avoid the social pressure to share their ethanol. Finally and probably most importantly, the observed differences in socializing between alcoholics and non-alcoholics could have resulted from the fact that, functionally, subjects were required to choose between drinking and socializing. Alcoholics worked about 6-8 hours per day on the operant task simply to earn ethanol. To purchase relief from isolation would have required either less drinking or more operant work in the private room. Thus, it is possible that the differences observed simply reflect the fact that the reinforcing efficacy of ethanol is much greater in alcoholics than in non-alcoholics.

The clinical literature concerning social factors and alcoholic drinking contains numerous speculative suggestions that alcoholics and non-alcoholics react differently to social influences. Interestingly, there is no uniformity concerning the direction of this alleged difference--at times alcoholics are described as uniquely sensitive to social influences, and at other times uniquely insensitive. The available data are compatible with there being no difference. It is, of course, possible that clear differences in responsivity to social influences will be documented in the future. It may be that the concept of social influence is too general for orderly differences to be observed. Perhaps differences will become apparent as we become more precise in our identifications of the qualitative nature of various social influence procedures.

ACKNOWLEDGMENT

Preparation of this paper was supported in part by NIAAA Grant No. AA-00179.

REFERENCES

Bigelow, G., Griffiths, R. R., & Liebson, I. Experimental models for the modification of human drug self-administration: Methodological developments in the study of ethanol self-administration by alcoholics. Federation Proceedings, 1975, 34, 1785-1792(a).

Bigelow, G., Griffiths, R. R., & Liebson, I. Experimental human drug self-administration: Methodology and application to the study of sedative abuse. Pharmacological Reviews, 1975, 26, 523-531(b).

Bigelow, G., Liebson, I., & Griffiths, R. R. Alcoholic drinking: Suppression by a brief time-out procedure. Behaviour Research and Therapy, 1974, 12, 107-115.

Caudill, B. D., & Marlatt, G. A. Modeling influences in social drinking: An experimental analogue. Journal of Consulting and Clinical Psychology, 1975, 43, 405-415.

Cohen, M., Liebson, I. A., Faillace, L. A., & Allen, R. P. Moderate drinking by chronic alcoholics. Journal of Nervous and Mental Disease, 1971, 153, 434-444.

Diethelm, O., & Barr, R. M. Psychotherapeutic interviews and alcohol intoxication. Quarterly Journal on Studies of Alcohol, 1962, 23, 243-251.

Docter, R. F., & Bernal, M. E. Immediate and prolonged psychophysiological effects of sustained alcohol intake in alcoholics. Quarterly Journal on Studies of Alcohol, 1964, 25, 438-450.

Goldman, M. S., Taylor, H. A., Carruth, M. L., & Nathan, P. E. Effects of group decision-making on group drinking by alcoholics. Quarterly Journal of Studies on Alcohol, 1973, 34, 807-822.

Garlington, W. R., & Dericco, D. A. The effect of modeling on drinking rate. Journal of Applied Behavioral Analysis, 1977, 2, 207-211.

Griffiths, R. R., Bigelow, G., & Liebson, I. Assessment of effects of ethanol self-administration on social interactions in alcoholics. Psychopharmacologia, 1974, 38, 105-110(a).

Griffiths, R. R., Bigelow, G., & Liebson, I. Suppression of ethanol self-administration in alcoholics by contingent time-out from social interactions. Behaviour Research and Therapy, 1974, 12, 327-334(b).

Griffiths, R. R., Bigelow, G., & Liebson, I. Effect of ethanol self-administration on choice behavior: Money vs. socializing. Pharmacology, Biochemistry and Behavior, 1975, 3, 443-446.

Griffiths, R. R., Bigelow, G., & Liebson, I. Facilitation of human tobacco self-administration by ethanol: A behavioral analysis. Journal of the Experimental Analysis of Behavior, 1976, 25, 279-292.

Griffiths, R. R., Bigelow, G., & Liebson, I. Comparison of social time-out and activity time-out procedures in suppressing ethanol self-administration in alcoholics. Behaviour Research and Therapy, in press.

Griffiths, R. R., Stitzer, M., Corker, K., Bigelow, G., & Liebson, I. Drug-produced changes in human social behavior: Facilitation by d-amphetamine. Pharmacology, Biochemistry and Behavior, in press.

Griffiths, R. R., Wurster, R. M., & Brady, J. V. Discrete-trial choice procedure: Effects of naloxone and methadone on choice between food and heroin. Pharmacological Reviews, 1975, 27, 357-365.

Higgins, R. L., & Marlatt, G. A. Fear of interpersonal evaluation as a determinant of alcohol consumption in male social drinkers. Journal of Abnormal Psychology, 1975, 84, 644-651.

Hunt, G., & Azrin, N. A community reinforcement approach to alcoholism. Behaviour Research and Therapy, 1973, 11, 91-104.

Marlatt, G. A., Kosturn, C., & Lang, A. Provocation to anger and opportunity for retaliation as determinants of alcohol consumption in social drinkers. Journal of Abnormal Psychology, 1975, 84, 652-659.

McNamee, H. B., Mello, N. K., & Mendelson, J. H. Experimental analysis of drinking patterns of alcoholics: Concurrent psychiatric observations. American Journal of Psychiatry, 1968, 124, 1063-1069.

Mendelson, J. H. (Ed.) Experimentally induced chronic intoxication and withdrawal in alcoholics. Quarterly Journal of Studies on Alcohol, Suppl No. 2, 1964.

Mendelson, J. H., & Mello, N. K. Experimental analysis of drinking behavior of chronic alcoholics. Annals of the New York Academy of Sciences, 1966, 133, 828-845.

Miller, P. M., Hersen, M., Eisler, R., & Hilsman, G. Effects of social stress on operant drinking of alcoholics and social drinkers. Behaviour Research and Therapy, 1974, 12, 67-72.

Miller, P. M. A behavioral intervention program for chronic public drunkenness offenders. Archives of General Psychiatry, 1975, 32, 915-918.

Mowrer, H. R., & Mowrer, E. R. Ecological and familial factors associated with inebriety. Quarterly Journal of Studies on Alcohol, 1945, 6, 36-44.

Nathan, P. E., Goldman, M. S., Lisman, S. A., & Taylor, H. A. Alcohol and alcoholics: A behavioral approach. Transactions of the New York Academy of Science, 1972, 34, 602-627.

Nathan, P. E., & O'Brien, J. S. An experimental analysis of the behavior of alcoholics and nonalcoholics during prolonged experimental drinking: A necessary precursor of behavior therapy? Behavior Therapy, 1971, 2, 455-476.

Nathan, P. E., O'Brien, J. S., & Lowenstein, L. M. Operant studies of chronic alcoholism: Interaction of alcohol and alcoholics. In M. K. Roach, W. M. McIsaac, and P. J. Creaven (Eds.), Biological Aspects of Alcohol. Austin: University of Texas, 1971.

O'Leary, K., O'Leary, S., & Donovan, D. M. Social skill acquisition and psychosocial development of alcoholics: A review. Addictive Behaviors, 1976, 1, 111-120.

Pickens, R., Bigelow, G., & Griffiths, R. R. An experimental approach to treating chronic alcoholism: A case study and one-year follow-up. Behaviour Research and Therapy, 1973, 11, 321-325.

Smith, R. C., Parker, E. A., & Noble, E. P. Alcohol's effect on some formal aspects of verbal social communication. Archives of General Psychiatry, 1975, 32, 1394-1398.

Steinglass, P. The simulated drinking gang: An experimental model for the study of a systems approach to alcoholism. Journal of Nervous and Mental Disease, 1975, 161, 110-122.

Steinglass, P., Weiner, S., & Mendelson, J. H. A systems approach to alcoholism: A model and its clinical application. Archives of General Psychiatry, 1971, 24, 401.

Weiner, S., Tamerin, J. S., Steinglass, P., & Mendelson, J. H. Familial patterns in chronic alcoholism: A study of a father and son during experimental intoxication. American Journal of Psychiatry, 1971, 127, 1646.

Zwerling, I., & Rosenbaum, M. Alcohol addiction and personality (non-psychotic conditions). In S. Arieti (Ed.), American Handbook of Psychiatry. New York: Basic Books, 1959.

A NONBEHAVIORIST'S VIEW OF THE BEHAVIORAL PROBLEM WITH ALCOHOLISM

Mark Keller

The Center of Alcohol Studies

Rutgers University

The program for the conference from which this book was derived called me Editor of the Journal of Studies on Alcohol. That
was true when the program was made up. Alas! At Rutgers, The
State University of New Jersey, where that Journal is published
and where editing it was one of my extra-academic functions, a
state law prevails under which, though I did not retire, "they"
retired me, almost two months before the conference began. For
consolation, the President of the University sent me a charming
letter in which he mentions flatteringly my services to my scholarly
discipline--that made me wonder what my discipline was (and if the
President knows, he knows more than I do)--and he concluded by
appointing me Professor Emeritus. Thereupon, I also became Editor
Emeritus of the Journal of Studies on Alcohol, no longer editor.

Another correction to the same program has to do with the
touchy question of my discipline--you can see now why I mentioned
it. I am to discuss "A Nonbehaviorist's View." Well, nonbehav-
iorism is a peculiarly negative discipline, but it's something.
Yet I wonder and wonder. Am I really a nonbehaviorist? Of course,
if a behaviorist has to be a member of a certain society, or if he
has to have a specific academic degree, like B.D.--I am not sure
whether that stands for Doctor of Behaviorism or Bachelor of
Divinity--or if he has to subscribe to a formal doctrine of faith
after Saint Watson or Saint Skinner or Saint Wolpe, then I could
be described as a nonbehaviorist. But the circumstances of my
life and my beliefs are not pure. For example, since about 30
years ago, I have believed that alcoholism is best explained as a
learned behavior, and I have dared to explain it that way in lec-
tures and in publications (Keller, 1969, 1975a, 1975b). Is it
possible that nevertheless I am a nonbehaviorist?

Now the organizers of this conference undoubtedly know a behaviorist when they see one. And if they have ruled me out, they must be right. Evidently, it's just I who am at fault--I don't know what I am, not even what I am not. The only sure thing is, apparently, that I am undisciplined. You can see why the President's reference to my scholarly discipline puzzled me.

From me, then, you cannot expect a behaviorist's viewpoint. Nor, I am afraid--at least not certainly--can you expect a nonbehaviorist's viewpoint. Only an undisciplined viewpoint. In the Proceedings volume of this conference, perhaps our editors will change my title from "Nonbehaviorist" to "Misbehaviorist." (Editorial Note: Sorry, Mark!) So that there should be no sex confusion, I would prefer "Dysbehaviorist," and that might be especially appropriate because long ago I suggested that alcoholism is a dysbehaviorism. I am also on record as proposing that alcoholism is a dysism (Keller, 1974, p. 200).

So what do I think about alcoholism?

As psychologists, you might find interesting the history of the development of a viewpoint. I started in this field--what was to become this field--over 40 years ago--not only ignorant about alcoholism, but uninterested. My interest was in the systematization and reporting of knowledge. I thought of scientific knowledge and, at that time especially, what I supposed to be medical knowledge. I would today call it biomedical science. It happened that where I got an opportunity to try out this work there were literally multitudes of alcoholic patients with every imaginable disease that people can acquire in consequence of alcoholism. In fact, one of the early things I did was to form a classification of all alcohol-related diseases. As a result, when in 1942 the American Medical Association published its Standard Nomenclature of Disease (Jordan, 1942), I was able to review it critically--that, by the way (more than 100 papers ago), was my first signed publication in this field (Keller, 1942)--and in that review, I was able to note that they had missed listing Marchiafava-Bignami's disease. And incidentally, in spite of an appreciative letter from the editor of that volume, with a promise to attend to my observations, subsequent editions of the A.M.A. Standard Nomenclature, under different editorships (Thompson and Hayden, 1961), have continued to omit that disease, though it is included in the International Classification of Diseases (WHO, 1972). I cannot imagine what the nosologists of the A.M.A. have against Marchiafava and Bignami and the degenerated corpus callosum.

Compiling a classified list of alcohol-related disorders was the least of my undertakings. Actually, without realizing it, I had begun to systematize the "scientific alcohol literature." At first it was strictly biomedical, but it was impossible to ignore

clinical and experimental psychology, social work, law, and eventually social science. I found myself mixed up with such branches
as anthropology, history, economics and religion, besides sociology.
Even belles lettres, even drinking songs and limericks, demanded
attention and inclusion. By the time of that first signed publication I mentioned, I had moved from Norman Jolliffe's dissolving
empire at New York University Medical School and the Psychiatric
Division of Bellevue Hospital to Howard W. Haggard's burgeoning
empire in the Laboratory of Applied Physiology at Yale University,
where the ingenious E. M. Jellinek was creating a Center of Alcohol
Studies. An important feature of the developing center was the
documentation of the alcohol literature (Keller, 1964). The systematic organization of that literature was by then well under way.
Besides the Journal of Studies on Alcohol, we had established the
Classified Abstract Archive of the Alcohol Literature, though it
existed in only one copy in our files, and the Master Catalog of
the Alcohol Literature. The International Bibliography was also
under way, and we had begun to create specialized and, what I considered superior, indexes. Without knowing it, I was becoming
what was later to be called an information scientist. Having ut-
tered this word, I must hasten to note that I have never called
myself an information scientist. It is a sensitive subject with
me. Though I have been a member of the American Society of Information Science from the first adoption of that name, and though I
have even given a paper at an international conference on information science (Keller, 1972), I hold that, so far, there isn't any
information science. So my discipline is certainly not information
science.

Now if you learned to read at age 3 and this caused a neurosis,
bibliophilia, and if then you undertake to gather and systematize
the entire scientific literature about alcohol, you can't help
becoming knowledgeable about alcohol--indeed, perhaps you become
an alcohologist as well as a knowledgist. But nobody is sure what
you are, so you get invited to give papers at meetings of physicians, nurses, social workers, lawyers, sociologists, biochemists,
anthropologists, educators, psychiatrists, alcoholics, and psychologists. You also get invited to write articles in the scientific-
professional journals--which not only don't pay you but expect you
to buy reprints of your own article at exorbitant prices; but
sometimes you get invited to write even in encyclopedias and popular magazines which do pay--sometimes rather well.

Now I am coming to the behavioral-psychology connection.

It is in the later 1940's, and I need urgently to buy a new
automobile. They want $800 for it--remember, this is about 30
years ago. I have only $50. So what does God do? A miracle, of
course. An editor of one of the popular national ladies' magazines
comes along and invites me to write an article for that magazine to

explain alcoholism--from A to Z; that is, from cause to cure. I
consent, of course. He has casually mentioned that the pay is
$750.

You have guessed that I set to work promptly, and the ink
flows magnificently. I have no problems. My head is as full of
facts as my pen is of vocabulary--and I have the best sources in
the world at my fingertips. I hasten to complete the work--and it
is nearly done. Here are all the complex causes, here are all the
convoluted courses, here are all the costly consequences, and
finally all the tried and tentative treatments, from Alcoholics
Anonymous to Zen, all of which report approximately the same rate
of success, about 33 percent. That is, about 33 percent become
permanent abstainers. Finished? But not quite. There is one em-
barrassing unresolved problem. It has nagged a bit at one's pre-
conscious before, but the unsuppressible superego has now forced
it up into the unveiled conscious. The question is, "Why aren't
some of the alcoholics really cured so that they can become normal
drinkers?"

Well, of course, having read just about everything, it is
unavoidable to recall that somewhere, here or there, one has seen
a report of some alcoholic returning to moderated, perhaps normal,
drinking. Of course one ignored that. In biology, certainly in
the constitution of the creature homo, there are no absolutes. How
often had I heard Dean John Wyckoff, at NYU Medical School, warning
the students, "But remember, it's never always." The assumption
that the only successful outcome of the treatment for alcoholism is
permanent total abstinence is valid, as I had written. It was an
experientially established pragmatic truth. But why was it so? Of
course, my article was not complete until I had explained why
sobered-up alcoholics can never drink again.

Now I had explained alcoholism in the article in terms which
today I might dare to call behavioral psychology. I had explained
it as learned behavior. From all my learning about it, that is
how I had come to understand the process, the development of an
addiction. And I thought I knew that what was learned could be
unlearned. Therefore, the alcoholismic behavior should be subject
to unlearning. Why did we have a law--for that was what it
amounted to--that of all behaviors, alcoholism could not be unlearned? Of course, until that was explained, the article dangled
incomplete--a lion without a roar.

I reexamined the alcohol literature. I delved into the psychological literature. I reread Pavlov, and I read Sherrington,
and I discussed it with a clever learning-and-thinking psychologist. To no resolution. I had no satisfactory answer. I abandoned the article--it was never finished. (I bought a second-hand
car.)

My thoughts were, I think, clear on the matter. In theory, it ought to be possible for an alcoholic, after appropriate treatment, to resume normal drinking--even to learn normal drinking as a new behavior, if he was one of those who had never originally learned to drink temperately. But in real life, everybody was agreed that alcoholics could never drink safely again, and it was especially the experts, the therapists, including the psychiatrists all over the world, who subscribed unanimously to this prudent principle. Also, it was not a new notion, not a recent invention of A.A. or Jellinek, but well entrenched in historic precedent and practice. Certainly throughout the 19th century the many physicians who took an interest in alcoholism and its treatment, who founded special societies and special journals in both England and America, and in other countries too, were unanimous in requiring alcoholic patients to become teetotalers. I even thought I tracked this principle down to early Bible times. In the First Book of Samuel, when the high priest Eli mistakenly diagnoses Hannah's condition as alcoholism, his prescription--let us remember that in those days the priests were also the doctors--his prescription was to give up drink altogether. That is how I read the original text. Historical unanimous agreement--but without a logical theory to explain it. And yet with a peculiar importance which was soon brought home to me.

One day, still in the latter 1940's, I was being interviewed in my office by a reporter. After I had told him what I thought about alcoholism he asked, very logically, "And why can't alcoholics ever drink again?" I replied, like a natural-born wise guy, "Well of course in theory there is no reason why an alcoholic shouldn't be able to drink again," and I proceeded to expatiate on that theme, intending finally to explain that in practice, however, the case is different. Then suddenly I noticed the machine that was recording every word--I had quite forgotten the machine--and I exclaimed, "But you can't quote me on this theory." The reporter said, "But this is not for the popular press, it's for a physician's journal." I said, "Well, one can express a scientific theory to physicians." How young and innocent I was! My theoretical viewpoint was published in that journal--one of those that goes free to physicians--and immediately it was picked up by the newspapers. They had a good headline at last: "Yale Professor Says Alcoholics Can Drink Again." And then the terrible letters began to arrive: "My husband hasn't had a drink in 6 months, and now he has read your article..." So I had learned something new--I learned to shut up.

E. M. Jellinek used to like to say, "Mark has a far look." I gave up talking about that theory, but I thought about it every now and then with anticipatory anxiety. By now, I was fairly in full charge of the Quarterly Journal of Studies on Alcohol--both Haggard and Jellinek were glad to leave the editing to me--and I used to wonder, every now and then, when I would receive "that" article for publication--an article reporting normal drinking in recovered alcoholics. I even thought I had reason to expect it, and I thought that when it was written it would naturally be submitted to us. And it came.

It was not quite what I had feared. And it was no incredible "Rand Report." It was a first-rate report of a single case, an alcoholic treated by a psychoanalyst for 3 years, and the patient had been able, for more than 5 years, to resume normal moderate drinking. It was a good article. It met the canons of scientific reporting. Not without anxiety and trepidation (I confess I wished this case hadn't happened, though I had known all along it was bound to come). I did my duty as a responsible editor of a scientific periodical. I published it (Shea, 1954). That was almost 25 years ago.

There was. no reaction at all. A couple of long-abstaining alcoholics on our staff--one in Alcoholics Anonymous, the other not-- shrugged it off. They said--and I suppose this was the reaction in the wider world--"A 3-year psychoanalysis?" In other words, the world thought, perhaps rightly, that this "unique" case was irrelevant. Some no doubt thought it was the exception that "proved" the rule. But it was not irrelevant to my theoretical position. And with my presumptive--and anxiety-provoking--far look, I awaited the next development.

It was dramatic when it came, after 7 years. Out of 93 reliably diagnosed alcohol addicts treated in one of the best psychiatric hospitals in the world--the deservedly famous Maudsley Institute--7 patients had returned to normal drinking. Not for any ridiculously insignificant 30-day period, but for a minimum of 7 years and ranging up to 11 years. The follow-up was as sound as the diagnosis and reporting, and the article was signed by the dean of the Institute.

I remember the words of one of my editorial referees--the clinical director of a great research foundation. He wrote, "Dear Mark, this is terrible, but you have to publish it." I had known, of course, that that's what I would be told, and of course I did my duty again and published it (Davies, 1962). I also published about 18 comments on the article (Tiebout et al., 1963). Some tried to explain it, some tried to explain it away. At any rate, the article by Dr. D. L. Davies opened up the question.

Not very long after, I published another article (Kendall, 1965) from the same Institute in which Dr. R. E. Kendell reported on four more alcoholics seen there, but not treated, who eventually nevertheless became controlled drinkers for at least 3 and up to 8 years.

   If you have kept track of the arithmetic, you know that it
adds up to a total of 12 cases of confirmed alcoholics--by which I
mean alcohol addicts, and nothing less--who became normal drinkers
for at least 3 years, most of them much longer. I later published
a paper (Pattison et al., 1968) reporting 11 cases with a mean
duration of 20 months but I don't count these as sure successes;
and because of brief durations or uncertain diagnoses I do not
count the cases in several other reports cited by Kendell and by
Pattison et al. I indulge a strong personal bias in this matter.
I have not forgotten, for example, the case of Mr. Clapp who, de-
fying the conventional wisdom, quit behaving like an alcoholic and
drank moderately under careful self-control. He was so successful
at it that he published a book about it (Clapp, 1942)--its title
was "Drunks are Square Pegs." After about 2 years Mr. Clapp had a
great fall. He then joined the more prudent company of abstainers
and wrote another book (Clapp, 1949) giving his fellow alcoholics
more conservative advice. That's one reason why I think 2 years
is too soon--by 2 years--to think the outcome secure. My own
"magic number"--for reasons too extraneous to be detailed here--is
4 years. At 4 years, I am willing to consider an alcoholic reasonably safe in abstinence--or reasonably safe in controlled drinking.
And so far, I know of only those 12 cases of controlled drinking,
along with 12^n cases of controlled abstinence.

   We are now in the new era. Several behavioral psychologists--
some of them psychiatrists too--relying on the principles of
learning psychology, have experimented in shaping the behavior of
a number of patients with drinking problems so that these patients
could become controlled drinkers rather than abstainers. And they
have reported some successes (e.g., Alterman et al., 1974; Cohen
et al., 1971; Lovibond and Caddy, 1970; Mills et al., 1971; Pattison
et al., 1968; Popham and Schmidt, 1976; Sobell and Sobell, 1975,
1976).

   I wish I had had the time and opportunity to review all those
cases, and many others that have been reported, in advance of this
occasion so that I could comment confidently on the characteristics
of the patients--for example, whether I think they are alcohol addicts or mere problem drinkers--and on the terms of the outcomes,
especially the duration of controlled drinking, and on the circumstances of the outcomes. From what I said earlier, it should be
obvious that I know of no reason why some alcoholics should not be
able to become controlled drinkers; and furthermore, it should be
obvious that I am inclined to think that treatment based on learning theory is the likely way to achieve it. I am, then, in principle, prepared to believe.

   It is particularly gratifying to me that the behavioral psychologists whom I know who have been experimenting along those
lines have worked and reported within the best tradition of

science, and within the humane and ethical constraints which all
of us expect from scientific experimenters--in the selection of
subjects, and in the care and after-care due to experimental subjects. It should not be necessary to allude to this fact, but all
of us are aware of the bitterness and enmity, and even some attempt
to sabotage the support of research that occurred recently following upon a much publicized claim of the restoration of a large
proportion of treated alcoholics to moderate drinking (Armor et al.,
1976). This part of my discussion of the possibility of such restoration would therefore be incomplete if I did not comment on
that report and its relation to the theory and the undertaking of
behavioral psychology in the treatment of alcoholism. As you will
see, eventually that will lead to an expression of my own view of
alcoholism.

The so-called Rand Report (1976) is an analysis of data from a
follow-up that was supposed to examine the outcome in a presumably
randomized sample consisting of more than 2,000 alcoholic patients
from 10 out of more than 40 clinics. Actually data from only 8 of
the 10 clinics were reported; and only 55.3 percent of the patients
selected as the sample from those 8 were interviewed. It is impossible to know how many of the people in the interviewed sample
were alcohol addicts--although it can safely be assumed that some
were not. The method of study is peculiar--almost beyond belief.
For example, the total amount of absolute alcohol consumed by a
patient during the month before admission is compared to the total
consumed during the month before the follow-up interview.· The total is then divided by the days in the month. If the resulting
daily average is less than 1.5 ounces of absolute alcohol, that is,
3 ounces a day of 100-U.S.-proof whiskey, the patient is recorded
as a normal drinker. A patient who drank a quart of 90-proof whiskey on each of 3 days during the month and got drunk three times
during the month before the follow-up interview, is nevertheless
counted as a normal drinker. I will not here discuss the other
naivetés of the Rand Report, obviously prepared by statistical
sophisticates lacking in clinical orientation. With that sort of
analysis, a substantial percentage of the patients--always called
"clients" in the Report--were reported as alcoholics who had successfully resumed normal drinking. The newspapers naturally
treated this as big headline news. And it was naturally very up-
setting to the people in what sometimes is called "the alcoholism
movement," or the "alcoholism establishment." From some sources,
I heard that attempts were made to stop the support of the re-
searches of some behavioral psychologists.

Popular reactions may be seen as irrelevant to the sacred
truths of science. The announcement that the earth is a spinning
ball aroused some negative reactions, too. Yet we ought to note
this fact: During the years when the Journal of Studies on Alcohol
was publishing several reports of alcohol addicts--positively

expertly diagnosed alcohol addicts--becoming normal drinkers, and
publishing theoretical papers by behavioral psychologists and psy-
chiatrists--some of whom are presenting papers at this meeting--
challenging the assumption that abstinence was the sole appropriate
goal in the treatment of alcoholism--during all those years and all
those papers, the Journal never--never--never--received one word of
criticism for publishing those papers. Some people, perhaps members of the so-called alcoholism establishment, wrote appropriate
and publishable critical comments on the papers. Only after the
questionably scientific Rand Report was popularized did the Journal
come under some criticism for publishing two papers dealing with
treatment for a non-abstinence goal. To add the comic note--of
these two papers, one (Pomerleau et al., 1976) was purely theoretical, suggesting that the alternative goal should be carefully investigated, and the other (Popham and Schmidt, 1976) reported a
controlled experiment in which treatment for abstinence gave better
results than treatment for moderated drinking.

It is, then, a fact of life that scientific work and scientific
publication can arouse public emotions and ire. It happened in
astronomy. We tend to forget it, but there are still a few people
who prefer to believe the earth is not ball-shaped, even in spite
of pictures taken from far-out space. It happened in biology, and
there are many more people who don't believe in the Darwin-based
theory of evolution, and some would like to suppress the teaching
of it to innocent youth. I think it is normal and expectable that
many who have witnessed and experienced the tragedies of failed
attempts by alcoholics to achieve permanent moderate drinking, and
have witnessed and experienced the happy effects when alcoholics
succeed in achieving permanent abstinence--I think it is normal
and expectable for them to be alarmed and upset by claims of multitudes of alcoholics being restored as normal drinkers. It is natural for them to fear that great harm may be done by the popularization of such "news."

I have been asked to talk about "the behavioral problem with
alcoholism." I don't see a behavioral problem. I think, as I
stated earlier, that the viewpoint of behavioral psychology about
the process in the development of alcoholism is correct as far as
it goes. I don't think it provides a sufficient account. There
is a question of etiology that learning theory does not answer. We
see that John learns to become an alcohol addict. And we see that
his brother--even his twin brother--James does not learn to be an
alcohol addict. Why John and not James? Perhaps learning psychology can explain that. Perhaps another psychology--it could even
be Freudian psychology--can explain it better. But when it comes
to the process, I see it adequately explained today by learning
psychology. So here I see no problem with the viewpoint of behavioral psychology. We come, then, to treatment. If the condition
is learned, should it not be most appropriately treated by the

principles of behavioral psychology, with the aim of unlearning the
maladaptive, harmful or undesired behavior?

In principle, yes.

But here I see a problem. In practice I see a problem.

It is a problem in the containment of hubris! The problem is
not with behavioral psychology but with the behavior of some psy-
chologists--that is, their reporting behavior. And it is a problem
in the realm of ethics--of all things, yes, ethics!

As an editor and a reader of the literature that reports on
therapeutic trials, I am familiar with the element of enthusiasm
that tends to glisten in early reports of new treatments. You are
all familiar with it. Just think of the last 99 cures of alcoholism
that were announced until five years ago and then try to think of
one that is generally being applied. I won't go so far back as the
chemical cure promised by a Dr. J. J. Smith away back in the 1940's
or 1950's which was going to make it possible for all alcoholics to
drink normally again. Most of you probably have never heard of it.
I may be the last man alive to remember it. But what about NAD--
nicotinamide adenine dinucleotide? Only a few years ago, this
treatment was so promising--to make alcoholics able to abstain--
that it was actually patented by a great corporation. The library
of the Rutgers Center of Alcohol Studies has a copy of the patent.
Or, let us get away from chemicals. What happened to thought capsules? They undoubtedly helped some alcoholics to become abstainers
But you can't get McGoldrick's thought capsules in any drug store--
neither over-the-counter nor by prescription. Perhaps it is because their inventor did not believe alcoholism is a disease. One
only has to wonder why, in that case, the thought capsules were designed to convert the alcoholic patients--excuse me, clients--into
abstainers, not normal drinkers. Does anybody remember the grape
treatment? The non-shock electric-current treatment? I'll spare
you the whole list. If you are interested, just consult the indexes
of the Journal of Studies on Alcohol under "Treatment of Alcoholism.

New treatments, new trials, new efforts, are reported enthu-
siastically. Surely that is normal. As an editor, I have often
gently advised my authors to moderate the confident tone of their
reports of new trials. I am sure no author who accepted this editorial advice ever had reason to regret it.

With respect to reports of the treatment of alcoholics by
modalities of behavioral psychology, with the aim of achieving
normal or controlled or moderated drinking, I feel justified in
urging the utmost conservatism in reporting. Not because I think
the aim cannot be achieved. I suppose that some day it may become
possible in many cases, and this justifies well-designed experiments

But I think that as yet there are too many unresolved mysteries in alcoholism for the achievement to be very likely in any substantial number of alcohol addicts.

One of the main problems that we ought to consider is the question of diagnosis. One sure mark of scientific process is the finding of distinctions, with the strict use of precise definitions. What, for example, is alcohol abuse? Here is a recently published definition.

> "Alcohol abuse: The intake of alcohol-containing beverage in a quantity or in a manner that evokes disapproval. Sometimes used pejoratively, sometimes ambiguously as a substitute for alcohol addiction, alcohol dependence, alcohol intoxication, alcohol misuse, alcoholism, drunkenness, excessive drinking, habitual excessive drinking, habitual drunkenness, problem drinking, and possibly with other meanings or with a . combination of these meanings either to avoid commitment to a specific meaning or from uncertainty about the nature of the behavior or the condition thus labelled." (Keller, 1977)

Now, if anybody tells me he has successfully treated alcohol abuse, or alcohol abusers, or abusive drinkers, with consequent moderated drinking, you can guess what I think of that. The kindest thing I can say about it is that I don't know whether he has been treating alcohol addicts.

The same thing applies to problem drinking.

I suggest that anyone who considers himself a professional, and undertakes to work as a therapist, ought to be a diagnostician. If he is going to treat people because they manifest the behavior of suspicion-arousing or distress-causing alcohol intake, he ought to be able to distinguish between alcoholism--that is, alcohol addiction--a disease--and less serious problems, perhaps prodromal symptoms, which may well be described as "problem drinking"--though I can accept many other terms, including habitual drunkenness.

I think I have arrived at the issue of the disease conception of alcoholism (Keller, 1976). I believe there is such a condition as a disease, addiction--more specifically to the present purpose, drug addiction, and still more specifically to the present topic, alcohol addiction, more commonly called until now alcoholism, and with proposals that it be called alcohol dependence or alcohol-dependence syndrome (Keller, 1977). I think it is an ethical imperative for the one who reports the results of any treatment to be able to tell us whether he has been treating alcohol addiction or problem-causing over-drinking. Alcohol addiction is marked by a disablement, an impairment of the ability consistently to refrain from drinking even if drinking is not desired or intended, and by a

further impairment of the ability consistently to stop if drinking
is started. This impairment of behavioral control is reasonably
diagnosable. I will not here detail the basis of diagnosing a drug
addiction, but I do not think membership in any association, or a
complaining spouse, or a hairline score on a paper-and-pencil test,
or being convicted of a traffic offense, to cite some examples,
constitutes a diagnosis.

Of course, there are those who do not believe that there is a
disease, alcoholism. Or they do not believe addiction is a disease.
Or they do not believe behavioral disorders--what I have termed
"dysbehaviorisms"--can be diseases. There are physicians who stand
firm in the 19th century definitions of disease which demand a manifest physical-anatomical pathology. But mostly it is social scientists who--with an antimedical predilection that Freud would have
enjoyed explaining--would deny medical doctors the right to deal
with behavioral disorders. They therefore oppose "medicalizing"
and "clinicizing" the dysbehaviorisms--especially alcoholism. It
is understood, by context, that "medicalizing" and "clinicizing"
are bad behaviors.

It may be that some behavioral psychologists agree with the
backward-looking physicians and the sideways-looking sociologists
in their conception of what constitutes disease, and define alcoholism as a non-disease. Some behavioral psychologists may join
some of the lay therapists--now often called "counselors"--in feeling embarrassed to seem to be imitation doctors when they deal
remedially with such people as alcoholics. They therefore call
these people "clients"--not patients--although they then sometimes
forget themselves and refer to what they are doing with their
"clients" as "treatment."

There is no question that the social scientists who conduct
surveys of drinking behaviors are right in seeing only a range of
drinking behaviors, from none, to a little, to a lot, to an awful
lot; and likewise a range of motivations, from antipathy to innocent sociality, to escapism; and a range of consequences, from
seemingly none, to all sorts and quantities of troubles, to death.

They are right because that is all that their discipline seeks
and all that their methodology allows. Some of the behavior encom-
passed by their descriptions and classifications could represent
behavioral disease. But they are not diagnosticians, and there is
no reason why they should distinguish it as disease. Nor should
they go out of their way and their competence to deny that there
may be disease in some of those behaviors.

To my understanding, sociology can be a branch, and should
even be a root, in medicine. Yet there are aspects of sociology
which are divorced, or can divorce themselves, from any connection

with medicine. The case of psychology is different. I am not sure any aspect of psychology can be irrelevant or indifferent to medicine--either basic biomedical science or clinical medicine. I think it inconceivable that a branch of psychology which studies self-harming human behavior, and even attempts to correct it, by modalities which inevitably constitute what we call treatments--it is inconceivable that such a branch of psychology should be divorced from medicine. The study cannot be distinguished from basic biomedical science. The treatments cannot be distinguished from clinical medicine. These are facts. If alcoholism were not a disease, they would not be facts.

The behavioral psychologists are in fact clinicians, whether they happen to be degreed M.D.'s or Ph.D.'s, and I hope this does not embarrass my friends, the behavioral psychologists, who treat alcoholic patients--whatever the goal of therapy.

And now I have come full circle and face to face with the problem set me by the Directors of this meeting--my view of the problem. You will recall that I have already said it is not a problem of behavioral psychology but of behavioral psychologists. It is the problem of how to proceed, as experimenters, as scientists, as clinicians, in exploring all the potentials of the treatment of alcoholism.

But I think I have already indicated the solution that represents my viewpoint of alcoholism. It invokes two rules.

The first rule needs to be painstaking diagnosis. With this goes the responsibility to use unambiguous terminology. We hear loose talk of "10 million alcoholics" in the U.S.A. That statistic is a fiction (Keller, 1975c). No matter how often you hear it, or read it, you will not be able to discover any evidential basis for it--certainly no basis in any branch of science that there are 10 million people in the United States of America who have a disease, alcohol addiction. There are believable indications that there may be a few million. And there are believable indications that there are also several million who are using alcohol in such a way and to such a degree that they sometimes experience or cause some problems, and they may be at some risk of becoming alcohol addicts. It may be convenient to call them "problem drinkers." If you are going to apply your experimental methods and therapeutic skills to any of these people, the canons of science and the ethics of your profession require that you should distinguish among your subjects and classify your patients.

Some of the people who are most vehement in denying that you can ever cure an alcoholic so that he can become a normal drinker are the same ones who are most eager to enlarge the number of alcoholics. For them, 10 million alcoholics are hardly enough. It is

tempting to say that they deserve the Rand Report. They have ama-
teurishly attached the label "alcoholic"--the diagnostic term,
"alcoholism"--to millions who are not alcohol addicts by strict
diagnostic criteria. Very likely a lot of those mislabeled and
misdiagnosed "alcoholics"--in quotation marks--have moderated their
over-drinking after getting into some trouble on account of it.

But psychologists cannot afford to behave like amateurs.
Strict diagnosis is of the essence in their professional and scientific behavior.

The second rule comes from me in the form of my gentle editorial advice. Be modest, cautious, conservative, in reporting the
results of your experiments. I suspect, I have always supposed,
that it ought to be possible to really cure at least some alcohol
addicts. What an unbelievable disease alcoholism would be if no
alcoholic could ever be cured! (Have I not already quoted Dean
John Wyckoff, that "It's never always?" Perhaps, if one is not
afraid of a paradox, one should add, with him, "and it's never,
never.") But conservatism in reporting miracles is a sound, prudent and ethical policy. We should never forget that many patients
with many serious diseases have remarkable--miraculous--recoveries
under new treatments, followed usually by relapse after a while.
We should remember that there is nothing new in alcoholics, like
people with other diseases, having a remission lasting a year or
two, only eventually to relapse. We should remember that in the
treatment of alcoholics, abstinence has worked for multitudes.
Whether moderated drink ing will work for very many is a question
not likely to be answered in a hurry. Therefore, at the very least,
we should attach a conservative timer to our claims. My arbitrary
period is a minimum of 4 years. If a carefully diagnosed alcohol
addict has been drinking moderately during 4 years, I would be willing to call it a recovery. Short of that, I would be content to
speak of remission--the same term that I would use in the case of
alcoholics who abstain.

My two suggested rules--strict diagnosis and conservative re-
porting--are appropriate for scientists. It is not easy to be a
scientist. One must be able to sacrifice certain sorts of reward--
for example, headlines in the so-called news media. But my friends
the behavioral psychologists are scientists, and that is why I am
proud that they let me bask in their company, and that is why I
dare to give them advice.

REFERENCES

Alterman, A. I., Gottheil, E., Skoloda, T. E. & Grasberger, J. C.
    Social modification of drinking in alcoholics. Quarterly
    Journal of Studies on Alcohol, 1974, 351, 917-924.

Armor, D. J., Polich, J. M. & Stambul, H. B. <u>Alcoholism and treat-ment</u>. Santa Monica, CA., Rand Corp., 1976.

Clapp, C., Jr. <u>Drunks are square pegs</u>. New York: Island Press, 1942.

Clapp, C., Jr. <u>Drinking's not the problem</u>. New York: Crowell, 1949.

Cohen, M., Liebson, I. A., Faillace, L. A. & Allen, R. P. Moderate drinking by chronic alcoholics: A schedule-dependent phenomenon. <u>Journal of Nervous and Mental Disorders</u>, 1971, <u>153</u>, 434-444.

Davies, D. L. Normal drinking in recovered alcohol addicts. <u>Quarterly Journal of Studies on Alcohol</u>, 1962, <u>23</u>, 94-104.

Jordan, E. P. (Ed.), <u>Standard nomenclature of disease</u>. Chicago: American Medical Association, 1942.

Keller, M. Nomenclature and classification of "alcoholic" diseases. <u>Quarterly Journal of Studies on Alcohol</u>, 1942, <u>3</u>, 518-523.

Keller, M. Documentation of the alcohol literature: A scheme for an interdisciplinary field of study. <u>Quarterly Journal of Studies on Alcohol</u>, 1964, <u>25</u>, 725-741.

Keller, M. Some views on the nature of addiction. Presented at the 15th International Institute on the Prevention and Treatment of Alcoholism (The E. M. Jellinek Memorial Lecture), Budapest, Hungary, June, 1969.

Keller, M. A special-library information-center model for a socie-tal-problem field. In L. Vilentchuk, & G. Haimovic (Eds.), <u>Proceedings of the ISLIC International Conference on Informa-tion Science</u>, Vol. 1. Tel Aviv: National Center of Scientific and Technical Information, 1972.

Keller, M. Final summing up. In N. Kessel, A. Hawker, & H. Chalke (Eds.), <u>Alcoholism: A medical profile</u>. London: Edsall, 1974.

Keller, M. Multidisciplinary perspectives on alcoholism and the need for integration: An historical and prospective note. <u>Quarterly Journal of Studies on Alcohol</u>, 1975, <u>36</u>, 133-147(a).

Keller, M. The nature of addiction: Some second thoughts. <u>Alcohol-ism</u>, 1975, <u>11</u>, 28-32(b).

Keller, M. Problems of epidemiology in alcohol problems. Journal of Studies on Alcohol, 1975, 36, 1442-1451(c).

Keller, M. The disease concept of alcoholism revisited. Journal of Studies on Alcohol, 1976, 37, 1694-1717.

Keller, M. A lexicon of disablements related to alcohol consumption In G. Edwards et al. (Eds.), Alcohol-related disabilities. Geneva: World Health Organization, 1977.

Kendell, R. E. Normal drinking by former alcohol addicts. Quarterly Journal of Studies on Alcohol, 1965, 26, 247-257.

Lovibond, S. H. & Caddy, G. Discriminated aversive control in the moderation of alcoholics' drinking behavior. Behavior Therapy, 1970, 11, 437-444.

Mills, K. C., Sobell, M. B. & Schaefer, H. H. Training social drinking as an alternative to abstinence for alcoholics. Behavior Therapy, 1971, 21, 18-27.

Pattison, E. M., Headley, E. B., Gleser, G. C. & Gottschalk, L. A. Abstinence and normal drinking: An assessment of changes in drinking patterns in alcoholics after treatment. Quarterly Journal of Studies on Alcohol, 1968, 29, 610-633.

Pomerleau, O., Pertschuk, M. & Stinnett, J. A critical examination of some current assumptions in the treatment of alcoholism. Journal of Studies on Alcohol, 1976, 37, 849-876.

Popham, R. E. & Schmidt, W. Some factors affecting the likelihood of moderate drinking by treated alcoholics. Journal of Studies on Alcohol, 1976, 37, 868-882.

Shea, J. E. Psychoanalytic therapy and alcoholism. Quarterly Journal of Studies on Alcohol, 1954, 15, 595-605.

Sobell, L. C. & Sobell, M. B. Legitimizing alternatives to abstinence: Implications now and for the future. Journal of Alcoholism, 1975, 10, 5-16.

Sobell, M. B. & Sobell, L. C. Second year treatment outcome of alcoholics treated by individualized behavior therapy: Results Behavior Research and Therapy, 1976, 14, 195-215.

Thompson, E. T. & Hayden, A. C. (Eds.), Standard nomenclature of diseases and operations. 5th Ed. New York: Blakiston (published for the American Medical Association), 1961.

Tiebout, H. M. and others. Normal drinking in recovered alcohol
     addicts: Comment on the article by D. L. Davies. Quarterly
     Journal of Studies on Alcohol, 1963, 24, 109-121.

World Health Organization. Manual of the international statistical
     classification of diseases, injuries and causes of death.
     Geneva, 1972.

INDEX

Abstinence, 19, 215-217, 264, 289, 298, 306, 326, 386
  Abstinence Violation Effect, 198, 296-299, 327
  alternatives to, 177-211 (see also Controlled Drinking)
  group therapy and, 144-146
  profile, 17
  self-monitoring and, 136
  social learning theory and, 73
  therapy goal of, 2, 198-201, 203, 232
Adjustment scales, 14
Adolescence
  drinking in, 24
Adoption studies, 46
Alcohol
  addiction to, 6, 16, 61-63 (see also Alcoholism)
  analgesia, 60
  and associated problems, 237, 246, 294-296
  dependence, 74, 196-197, 238, 255, 277, 348
  expectancies about, 293-296
    behavioral effects and, 317-325
    sex differences in, 324
  physiological effects of, 42-44, 295
  use of, 6, 85
    scales to measure, 9-13
    surveys on, 236-243
Alcohol abuse (see also Alcoholism)
  and adolescence, 24
  behavior therapy and, 119-122

cognitive processes in, 315-333
composite picture of, 59, 265
definitiion of, 256, 391
multivariate analysis of, 71-119
and other drug use, 32
theories of, 87-88
Alcohol Intake Sheet, 136
Alcoholics Anonymous, 2, 73, 183-184, 212, 278, 290, 298, 300, 385
  and disease concept of alcoholism, 72
Alcoholism
  age differences and phase theories of, 20-25
  blood alcohol level discrimination and, 161-177
  children at risk for, 46
  cognitive variables in, 92-97, 326, 315-341
  craving and, 61-63, 72, 135, 271-310, 325
  criteria of, 74-76
  differential diagnosis of, 1, 20, 76, 393-394
  escape-avoidance paradigms of, 88, 321-325
  etiology of, 5, 41-71, 93, 256
  Jellinek's system of, 20-21, 25, 71-73, 246, 259-260
  loss of control concept of, 95, 161, 176, 271, 278-280, 294-295, 299-301, 325
  and craving, 341-349